Interactive spatial data analysis

Interactive spatial data analysis

Trevor C. Bailey
Department of Mathematical Statistics
and Operational Research
University of Exeter

Anthony C. Gatrell
Department of Geography
Lancaster University

Longman Scientific & Technical
Longman Group Limited
Longman House, Burnt Mill, Harlow
Essex CM20 2JE, England
(and associated companies throughout the world)

Copublished in the United States with
John Wiley & Sons, Inc., 605 Third Avenue, New York NY 10158

© Longman Group Limited 1995

All rights reserved; no part of this publication may be reproduced, stored in a retrieval system, or transmitted, in any form or by any means, electronic, mechanical, photocopying, recording, or otherwise without either the prior written permission of the Publishers or a licence permitting restricted copying in the United Kingdom issued by the Copyright Licensing Agency Ltd, 90 Tottenham Court Road, London W1P 9HE.

The software package INFO-MAP distributed with this book was originally and wholly written by one of the authors. It has been especially modified to accompany the book and is distributed by the publishers with the full agreement of Claymore Services Limited, who hold copyright to the package, and Media Cybernetics Inc., whose 'Halo graphics library' is incorporated into the software.

Notice of liability
The authors of this book, the publishers, and Claymore Services Limited do not accept any responsibility for loss occasioned to any person using the disk supplied with this book or acting or refraining from acting as a result of material contained in the disk.

Trademarks
Throughout this book trademarks are used. Rather than put a trademark symbol in every occurrence of a trademarked name, we state that we are using the names only in an editorial fashion and to the benefit of the trademark owner with no intention of infringement of the trademark.

First published 1995

British Library Cataloguing in Publication Data
A catalogue entry for this title is available from the British Library.

ISBN 0-582-24493-5

Library of Congress Cataloging-in-Publication data
A catalog entry for this title is available from the Library of Congress

ISBN 0–470–23502–0 (USA only)

Typeset by 22 in 10/12 Times
Printed in Malaysia

To Alison, Sarah and Simon *(Trevor Bailey)*
To Caroline, and Richard, Anthony and Laura *(Anthony Gatrell)*

Contents

Preface

The past decade has seen considerable interest in the analysis and modelling of spatial data. By 'spatial data' we mean data where, in addition to values relating to the primary phenomenon, or phenomena, of interest, the relative spatial locations of observations are also recorded, because these may be of possible importance in interpreting the data. Not surprisingly, such interest is particularly strong in the fields of geography and geology, but similar kinds of data are encountered in many other disciplines concerned with spatial or geographical variations and patterns—for example in environmental science, epidemiology, ecology and economics. At one extreme, the spatial dimension may involve the microscopic scales that arise in medical or biological research; at the other, the huge distances encountered in astronomical studies.

The existence of this wide audience for spatial data analysis has generated a number of excellent texts on the subject over the same period. However, the majority of these have been addressed to either the researcher, or to the graduate student. In general, spatial data analysis has tended to be treated as a somewhat specialist topic, requiring a level of statistical background too advanced for an undergraduate audience.

This provided our first objective in producing this book. Undergraduates in most of the fields mentioned above are exposed to courses in statistical techniques on the basis that these quantitative skills are important in their studies and perhaps in their careers. However, few such courses involve much spatial data analysis, even though the analysis of spatial patterns and relationships is important in many of these disciplines. Indeed, there are few courses on spatial data analysis even in undergraduate degrees specialising in statistics. Although we take the view that some aspects of spatially referenced data can be usefully analysed by the routine, non-spatial, methods which are 'standard fare' in most courses in statistical analysis, we also believe strongly that the analysis of spatial data can provide particular problems where special methods are valuable and sometimes essential if the analyst is to avoid mistaken conclusions. We think the undergraduate audience ought to know about at least some of these methods. We wished therefore, to provide a text directed primarily at introducing spatial data analysis to the undergraduate interested in applied statistical methods. Of course we hope that it will also be useful to others, such as the researcher requiring an overview of the subject, which may then be subsequently refined by reference to other more advanced texts.

With this overall objective in mind, we have not attempted to discuss exhaustively the whole of spatial statistics, but have restricted ourselves to a subset of methods which we believe are both relatively accessible and of real practical use. We have included a mixture of both informal, visual or exploratory techniques, and formal statistical modelling, parameter estimation and hypothesis testing. We wished to emphasise the way in which these approaches complement and reinforce each other in the analysis of real data. We have tended not to present methods in their full generality, preferring to provide their basic forms, with guides to further reading at the end of each chapter. To avoid presenting the reader with 'statistical black boxes', we have tried to explain clearly the rationale and principles behind the methods discussed. We have therefore not held back from mathematics where we have felt it necessary; but at the same time we have not stressed mathematical or statistical theory, rigour, proof or generality—we have tried to provide a 'practitioner's guide' to spatial analysis, rather than a reference text. However, in doing so within a manageable number of pages, we have had to presume a moderate level of general mathematical background, including a familiarity with elementary vector and matrix algebra. We also expect our reader to have already taken a first course in statistical methods and met ideas of probability distributions, expectation, variance, parameter estimation and significance tests. We further assume that the reader has encountered simple forms of statistical modelling, such as multiple regression.

Our second overall objective was to provide a text that was 'data driven' rather than 'theory led'. Our organisation of material is therefore firmly based on the type of spatial problem and spatial data concerned. Following on from the introductory Part A, each of the subsequent four Parts B–E of the book relates to a different class of spatial problem and type of spatial data. Each of these parts is as self contained as possible, consisting of two chapters (except for Part E), the first of which contains 'core material', the second involving more specialised topics. This structure should allow the reader to 'dip into' the book without necessarily having to read it from the beginning. We felt it to be particularly important to motivate methods by reference to examples and case studies which involved real data sets and reflected a range of applications across disciplines. We have therefore included a case study section in each part of the book, which introduces and describes a number of data sets; and we have distributed these with the book, on diskette. We use them in each part of the book, as appropriate, to illustrate the methods discussed there. Providing them on diskette also means they are available for readers to work on in their own time.

Like many others, we feel that much of the learning in a subject like data analysis comes from trial and error and involves a considerable amount of experience. Our third objective was therefore to try and make our example data sets come 'alive' for the reader, rather than just providing a file of the data on disk. We wanted readers to understand that statistical analysis is rarely 'right', but perhaps at best 'not wrong' and to encourage them to experiment with methods and emphasise the interpretation of results and the validity of the assumptions upon which they depend. The teaching of statistics has been transformed in recent years by interactive computer packages which not only permit routine analyses to be performed quickly and easily, but also allow creative use of plots, simulations and transformations of the data. Indeed, many recent methodological developments in statistics itself have been heavily influenced by dynamic graphical computing environments. Recent major advances in digital mapping, computer cartography and Geographical Information Systems (GIS) potentially provide similar computer environments within which to visualise spatial data and explore spatial patterns and relationships interactively. Unfortunately, such technology is not currently as widely available as conventional statistical packages, few of which are designed to handle spatial data structures. Excellent spatial analysis software does exist, but costs, hardware requirements and steep user learning curves, can prove prohibitive.

We therefore felt it necessary to develop a special purpose computer package to accompany the book, incorporating the data sets discussed in the text. We make no great claims for this package; it is a relatively simple application that will run on any typical PC with a standard colour graphics screen. By no means all of the methods discussed in the text are implemented in the software, and for those that are, only fairly simple variants have been made available. Nevertheless, we hope the software will allow readers to use the data sets and experience the 'look and feel' of interactive spatial data analysis. In particular, we would like to think that this will enable and encourage students of spatial analysis to do a considerable amount of self instruction outside the 'chalk and talk' environment, working on the computer exercises that we refer to in the text, or possibly creating their own maps and data sets. At the same time, although the book is certainly not a GIS text, and the software package provided is far from qualifying as a GIS, we hope that by using it readers will encounter cartographic and visualisation issues and the linking together of data sets, as well as spatial analysis—all issues at the heart of GIS.

Whether we have achieved any of the objectives discussed above is, of course, for our readers to judge; but if we have not then it is certainly our failing and not that of the others who have contributed to the book and to whom we are indebted.

At an organisational level, we are grateful to The Department of Mathematical Statistics and Operational Research, Exeter University, for providing funds to allow research visits that aided 'spatial interaction' between the authors. In addition, Lancaster University is thanked for granting a period of study leave that allowed one of us to devote time to the book, while the

Department of Geography, McMaster University, also provided a valuable environment for reflection and research. At the same time we would like to acknowledge the help and co-operation we have received from Mike Butts and Chris Whitt of Claymore Services Ltd, who kindly allowed us to modify and distribute the software that accompanies the book.

We also owe a considerable debt to those individuals and organisations that provided us with, or gave permission to use, various of the data sets and diagrams that we have included. In particular we acknowledge the following sources for permission to reproduce or modify diagrams: British Medical Association (Figure 1.1); Institute of British Geographers (Figures 1.3 and 1.5); Royal Geographical Society (Figures 1.6 and 1.7); Wolf Dieter Rase (Figure 2.1); Ian Bracken and Chris Webster (Figure 2.3); American Geophysical Union (Figures 5.2, 5.21 and 5.22); Daniel Dorling (Figure 7.2); David Martin (Figure 7.5); Waldo Tobler (Figure 9.5); Ohio State University Press (Figure 9.3) and Elsevier Science Ltd (Figure 9.6). We also thank all those concerned with the sources of our data sets and hope that due credit has been given to them in the references we provide in the 'Further reading' sections, throughout the book.

We would also like to thank more specifically, a number of individuals who have helped in various other ways. Assistance in preparing data and carrying out analyses was provided by: Barry Rowlingson, Isobel Naumann, and David Howes (Lancaster University); Peter Fryers and in particular Ken Powell (Exeter University). We are most grateful to them for assisting us. Many of the figures were drawn by Nicky Shadbolt of the Department of Geography, Lancaster University, and we thank her for her efforts on our behalf. Finally, we are grateful to various colleagues for other assistance, either in the form of suggestions, helping with theoretical and computer problems, reading sections of the book, or as sources of intellectual stimulation (or all of these!). In particular, we thank Robin Flowerdew and Peter Diggle (Lancaster University); John Hinde, Wojtek Krzanowski, Alan Munford and David Smith (Exeter University).

Part A

Introduction

1

Spatial data analysis

Here we give an introduction to spatial data analysis and distinguish it from other forms of data analysis. We identify different classes of spatial problem and types of spatial data, and draw a loose distinction between techniques for visualising, exploring and modelling of such data. We then explain the scope and organisation of the book in terms of these distinctions and introduce some fundamental issues related to spatial analysis and modelling. We conclude with a short discussion of some practical problems which confront the spatial analyst and which we urge our reader to bear in mind throughout subsequent chapters.

1.1 Introduction

Imagine you work for the research department of a major political party and have been asked to look at the relationship between support for that party and levels of personal income. You have survey data from a sample of electoral districts which includes the percentage support for your political party and corresponding average per capita personal income. Since you are the kind of person that we would expect to read this book—that is, you have at some time taken an introductory course in statistical methods, even though you have not necessarily specialised in the subject—we hope that you would feel fairly confident about assessing the nature and strength of any relationship between these two variables. You remember something about relevant basic techniques that might be of use. For example, perhaps you might produce a scatter plot of the two variables against each other and, if there appears to be a reasonable degree of association, you might go on to calculate a correlation coefficient between them and formally test that its value is statistically significant. You may wish to model more precisely the quantitative nature of the relationship, maybe by fitting a linear regression between the two variables, and so on.

These kinds of general purpose data analyses will undoubtedly give some insight into the strength of the relationship in which you are interested; but you will probably be rightly wary of placing too much confidence in the results. You realise that there will be a whole host of other factors affecting voting preference, which you have not accounted for, and which may be confounding any relationship that you have established. Some of these you may be able to address by straightforward extensions to your earlier analysis, for example by including additional 'explanatory' variables into your regression model. However, one factor which may not immediately strike you as a problem, but in which we are very interested in this book, is that the observational units you have been analysing (electoral districts) have a spatial configuration—some may neighbour others, some may be a long way apart—and this may be relevant to your analysis.

It is very likely, for example, that both party support and per capita income in one district are related to those in neighbouring districts, in addition to any relationship that they may have with each other. The reasons for this may be too complex to disentangle and account for separately in your model except by reference to the geographical proximity of the districts as a surrogate measure. Even though you are not interested in a spatial question, one may have entered through the 'back door'. Most elementary general purpose data analysis makes the fundamental assumption that the observational units analysed represent *independent* pieces of evidence about the relationship under study. In your case, depending on your particular sample of electoral districts, the values of the variables which you are analysing may well be *spatially correlated* across observational units. In essence you may have less independent evidence about the relationship than you think and somehow you need to allow for this in your analysis.

In fact, you are in a similar, although possibly more complex, situation to a colleague of yours who has been asked to explore the way in which overall support for the party is related to changes in national economic growth. Her problem is that she is, by necessity, forced to examine any possible relationship in these variables by reference to data for successive time periods. However, it is very likely that party support in any time period will be partly a function of what happened in preceding time periods, in addition to any relationship with economic growth in the current period—data for her observational units (time periods) are serially or *temporally correlated.* If she were to analyse such data using standard methods which are intended for independent observations (such as the views of different, randomly selected individuals in an opinion poll), then she may well come to misleading conclusions—similarly, for you and your analysis. In her case the possibilities for the structure of such temporal correlation are relatively straightforward—time is one dimensional and the direction of correlation will be backwards in time, the only question being how far. Your case is possibly more complex—any spatial correlation in your data occurs in two dimensions and no particular direction is ruled out *a priori.*

She needs to consult a book on *time series analysis*. You, we presume, have already purchased (or borrowed!) this one on *spatial data analysis* and commenced reading it. We hope you will persevere, because this book should not only prove useful to you in modifying your analysis of non-spatial hypotheses to account for the use of data which are spatially or geographically referenced, but will also help you to address explicitly spatial questions which are not discussed at all in many elementary texts on data analysis—such as forecasting the effect that forthcoming changes in the boundaries of electoral districts might have on the party's fortunes in future elections.

Furthermore, if your area of work or interest is outside party politics, you will be pleased to know that we have not written this book solely to enhance the promotional prospects of political party researchers! We will be concerned generally with ways of analysing all varieties of data in a spatial context. Here are some other examples of the kinds of problem we are interested in, just to set the scene. We hope that you will agree that they are all important, practical, problems.

- Seismologists collect data on the regional distribution of earthquakes. Does this distribution show any pattern or predictability over space?
- Public health specialists (epidemiologists) collect data on the occurrence of diseases. Does the distribution of cases of a disease form a pattern in space? Is there some association with possible sources of environmental pollution? Is there any evidence that a particular disease is passed on from one individual to another?
- Police wish to investigate if there is any spatial pattern to the distribution of burglaries. Do the rates of burglaries in particular areas correlate with socio-economic characteristics of those areas?
- Environmental scientists (but also others concerned with 'image' data) collect data obtained from satellites. Such data may be very 'noisy'. Can we filter out the noise to reveal underlying pattern?
- Geologists wish to estimate the extent of a mineral deposit over a particular region, given data on borehole samples taken from locations scattered across the area. How can we make sensible estimates?
- A groundwater hydrologist collects data on the concentration of a toxic chemical in samples collected from a series of wells. Can we use these samples to construct a regional map of likely contamination?
- Retailers wish to use socio-economic data, available for small areas from the population Census, to assess the likely demand for their products if they open, or expand, an outlet. How are we to classify such areas?
- The same retailers collect information on movements of shoppers from residential 'zones' to stores. Can we build models of such flows? Can we predict changes in such flows if we expand an outlet or open a new one?

These examples illustrate the breadth of application of spatial data analysis, and suggest that the subject is of relevance in many different fields. Whether we are geographers, statisticians, economists, sociologists, epidemiologists, planners, biologists or environmental scientists, we are often faced with problems of a spatial nature, or with problems that involve spatial data. Our aim in this book is to show how the spatial nature of the data can be recognised explicitly and, if necessary, properly incorporated into our analyses.

Clearly, some of these disciplines will find a spatial dimension to their analyses more important than others. For example, geologists interested in searching for mineral deposits and predicting the volume of ore in rock have seen a major branch of their subject (and of spatial data analysis), 'geostatistics', emerge around this problem. Similarly, geographers have an obvious interest in spatial analysis techniques. For other researchers, such as sociologists and botanists, the spatial dimension, although often a necessary consideration in analysis, may be of secondary rather than primary importance in the sorts of problems they study.

This wide and varying interest in spatial data analysis has meant some disciplines have tended, independently of others, to develop their own distinctive styles of analysis. As a result, terminology and notation differ from discipline to discipline and inevitably a fair amount of 're-inventing the wheel' has gone on. This means that the subject of spatial data analysis is somewhat scattered throughout the literature of various disciplines and can appear, particularly to those coming to the field for the first time, as a collection of *ad hoc* techniques, with little sense of coherence or underlying theory, and can consequently seem rather difficult to get to grips with. This is not really so, as is well demonstrated by a number of recent unifying texts on the subject. However, such texts tend to require a fairly advanced statistical background. We hope in this book to follow their example and emphasise the coherence of the subject by structuring our discussion accordingly, but to do so at a more introductory level, addressing an audience which may not be so statistically specialised.

In the remainder of this chapter, we commence this task by first defining more precisely what we mean by spatial data analysis and what distinguishes it from its non-spatial counterpart. We try to convince our reader that 'space can make a difference'; in other words, that the recognition of a spatial dimension in analysis may yield different, and more meaningful, results, than an analysis which ignores it. From this we are able to move on to consider a set of illustrative case studies, which set the scene for the different classes of spatial problem addressed in the various subsequent parts of the book; each of which is related to one of these particular types of spatial problem. We follow this by a general discussion of the different types of spatial phenomena and spatial relationships that may arise in analysing any of these classes of problem. We then introduce some basic general concepts which we will apply in each part of the book to further structure the spatial analysis methods presented. In particular, we draw a distinction, which is useful, although not cast in tablets of stone, between methods which allow us to *visualise* spatial data, those which allow us to *explore* any structure and suggest possible hypotheses and finally, those that involve more formal statistical *models* of such data. Discussion of the latter gives us an opportunity to introduce some important ideas concerning the statistical modelling of spatial phenomena, which arise in different contexts throughout the book. We conclude the chapter with a short discussion of a number of important practical problems which may confront the spatial analyst, acknowledging that many of these will never be resolved by analytical techniques alone, but will ultimately involve the insight, experience and more subjective judgement of the analyst in the particular case in question. We encourage the reader to bear these in mind in relation to any of the subsequent methods presented in the book.

1.2 Spatial versus non-spatial data analysis

In broad terms one might define spatial analysis as the quantitative study of phenomena that are located in space. However, it would indeed be ambitious to attempt to cover such a broad field in one book! We have therefore decided to limit our discussion in the chapters which follow, to that important subset of the subject which we shall refer to as *spatial data analysis*.

The particular distinction which we draw between *spatial analysis* and *spatial data analysis* needs some clarification. We are concerned in this book with the situation where observational data are available on some process operating in space and methods are sought to describe or explain the behaviour of this process and its possible relationship to other spatial phenomena. The object of analysis is to increase our basic understanding of the process, assess the evidence in favour of various hypotheses concerning it, or possibly to predict values in areas where observations have not been made. The data with which we are concerned constitute a sample of observations on the process from which we attempt to infer its overall behaviour.

By defining spatial data analysis in this way we place ourselves firmly in the area of statistical description and modelling of spatial data and accordingly restrict attention in the book to a particular set of methods. In doing so we have to exclude from our coverage some important quantitative methods which would be included under the more general heading of spatial analysis. For example, although we discuss very briefly various forms of network analysis, such as those concerned with routing problems, minimisation of transportation costs or the optimal siting of facilities, we do not go into great detail. These classes of problem do of course involve spatial data, but not primarily observational data of the type discussed above, and understanding, explaining, or predicting the data is not the objective of the analysis; these problems instead involve mathematical optimisation techniques. There are clearly areas where such methods do interact with those with which we are concerned and have defined as spatial data analysis. For example, network analysis may be useful to compute travelling times, which are then used as measures of spatial proximity in the modelling of some observed spatial process of interest. Alternatively, models of observational data on movement of people or goods between locations, may provide necessary input to methods concerned with the optimal siting of facilities. We shall endeavour to point out such links and, where appropriate, will refer the reader to references on relevant spatial analysis methods beyond the scope of our definition of spatial data analysis.

Having clarified what we mean by spatial data analysis as opposed to spatial analysis more broadly conceived, we need to make explicit a further distinction at the outset—that between *spatial* data analysis and *non-spatial* data analysis. We do not intend in this book to discuss all forms of statistical analysis that may be useful in relation to data that happen simply to be located in space. This would require us to present (yet again!) many of the standard techniques found in numerous general purpose texts on statistical data analysis. We have no wish to do this. In fact, we assume that our reader already has some familiarity with these sorts of techniques. Rather, we wish to focus on modifications, extensions and additions to such techniques which consider explicitly the importance of the locations, or the spatial arrangement of the objects being analysed; in other words, on *spatial* as opposed to *non-spatial* data analysis.

It is unnecessary for us to become too pedantic about precisely where this dividing line between spatial and non-spatial data analysis actually lies. We have no wish to become involved in theoretical 'hair splitting' between what is, and what is not, a spatial statistical method. For our purposes it will suffice to say that spatial data analysis is involved when data are spatially located *and* explicit consideration is given to the possible importance of their spatial arrangement in the analysis or in the interpretation of results.

A good example of the distinction we are trying to convey comes from the field of research known as island biogeography. Consider the relationship between number of plant species and geographical area for a set of small islands. There is a wealth of empirical evidence to suggest that the logarithm of the number of species is related to the logarithm of the area of the island. One explanation of this kind of empirical relationship is simply that as area increases there is a greater probability of a range of available habitats. Spatial data analysis is not necessarily involved at this stage; the mere fact that the units of observation (islands) are locational, or that one of the variables involved is geographical area, does not itself make the analysis a spatial one. However, other theories are more explicitly 'spatial'. We might expect, *a priori*, that the isolation of an island is an important factor, in terms of its distance from other islands or from a continental area. Indeed, some authors have considered modifications to estimation of the relationship between species number and area which allow for such factors. Work on the distribution of bird species in Pacific islands, for example, shows that isolation, in terms of distance from New Guinea, does indeed reduce the number of species. Such analysis is more clearly spatial, since the relative locations of the spatial units (their proximity to the mainland) are being exploited in the analysis.

The reason for our intended focus in this book on spatial, as opposed to non-spatial, forms of data analysis, is that when spatial data are involved the former often yield different and more meaningful results than analyses which ignore the spatial dimension. We tried to convince our imaginary political researcher of this in the introduction to this chapter. Let us now consider in more detail a further example to consolidate the point. We do not set this up as an exemplar of 'ideal' spatial data analysis; indeed, there are shortcomings. But it does illustrate well the importance of a spatial perspective when dealing with spatial data.

Consider the very geographical problem of trying to model spatial variation in precipitation in California. Suppose we take a set of 30 monitoring stations, distributed across the state as in Figure 1.1. For each of these we have recordings of: average annual precipitation (the 'response' variable of interest); and altitude, latitude, and distance from the coast, each of which is a possible covariate that might explain the variation in precipitation.

We include the data, along with a simple map, on the disk we supply with this book, as we do with other data sets we will be using in later chapters. Of course, to 'look' at this data you will need some software. You may already be familiar with simple mapping packages, or even more advanced software for handling geographical data. If so, you can 'import' the data we supply into such software. If not, then you can use the software that we have provided with this book to 'view' the data sets. In Chapter 2 we shall be looking in detail at using computers to view, manipulate and analyse spatial data, and at that time

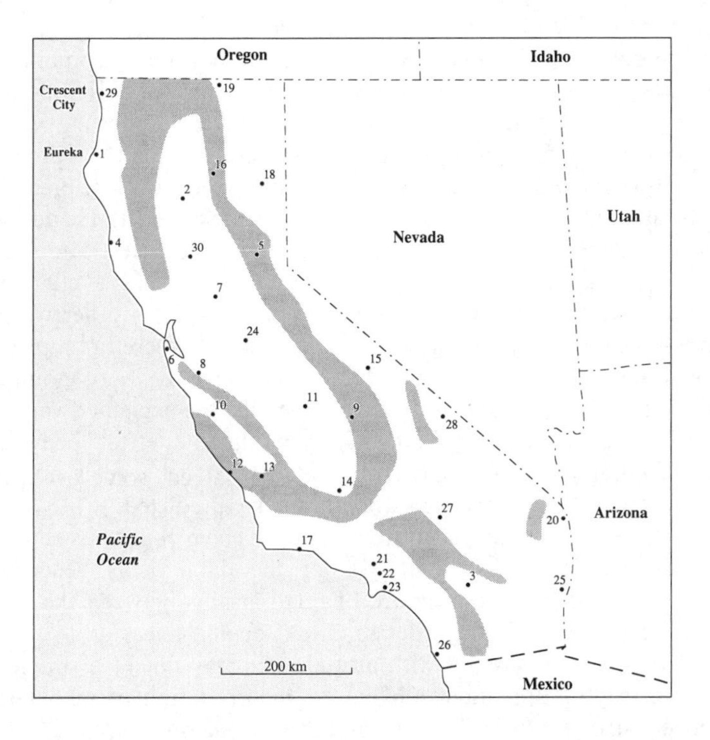

Fig. 1.1 Locations of rainfall measurement sites in California (shaded areas are mountains)

you can place the software package that we have provided into the wider context of what is currently possible and potentially desirable in terms of computing developments related to our subject. It will become apparent there that this is an area of rapid technological change and in no way do we wish to tie our text to any specific computer package. Therefore, throughout the book we avoid specific reference to use of our software, except in the series of computer exercises that accompany chapters. This is a book about spatial data analysis, and not about any particular software package. However, in the computer exercises we do encourage the reader to try out some of the ideas that we have described in the text, using our software and the data sets and maps that we have provided with it. In the course of discussing these exercises, we do become involved in specific details of using our package. At points in the text where there is a relevant computer exercise for the reader to refer to, we simply place a symbol in the margin to indicate this, such as the one which you see here. If you wish to follow our scheme you may like to refer to our first exercise now. If you are the kind of reader who prefers not to be distracted with our practical computer exercises, or wishes to use alternative computer software for spatial analysis and mapping, you can simply ignore the symbols in the margin, and rest assured that we will not worry you with details of the exercises in the main text.

Returning to our main discussion, in the paper that originally reported the Californian rainfall data, Peter Taylor fits a standard multiple regression model to the data. This involves trying to find the best linear combination of the covariates that explains variation in precipitation. Taylor finds that all three covariates are significant predictors of rainfall and about 60 per cent of the variation in precipitation is explained by them. Taylor then proceeds to map the residuals (the differences between the observed values of precipitation at the stations and those predicted by his three-variable model). He does this in order to see if there is any spatial pattern to these differences. The mountain ranges in the area are also indicated in Figure 1.1 and he finds a 'clustering' of negative residuals on the leeward side of the mountains. In other words, the model over-predicts precipitation at these locations. He therefore introduces a fourth covariate, where stations are coded 1 if they lie in the lee of mountains, 0 otherwise. When this variable is added to the model the explained variation in precipitation rises to 74 per cent, the rain shadow effect being highly significant. Mapping the residuals from this new model reveals no obvious spatial patterning, though two sites are still poorly predicted.

In an extension to the analysis of these data, Kelvyn Jones has shown how further, graphical, exploratory analysis of the data yields benefits. One modification is to examine the nature of relationships between the response variable and each covariate, in order to check for non-linearity; for example, although precipitation increases with altitude this may not be a simple linear increase with height. But, interestingly, the incorporation of an interaction effect, between distance from the coast and the rain shadow variable, generates a particularly good fit. The final model suggests that precipitation will be lowest when a station is both well inland *and* within the rain shadow.

Although there appears to be no obvious spatial relationship in the final residuals, missing from the analysis reported so far is any formal check that no *spatial correlation* or 'spatial persistence', remains in these unexplained variations in precipitation. We mentioned this consideration in our earlier example relating to support for a particular political party and levels of per capita income. In this case it may be that the amount of rainfall observed at one site will be similar to that measured close by, for reasons other than those explicitly incorporated into the model presented so far. If so then we may be able to improve our model by taking this into account in some way. We shall see in later chapters exactly how to employ methods which seek to allow for such residual spatial dependence. At this stage the point we wish to make is simply that there are possibly yet further adjustments that we might have to make to our precipitation model to allow for residual rainfall levels at one site being similar to those nearby.

Ex 1.4

This seems to us to be a useful example of the importance of an 'awareness of space' when analysing spatial data. There is an emphasis on spatial exploration of both data and results, in the form of both mapping and graphical plots. This then informs the more formal model-fitting. Disparities between model and reality are mapped and interpreted in terms of the spatial arrangement of the observations and their geographical context. We could then go on to use the model in a spatially predictive sense, by generating predictions of precipitation at locations which are not themselves monitoring sites. In this way, there is a close interaction between the statistical modelling and spatial interpolation.

As our earlier illustrations have indicated there are several important classes of problem for which such an explicitly spatial perspective is useful in data analysis. In the next section we shall look in more detail at a set of four empirical studies, and relate each of these to the problem areas dealt with in the four subsequent parts of the book.

1.3 Classes of problem in spatial data analysis

We hope that the discussion so far has given our reader a clear idea of the kind of methods that we will be concerned with in subsequent chapters; in other words, how we have defined the subject of spatial data analysis for the purposes of this book. The next obvious question is how we intend to structure our discussion of the various techniques which fall under this heading. We want now to explain this by examining a range of particular case studies, relating each of these examples to particular parts of the book. As we said in the Preface, this is an applied rather than a theoretical book and so we will start as we mean to go on, letting the structure of the book be dictated by practical applications.

In each case we believe that the applications considered deal with quite important contemporary problems. We hope that this will have the additional effect of convincing our reader that spatial data analysis is very far from being a collection of 'techniques in search of problems'. As in many areas of science, there have been plenty of perhaps rather narrow or far-fetched applications of spatial data analysis. This is unfortunate, but it would be unfair to tar all spatial data analysis with a 'too technique-driven' brush on the basis of these. We hope the examples we give will clearly demonstrate this.

We start by considering the kind of problem dealt with in Part B of the book, the first of our four methodological parts. In July 1976 over 180 cases of a severe form of pneumonia were diagnosed in Philadelphia, USA. The incidence was particularly high among men attending a conference of the American Legion, many of whom were staying in the same hotel. Other cases tended to be among those who had watched a parade of Legionnaires from outside the hotel. The risk factor was eventually thought to be exposure to droplets of water from air-conditioning and water-cooling systems. A considerable body of research into the disease (now known as Legionnaire's Disease), including some of a spatial nature, has since been undertaken. In particular, some writers have suggested that cases tend to occur sporadically and that 'clusters', such as that in Philadelphia, are rather rare. How are we to assess whether cases are sporadic, that is, single cases with no association in time or space with other cases, or part of a cluster of cases in time and space? This was a question addressed by epidemiologists and statisticians in a study of the disease incidence in Scotland. The epidemiologist Raj Bhopal collected data on the residential locations of over 400 cases, mostly in Glasgow and Edinburgh. Month of disease onset was also recorded. Working with the spatial statistician Peter Diggle, a statistical test was developed to determine whether pairs of cases were closer in time and space than might be expected on a purely random basis. Clear evidence of such 'space–time interaction' was uncovered, suggesting that cases were not sporadic but part of clusters (Figure 1.2).

The question that then naturally arises is that of the risk factors with which such clusters might be associated. As noted above, it has been suggested that the locations of cooling and air-conditioning systems infected with the *Legionella* bacterium are significant. But before any laboratory analyses of such water supplies can be carried out, the spatial analyst has to pinpoint the existence of sets of cases that are linked to possible point sources of contamination. The sorts of methods that are used in cluster detection, both in space and in space–time settings, are, as we have said, the subject of Part B of our book. There we are concerned with data comprising a *point pattern*, the locations of particular point events, and we want to investigate whether the proximity of the events, their locations in relation to each other, represents a 'significant' pattern. If we find clustering, say, we may then be interested in the question of over what spatial and/or temporal scales this occurs, and in whether particular spatial aggregations, or clusters, are associated with proximity to particular sources of some other factor.

Ex 1.5

In the previous example the analysis was carried out on data that comprised a set of points (locations of disease incidence), to each of which a simple attribute (month of disease onset) was attached. Consider now a different form

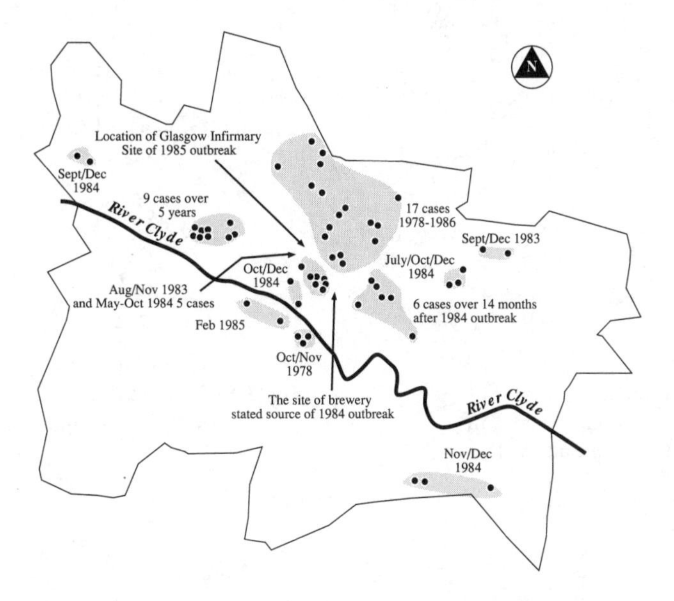

Fig. 1.2 Locations of cases of Legionnaires' disease in Glasgow

of spatial problem, typical of that dealt with in Part C of the book. Here, again, the data comprise a set of point locations, but attached to each is a continuous measurement, such as the volume of rainfall measured at each site over a given time interval. This is very similar to the Californian example we discussed in detail earlier. Our problem is not one of whether there is a pattern in the locations themselves; these are now simply the points at which sample measurements were taken. Rather, we are interested in understanding the pattern in the values at these sample locations. We can then perhaps use this understanding, to 'model' or estimate values of the variable of interest at locations that were not part of the spatial sample. This kind of problem is very common in the environmental sciences, where variables that describe soil, climate, geology, and so on, are inherently *spatially continuous*; in other words, air pressure, temperature, and soil acidity, to name but a few variables, are measurable at any location on the earth's (or perhaps only land) surface. But given that it is expensive, and ultimately absurd, to take measurements at vast numbers of locations, can we produce good estimates at locations that are not sampled?

A particular example of this is in attempting to derive rainfall maps for England and Wales using a body of techniques known collectively as *geostatistics*. As Margaret Oliver and colleagues report, annual rainfall measurements are sampled at 159 stations (Figure 1.3). Rainfall is an example of a variable that shows what hydrologists call spatial 'persistence' or what we

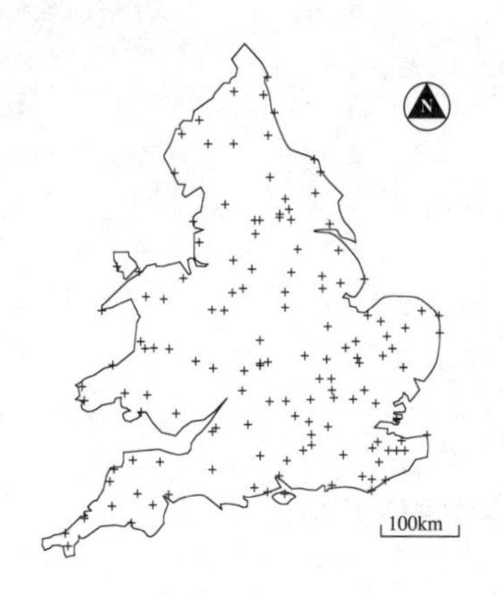

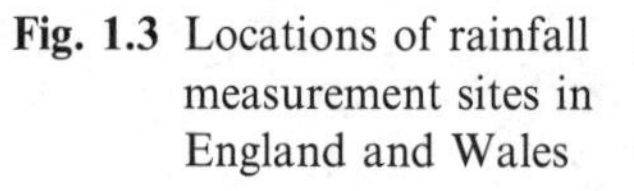

Fig. 1.3 Locations of rainfall measurement sites in England and Wales

Fig. 1.4 Contoured precipitation levels (mm) in England and Wales

Fig. 1.5 Prediction errors (mm) of precipitation in England and Wales

have already referred to as spatial correlation; in other words, the monitored rainfall at one station is likely to be similar to that 20 kilometres away, but probably less similar to the rainfall 200 kilometres distant. This pattern of correlation can be measured and used (possibly in conjunction with other covariates) in the estimation of rainfall elsewhere on the map. The resulting estimates can be contoured (Figure 1.4), revealing that annual rainfall is generally higher in the west. However, parts of Wales and the north-west (near the Lake District) are shown as rather drier than local experience suggests. This is because the estimates there are rather more uncertain; reference back to the distribution of weather stations reveals a paucity of stations in these areas, simply because the stations are used primarily for agricultural purposes and mountainous areas are under-sampled. The important point about the kind of analysis undertaken here is that it also generates an accompanying map of prediction errors (Figure 1.5), a map which highlights the confidence we may have in the predictions in different areas. It is these and related questions, in respect of spatially continuous data in general, rather than rainfall in particular, which form the subject of Part C of the book.

Ex 1.6

Let us turn now to a study that typifies the kind of spatial problem considered in Part D, where we deal with *area data*, spatial data that has been aggregated to areal units such as districts or census zones. We are all aware of the devastation that HIV infection and AIDS is wreaking in some parts of the world. The pool of HIV infection in parts of central Africa is huge. By the end of 1989 it was estimated that there were at least one million carriers of HIV in Uganda. There were over 12 000 reported cases, but this was thought to be a gross underestimate. Can we explore the distribution within a country such as Uganda, perhaps as a first step towards encouraging the implementation of policies to halt the spread? This is one of the tasks attempted by the geographers Andrew Cliff and Matthew Smallman-Raynor. Acknowledging the difficulties of putting together a data set to study as complex a topic as this, they mapped the incidence rates of AIDS in 34 districts of Uganda (Figure 1.6),

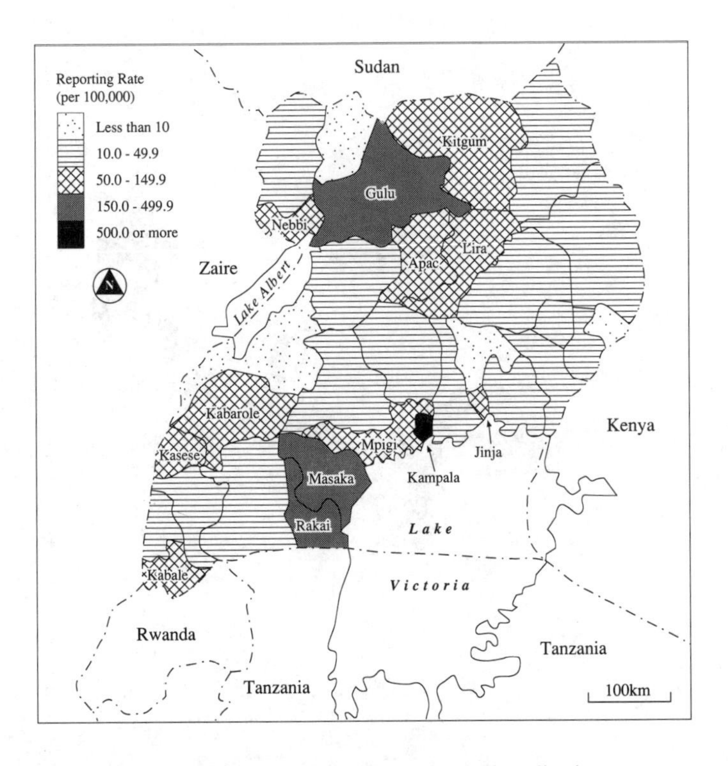

Fig. 1.6 Incidence of AIDS in Ugandan districts

a map that points to two foci of incidence. They then considered what hypotheses might explain the observed variation in this map.

One hypothesis is that the infection is acquired by those working in major urban areas, who then return home to infect partners in rural areas. Another hypothesis is that the infection is passed along major routes or 'corridors'. A third hypothesis is that involving contacts between soldiers and prostitutes; given that Uganda has been wracked by civil war during much of its recent history we surely cannot discount such factors? Cliff and Smallman-Raynor set out to test these various hypotheses, assembling an admittedly imperfect data set, but one which is probably as good as can be constructed. This involves explanatory variables such as accessibility to roads, in and out-migration rates, and army recruitment. Only the last of these variables can explain a significant proportion of variation in AIDS incidence, offering confirmation of the third hypothesis. However, the level of explanation is not high; there is considerable residual variation. This can be mapped (Figure 1.7) in order to see if there are clusters of districts with more AIDS cases than the model predicts. The residuals seem to be spatially quite random and this suggests the existence of non-spatial local factors that account for higher than expected rates in these areas. At this stage, spatial data analysis hands over to field workers, such as public health specialists, who can begin to interpret such residual variation. No-one, least of all the present authors, would be so naive as to suggest that

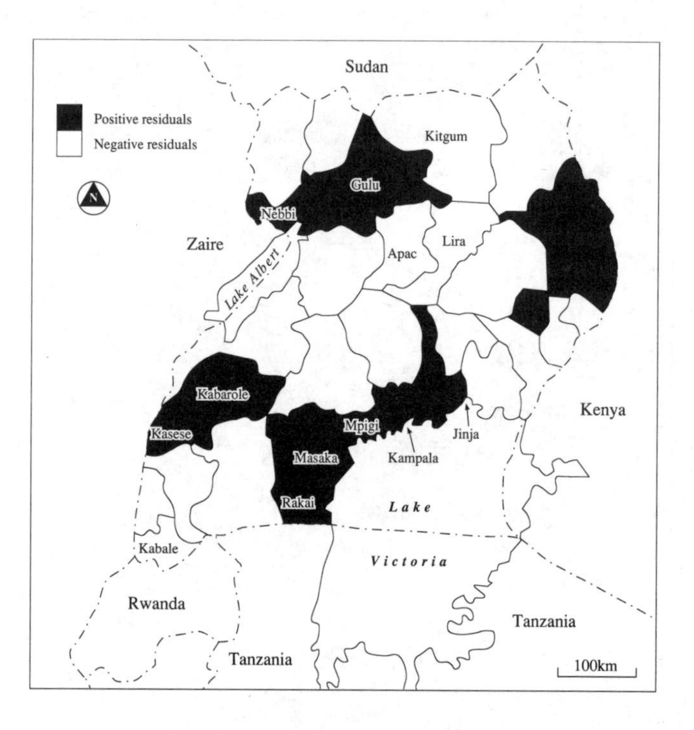

Fig. 1.7 AIDS residuals in Uganda

this kind of spatial data analysis can provide any definitive answers to the AIDS problem in Uganda; but it does offer some prospect of suggesting policies for controlling the pandemic. It is problems such as these and many others related to the analysis of area data, which form the subject of Part D.

Finally, we outline the kind of problem considered in Part E of the book. Planners, both in the private and public sectors, are often faced with problems of finding suitable locations for new investment. Whether dealing with sites for new supermarkets, or sites for such public facilities as hospitals, libraries or sports centres, planners need information about usage of existing facilities in order to make sensible investment decisions. One way, though rather crude, of seeing whether existing coverage is adequate is simply to draw circles of fixed radius around the sites and then to inspect the map for 'gaps'. But this takes no account of current human spatial behaviour or trip-making patterns. In other words, an important piece of information is the patterns of flows from residential locations to facilities. We can envisage this as a table, or matrix, of flows, from residential zones (rows of the matrix) to locations of existing facilities (columns). Can we model the observed pattern of flows?

This kind of problem goes under the general name of the modelling of *spatial interaction data* and has spawned a huge literature over the past thirty years. An interesting application is to the problem of modelling the demand for sports facilities. A specific example is work done by the geographers Goodchild and Booth on the demand for swimming pools in the city of London, Ontario. A set of 460 Census enumeration districts were used as residential zones, and a 10 per cent sample of visitors to 11 swimming pools was used to draw up a (460×11) matrix of flows. The problem is to construct a simple model of these flows; if this is possible we can then assess the effect of opening a new pool somewhere else, or the impact on existing flows if new housing is built in one or more zones. Goodchild and Booth show what the impact on flows might be if two new pools were constructed.

Such models are typically predicated on the assumption that the flow from a residential zone to a facility is directly related to the demand from that zone (such as population size) and the 'attractiveness' of the facility. In the current example, demand was weighted by the age distribution of the population, so that zones with the same total population generated different demands if their age structures differed. Spatial interaction models also assume that such interaction declines with distance, so that people are more likely to patronise a pool which is closer than one which is further away. This 'distance decay effect' can be built into the model in a variety of ways. When the model is fitted to observed data the importance of the distance decay effect can be estimated, as can the differential attractiveness of the swimming pools. Precisely how such models are constructed, and estimated statistically, we shall explore in Part E of the book.

Ex
1.8

These, then, are outlines of four case studies that deal with problems involving spatial data analysis. They each represent one of the classes of problem with which we deal in the book. To repeat, the first class of problem (Part B) deals with data for a set of point events, or a *point pattern*; sometimes these points will have simple attributes associated with them distinguishing one kind of event from another, but the important issue is that it is the spatial configuration or arrangement of the events themselves that is of interest. The second class of problem (Part C) arises where again we have a set of point sites, but the pattern of these locations is not itself the subject of analysis. Rather, there is a variable (or more than one variable) measured at these sites, and the problem is to understand the process generating these values and possibly then to use this information to make predictions elsewhere on the map. We refer to this as *spatially continuous data*. The third class of problem (Part D) concerns *area data*, data that have been aggregated to a set of areal units, such as counties, districts, census zones, and so on. We have one or more variables whose values are measured over this set of zones. The problem is to understand the spatial arrangement of these values, to detect patterns, and to examine relationships among the set of variables. The final class of problems we examine (Part E) are those involving *spatial interaction data*, data on flows that link a set of locations (either areas or points). We wish to understand the arrangement of flows, to build models of such flows, and maybe to use this information in making predictions about how the flows may change under certain scenarios.

We hope that this explains the structure of our subsequent chapters and that by motivating this in terms of 'real life' case studies we have convinced the reader that the methods we will present are not abstract 'spatial science', but, rather, a set of techniques that can help to illuminate the world and solve real problems. Each of the parts of the book that we have mentioned is relatively self contained, but there are general ideas which are common to all of them and we devote the rest of this chapter to introducing these.

1.4 Types of spatial phenomena and relationships

As we are writing a book on spatial data analysis we need to say something generally about the different types of spatial phenomena and spatial relationships that may be involved in our analyses.

Some authors have drawn a distinction between a *discrete* and *continuous* view of spatial phenomena. That is to say, a distinction is made between a conception of space as something filled with 'objects', and a view of space as covered with essentially continuous 'surfaces'. The former has been labelled an *entity* view of space, the latter a *field* view.

In the former the sorts of spatial phenomena being analysed are usually conceptualised as *points, lines* or *areas*. This means, for example, that we might think of things such as plants, people, shops, soil pits, the epicentres of earthquakes, and so on, as being located at points on the earth's surface. Of course, how appropriate this is depends upon the scale of investigation. If we are looking at the distribution of urban settlements at a national scale it is reasonable to treat them as a distribution of points; at the scale of a small region it becomes less sensible. Phenomena such as roads, streams, fault lines, can be treated as lines, though again there is scale dependence. On large scale maps of urban areas roads have detectable width, while the Amazon is hardly a 'line' where it flows into the Atlantic. Lines also mark the boundaries of areas and they represent points of equal height on maps of surfaces such as relief. By areas we understand those entities that are administratively or legally defined, such as counties, voting areas, health regions, and so on, as well as natural areas such as soil or vegetation zones on a map.

In a field view the emphasis is on the continuity of spatial phenomena. If we think of phenomena in the natural environment, such as temperature, relief, atmospheric pressure, soil characteristics, and so on, then such variables or data 'layers' can, in principle, be observed and measured anywhere on the earth's (or at least the land) surface. In practice, of course, such variables are 'discretised'. Thus relief might be sampled at a collection of sample points and represented as a series of contours, while temperature is sampled at a set of sites and represented also as a collection of lines (isotherms). Soil characteristics might also be sampled at a set of discrete locations and represented as a continuously varying field. In all cases, from discrete sampling an attempt is made to represent underlying continuity.

Whilst it is conceptually useful to keep in mind this distinction between an entity-oriented as opposed to a field-oriented view of spatial phenomena, we will not find it necessary to place a great deal of formal emphasis on which view we adopt in the various parts of this book. Certainly, in Part B we adopt an 'entity' view in that the objects we are interested in are viewed as occupying discrete point locations on the earth's surface. However, in Part C we switch to what might be called a 'field' view, where measurements at a set of sampled locations are used to construct a continuous spatial surface. Then, in Part D, we deal with a fixed set of areal units. Although some of the variables that might be measured in such units are inherently continuous (such as population density) for our purposes we shall mostly be treating the zones as fixed and discrete, again an 'entity' view. The same convention is adopted in Part E, where a discrete set of locations (typically areal units, but often regarded as points for the sake of computing distances between locations) is used as a framework for analysis.

In the 'entity' view spatial objects have features or *attributes* attached to them; we may also speak of the feature associated with a field as an attribute which varies continuously over space. Such attributes may be measured according to one of the classic measurement scales: nominal, ordinal, or interval. Points may be distinguished from each other in terms of type (e.g. tree species), ordered class (e.g. hamlets, villages, towns) or along a continuous measurement scale (e.g. depth of borehole). The same is true of lines (a class of object that we do not explicitly treat in the book). Streams may be clean or polluted; or ordered as first, second, or higher order streams according to one of the conventional ordering schemes; or have continuous data attached to them, such as volume of flow. Areas of land have 'land use' classes assigned to them (nominal data), but the land may be classified according to quality (ordered data). Other areas, such as census tracts, will typically have continuous data, such as population density or levels of economic activity, attached to them.

As we noted earlier in the chapter, if we deal with attributes alone, ignoring the spatial relationships between sample locations, we cannot claim to be doing spatial data analysis, even though our observational units may themselves be spatially defined. Thus, although the attribute data are of fundamental importance, divorced from their spatial context they lose value and meaning. In order to undertake spatial analysis we require, as a minimum, information on location and usually information for both locations and attributes, regardless of how these attributes are measured. If we wish only to study the spatial arrangement or pattern of entities then this is essentially a geometric question and we need only collect data on the locations of the entities. If, however, we wish to compare the arrangements of different types of entities or to study spatial pattern in measurements taken at locations, then we shall need to make use of the attribute information as well.

It is important to note that while we make a distinction between different types of entities such as points and areas, it is sometimes sensible, and often necessary, to transform one class of entity into another. This notion of 'transforming' one class of objects into another assumes particular significance in the field of Geographical Information Systems (GIS), which we discuss in Chapter 2. For example, some use is made in spatial data analysis of *Thiessen polygons*, which are areal units created around a set of points, such that all locations within a polygon are closer to the point object used to define that polygon than to any other such point object. Conversely, we might have a set of areal units and wish to represent these very simply by a set of points, one for each polygon. Typically, the geographical centre or *centroid* of an areal unit is used. This is useful if one wishes to measure distances between areal units, as may be required for example in applications of spatial interaction modelling.

More generally, this notion of 'new objects for old' relates to the subject of 'relations' between classes of map object. Such relationships can be of many different types. Consider, for example, relations between point objects. As we have seen, one branch of spatial data analysis seeks to describe and analyse the spatial arrangement of points. This involves measuring distances between points; distance is a spatial relation. As a second example, we might wish to compare two point distributions, such as the distribution of a set of disease 'cases' with a set of healthy 'controls'; this too will involve distance measurements.

Areal data give rise to numerous examples of spatial relationships. For example, simple information about spatial adjacency may be of interest. More commonly, there are many instances when knowledge of spatial proximity is linked to attribute information. Many times we wish to assess whether areas that are close together on the ground have similar values on one or more attributes. For example, do sets of neighbouring health districts tend to have the same mortality experience? Do adjacent 'pixels', the small areal units used as data-recording units in remote sensing, tend to have similar electromagnetic reflectances?

In addition, objects of one type may be 'related' to a different class of objects. We might, for instance, wish to see if there is an association between a set of points and a set of lines. One example is to test for the association between the occurrence of mineral deposits and configurations of geological lineaments. Another might be to test the hypothesis that there is a link between childhood leukaemia and proximity to high voltage power lines. In other contexts, a test of the association between a set of points and a set of areas might be required. For instance, is there a relationship between a set of plants of one species and soil type? Is there any relationship between the incidence of Alzheimer's disease (given by a set of residential locations of cases: a point pattern) and the presence of aluminium in water sampled in a set of water supply zones (areal units)?

We will not be concerned in subsequent chapters with explicit further development of the kind of concepts and relationships introduced in this section. However, they are important, since they form a fundamental backdrop to spatial data analysis. Such issues are best considered in depth under the heading of Geographical Information Systems (GIS). This is not a book about GIS, even though we see a close partnership currently developing between GIS and interactive spatial data analysis. The nature of this partnership is explored in more detail in Chapter 2.

1.5 General concepts in spatial data analysis

So far in this chapter we have attempted to convey what we mean by spatial data analysis and to explain what classes of problem we will be concerned with in each of the subsequent parts of the book. We now turn our attention to how we intend to organise the material presented *within* each of these parts.

Spatial data analysis, as we have defined it, involves the accurate description of data relating to a process operating in space, the exploration of patterns and relationships in such data, and the search for explanations of such patterns and relationships. As such, the subject involves a collection of analysis methods, some of which are oriented towards one end of this spectrum of activities, some towards the other. We have found this a useful basis upon which to structure the techniques and methods presented in each of the parts of the book.

We therefore draw a distinction between methods that are essentially concerned with *visualising* spatial data, those which are *exploratory*, concerned with summarising and investigating map pattern and relationships, and those which rely on the specification of a statistical *model* and the estimation of parameters. In this section we look briefly at each of these classes of method, setting out in broad terms what each involves and introducing some general ideas and concepts which relate to them.

We emphasise at the outset that the distinction between methods for visualising, exploring and modelling spatial data is useful, but not clear-cut. In particular, there is usually a close interplay of the three, with data being visualised initially, interesting aspects being explored, which then possibly leads to some modelling. The results of modelling may then be displayed again, further explored and perhaps the model refined as a result. The three approaches should therefore be seen not as 'either–or', but very much as interlinked in an interactive fashion. As is clear from the title of this book, we feel that the interaction between the analyst, the data, and possible models, is very important.

1.5.1 Visualising spatial data

An essential requirement in any data analysis is the ability to be able to 'see' the data being analysed. Plots of data and other graphical displays of various descriptions are the fundamental tools of the analyst concerned with seeking patterns in data, generating hypotheses and assessing the fit of proposed models, or the validity of predictions derived from them. The way in which data analysis is performed has been transformed in recent years by interactive computer packages which permit such views to be obtained quickly, easily and flexibly. Indeed many recent methodological developments in data analysis itself have been heavily influenced by these kinds of dynamic, graphical computing environments.

Spatial data analysis is no exception to this general rule and visualising spatial data means mapping. In a sense, the map is the spatial analyst's equivalent to the invaluable scatter plot in non-spatial analysis. As we shall see in more detail in Chapter 2, recent major advances in digital mapping, computer cartography and Geographical Information Systems now provide us with an environment within which to create such maps and explore spatial patterns and relationships quickly, easily and interactively.

Like any graphical techniques for displaying data, maps come in various forms, each useful in relation to different kinds of spatial data. For example, a simple dot map may be the natural tool for visualising a point pattern, whilst a zonally shaded, or choropleth, map may fulfil the same function in respect of area data. Some of these mapping techniques are so intuitive as to require little discussion, others may merit further explanation. In each part of the book we will devote a brief section to visualisation methods appropriate to the kind of data under discussion in that part. We emphasise that such techniques are not only useful in the spatial display of raw data, but can also be used, where appropriate, with transformed data or with other analytical results.

The demands made of the map in spatial data analysis do not necessarily require a polished product worthy of the cartographer's art; but cartographic considerations are not unimportant in using maps in spatial data analysis. Bad choices of map type, or the scaling used for data values, can lead to misleading conclusions being drawn from the display and can suggest inappropriate models for the process under study. We shall refer to such issues as we introduce particular visualisation methods.

1.5.2 Exploring spatial data

Exploratory methods in data analysis involve seeking good descriptions of data, thus helping the analyst to develop hypotheses about, and appropriate models for, such data. Such techniques are characterised by making few *a priori* assumptions about the data and many are designed specifically to be *robust*, that is, to be as resistant as possible to the effect of extreme data values or *outliers*. It comes as no surprise to learn, therefore, that many such methods emphasise graphical views of the data which are designed to highlight particular features and allow the analyst to detect pattern, relationships, unusual values, and so on.

In the context of spatial data analysis, the views that result from exploratory methods may be in the form of maps; others may involve more conventional plots. For example, as we discuss in Part B, some exploratory techniques, when applied to an observed pattern of point events, result in a contour map of the estimated intensity of occurrence of events over the whole study area; other exploratory techniques applied to the same set of events result in a graph designed to throw light on the degree of spatial dependence between the event locations. In each part of the book we will devote a section to exploratory methods which are appropriate to investigate various aspects of the kind of data under discussion in that part.

At this point the reader may feel that the distinction between exploratory methods and the visualisation techniques we discussed earlier is becoming rather blurred. If many exploratory spatial techniques result in different forms of maps, then how do they really differ from visualisation techniques? The dividing line is obviously somewhat artificial; we are really talking here about a continuum of methods of varying sophistication. In our mind the distinction is a question of the degree of data manipulation which the method involves. Suppose for example that we have age-standardised, cause-specific death rates in a number of administrative zones. A map of these raw rates, or simple transformations of them, is a visualisation technique. If we wish to smooth out local variations and see more clearly global trends we might produce a map of a 'spatial moving average' of the rates (where each rate is replaced by the average of itself and those in 'neighbouring' districts); we would class this as an exploratory technique, because it involves a significant degree of 'value-added' data manipulation.

A similar blurring occurs at the division between exploratory techniques and spatial modelling techniques, which we discuss in the next section. It would be convenient if one could simply say that exploratory techniques were distinguished by not involving any explicit model for the data under study. This is largely, but not wholly, the case. Several exploratory techniques involve informal comparison of some summary of the data with what one might expect this to look like if the data followed a particular model. So models do enter into exploratory techniques. Again, it is a question of degree, in this case that of how specific or restrictive the model involved may be and to what extent any comparisons made between data and model depend formally upon particular (and possibly incorrect) statistical assumptions. An illustration of the point is provided by the 'Geographical Analysis Machine' or GAM, developed by the quantitative geographer Stan Openshaw to detect clusters in point distributions of the incidence of diseases such as childhood leukaemia. This incorporates a technique which exhaustively compares the observed intensity of events in circles of varying radius, centred on a fine grid imposed over the study area, to that which would be expected in such regions if cases were distributed at random. Circles with 'significant' discrepancies are identified and retained for later display and investigation. This technique does involve a model—that for 'distributed at random'—and it performs repeated formal statistical comparisons with this model; but the validity of such comparisons does not depend upon the assumption of any specific alternative model. The technique is detecting 'clusters', not seeking to explain the process by which such 'clusters' may arise. Moreover, the use of exhaustive search cleverly avoids the need to specify over what spatial scale we expect clusters to occur. As such, this form of analysis makes very few *a priori* assumptions about the data and is fully in line with what we consider to be the spirit of an exploratory method.

1.5.3 Modelling spatial data

In some cases a judicious choice of exploratory analyses, combined with appropriate visualisation methods, may suffice to answer the questions which a researcher wishes to ask of a particular set of spatial data (or those which one is justified in asking given the 'quality' of those data). In other instances, however, there may be a requirement to formally test certain hypotheses, or to estimate with some precision the extent and form of relationships of interest. We are then forced to consider explicit statistical models for the data. Some readers may be unfamiliar with the idea of a statistical model and so we devote this section to introducing some basic ideas and terminology concerned with statistical models in general and then progress to focus on concepts which relate particularly to the modelling of spatial data. The ideas which we discuss here are fundamental to many of the analysis methods introduced in subsequent parts of the book, especially, of course, those in the 'modelling' section of each of these. Their importance is perhaps emphasised by noting that the words model or modelling have already been used over fifty times in this chapter.

Statistical models of one description or another are implicit in all formal statistical inference and hypothesis testing, even though the term 'model' may not be used explicitly in many elementary statistical texts. If you have not formally come across the idea before, but have experience with some basic statistical techniques such as 't' tests or regression, then it may be comforting to know that you have been involved in statistical modelling all along. We need to introduce the fundamental ideas and principles explicitly. We shall try not to be unduly formal about this, but we do need to explain them before we can go on to discuss important concepts concerned more particularly with statistical models of spatial data.

Since statistical models are concerned with phenomena which are *stochastic*, that is to say phenomena which are subject to uncertainty, or governed by the laws of probability, we will need a 'language' which allows us to represent such uncertainty mathematically. This is provided by the concept of a *random variable* and its associated *probability distribution*.

Random variables are used to represent stochastic phenomena mathematically; informally, we may think of them as simply being variables whose values are subject to uncertainty. Typically, we might represent the result of a single throw of a die, or the measurement of precipitation at a particular site in the UK, by a random variable say, Y. The name is somewhat unfortunate, since 'random' tends to imply that such variables are equally likely to take any of their possible values, whereas all that is really meant is that the values are subject to *some* form of uncertainty. One value or set of values may be more likely than others, but we still refer to the variable concerned as 'random'.

The relative chance of a random variable taking different possible values, is characterised by its associated probability distribution, so that we might refer to the random variable Y as having a probability distribution $f_Y(y)$. $f_Y(y)$ is a mathematical function which specifies the probability that particular values or ranges of values of Y will occur. In general, throughout this book we will attempt to adhere to a convention of representing random variables by upper case letters, and specific values of a random variable by lower case letters; so that Y may be a random variable that conceptually represents the result of throwing a die, but when it is actually thrown we get a particular observed value, y, of this random variable.

In general, random variables may be *discrete*, that is, only able to take a finite number of values (or an infinite set that can be put into a one-to-one correspondence with the positive integers); in this case $f_Y(y)$ is the probability that Y takes the specific value y. They may also be *continuous*, able to take any value within a continuous range, in which case $f_Y(y)$ is the *probability density* at the value y. The probability that a random variable Y takes values in some range (a, b) is therefore:

$$\sum_{y=a}^{b} f_Y(y) \qquad \text{if } Y \text{ is discrete}$$

$$\int_a^b f_Y(y)\mathrm{d}y \quad \text{if } Y \text{ is continuous}$$

We may also occasionally be interested in the *cumulative probability distribution* $F_Y(y)$ of a random variable Y, sometimes referred to as its *distribution function.*

This is simply a mathematical function which specifies the probability that Y takes any value less than or equal to the specific value y. Thus:

$$F_Y(y) = \begin{cases} \sum_{u=-\infty}^{y} f_Y(u) & \text{if } Y \text{ is discrete} \\ \int_{-\infty}^{y} f_Y(u)\mathrm{d}u & \text{if } Y \text{ is continuous} \end{cases}$$

We shall also be interested in the *expected value* or *mean*, of a random variable Y, or perhaps some function, $g(Y)$, of that random variable. As its name suggests, this is simply the 'average' value that we would expect Y or $g(Y)$ to take. It is therefore a weighted sum of the possible values, where the weights used for each value are the probability associated with that value. Therefore:

$$E(Y) = \begin{cases} \sum_{y=-\infty}^{\infty} y f_Y(y) & \text{if } Y \text{ is discrete} \\ \int_{-\infty}^{\infty} y f_Y(y)\mathrm{d}y & \text{if } Y \text{ is continuous} \end{cases}$$

or:

$$E(g(Y)) = \begin{cases} \sum_{y=-\infty}^{\infty} g(y) f_Y(y) & \text{if } Y \text{ is discrete} \\ \int_{-\infty}^{\infty} g(y) f_Y(u)\mathrm{d}y & \text{if } Y \text{ is continuous} \end{cases}$$

The expected value of one particular function of a random variable is often of particular interest; this is the expected squared deviation of the random variable from its mean. Clearly, this is a broad measure of how much its values tend to vary around their 'average'. It is therefore referred to as the *variance* of the random variable. Formally, $VAR(Y) = E((Y - E(Y))^2)$. The positive square root of this quantity is referred to as the *standard deviation* of the random variable.

All of these ideas generalise to the case where we are interested not just in one random variable but in more than one. If we have two random variables (X, Y) then we can speak of their *joint probability distribution*, $f_{XY}(x, y)$, which specifies the probability, or probability density, associated with X taking the specific value x, at the same time that Y takes the specific value y. As well as the mean and variance of Y or X, we may then also be interested in the expected tendency for values of X to be 'similar' to values of Y. A broad measure of this is the *covariance* of the two random values, defined as $COV(X, Y) = E((X - E(X))(Y - E(Y))))$. The covariance of two random variables divided by the product of their standard deviations is referred to as the *correlation* between them.

Informally, two random variables are said to be *independent* if the probabilistic behaviour of either one remains the same, no matter what values the other might take. In that case their joint probability distribution is simply the product of their individual probability distributions, so that $f_{XY}(x, y) = f_X(x)f_Y(y)$. If this is so then it follows that their covariance and correlation will be zero.

So much then for a very brief review of a 'language' with which to describe stochastic phenomena. We excuse our rather scanty coverage on the basis that we hope our reader already has some acquaintance with these elementary statistical concepts. Having established these ideas we can now turn to our primary concern in this section, that of statistical models.

A statistical model for a stochastic phenomenon consists of specifying a probability distribution for the random variable (or variables) that represent the phenomenon. Once this probability distribution is fully specified there is effectively nothing further that can be said concerning the behaviour of the phenomenon—by definition it cannot be specified to any greater degree of precision than the probability that particular values may occur.

In the case of modelling something simple, like throwing a die, which involves a single random variable, say Y, this means specifying a corresponding probability distribution, $f_Y(y)$. However, for more complex phenomena rather more can be involved. Consider, for example modelling levels of a photochemical oxidant such as ozone in a large rural region. The ozone level at each location, $\mathbf{s}$, in the region, $\mathcal{R}$, will vary during the day and from day to day according to some probability distribution, as illustrated in Figure 1.8.

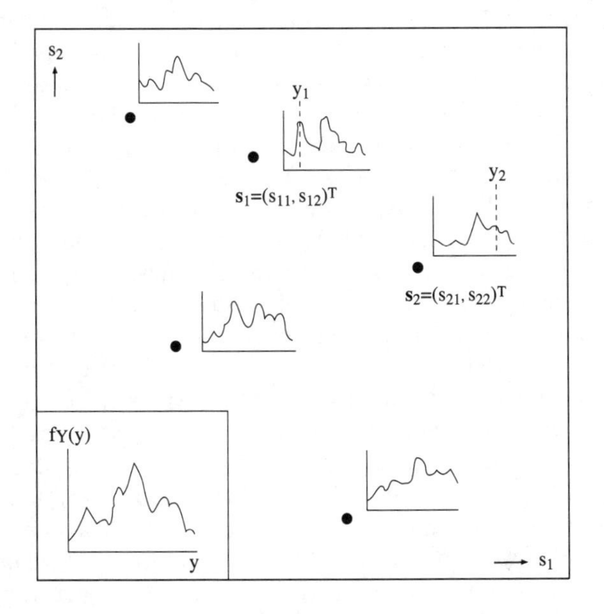

Fig. 1.8 Illustration of a spatial stochastic process

Note here in passing, the use of the (2×1) vector $\boldsymbol{s} = (s_1, s_2)^T$ to refer to a point location. This is simply a shorthand way of referring to the x-coordinate, s_1, and the y-coordinate, s_2, of a point. We will use this kind of notation throughout the book to denote point locations. In general, we will adhere to the convention of distinguishing vectors by bold typeface. Thus if we have two point locations we will refer to them by the two vectors $\boldsymbol{s}_1$ and $\boldsymbol{s}_2$, where, in terms of coordinates we imply, $\boldsymbol{s}_1 = (s_{11}, s_{12})^T$ and $\boldsymbol{s}_2 = (s_{21}, s_{22})^T$. For matrices we shall generally use upper case bold typeface. All vectors that we use will be column vectors unless we explicitly specify otherwise; this explains our use of the transpose operator '^{T}', in the expression $\boldsymbol{s} = (s_1, s_2)^T$. For the reader who is unfamiliar with vector or matrix notation and simple matrix operations, we list a suitable introductory reference at the end of this chapter.

To return to our main theme, the particular form of the ozone level probability distribution may well differ from location to location in $\mathcal{R}$, and furthermore the ozone levels at neighbouring locations may well be related in some way. For example, ozone levels at sites separated by 5 kilometres are probably quite similar, while those separated by 50 kilometres may be very different. Thus, to represent ozone levels in $\mathcal{R}$ we require a set of possibly non-independent random variables, $\{Y(\boldsymbol{s}), \boldsymbol{s} \in \mathcal{R}\}$. Such a set is often referred to as a *spatial stochastic process.* A complete statistical model for the ozone level in $\mathcal{R}$ involves specifying the joint probability distribution of every possible subset of these random variables. To say the least, this is a somewhat formidable task!

In general, then, how do we go about formulating a statistical model? In rare cases, like that of a throw of a die, the appropriate statistical model may be apparent purely from our theoretical knowledge about the phenomenon. For example, an obvious model for a fair die is: $f_Y(y) = \frac{1}{6}, \quad y = 1, \ldots, 6$. However, when we attempt to model phenomena more interesting than dice, like ozone levels, we are unlikely to know enough about the particular phenomenon to be able to fully specify a model in such a way. We must therefore rely on observational data to help us to arrive at an appropriate model. However, it should be appreciated that whilst data will help, alone they can rarely provide a definitive model. A typical data set in the ozone level example would consist of a set of observations $(y_1, y_2, \ldots)$, each at specific sites, $(\boldsymbol{s}_1, \boldsymbol{s}_2, \ldots)$ in $\mathcal{R}$. Note that strictly we should probably refer to our observed data values as $y(\boldsymbol{s}_i)$, since they are observations on the random variables $Y(\boldsymbol{s}_i)$; however, we shall mostly use the simpler notation y_i. Where it is convenient we may also refer to the random variable $Y(\boldsymbol{s}_i)$ simply as Y_i. This data set, $(y_1, y_2, \ldots)$, is often referred to as a *realisation* of the spatial process. It is just one observation (see Figure 1.8) from the joint probability distribution of the random variables $\{Y(\boldsymbol{s}_1), Y(\boldsymbol{s}_2), \ldots\}$, or in our simpler notation $\{Y_1, Y_2, \ldots\}$. Unfortunately, one observation does not give much information about a joint probability distribution, even if we were prepared to accept that this particular set of sites was 'typical' of sites in general in $\mathcal{R}$, which may itself be open to question.

In general, then, the specification of a model will involve using a combination of both data and 'reasonable' assumptions about the nature of phenomena. Such assumptions may arise, for example, from background theoretical knowledge about how one expects a phenomenon to behave, the results of previous analyses on the same, or a similar, phenomenon, or alternatively, from the judgement and intuition of the modeller. How 'reasonable' they are, in certain cases, can be assessed by exploratory analyses of aspects of the observed data appropriate to those particular assumptions. Once specified, they provide a basic 'framework' for a model. Usually this framework will amount to the specification of a general mathematical form for the probability distribution appropriate to the phenomenon, but one which involves certain *parameters* whose values are left unspecified. This general form is then further refined, or *fitted* (that is the values of the unknown parameters are *estimated*), by reference back to the observed data. The fitted model can then be evaluated, which may lead to modified assumptions and a different model being fitted and so on.

An illustration of this approach would be to apply a simple form of the familiar standard linear regression model, widely used in non-spatial analysis, to our ozone level example. An example of a model of this type would involve, firstly, the assumption that the random variables, $\{Y(\mathbf{s}), \mathbf{s} \in \mathcal{R}\}$, are independent; secondly, that their probability distributions only differ in their mean value, all other aspects being the same; thirdly, that this mean value is a simple linear function of location, say $E(Y(\mathbf{s})) = \beta_0 + \beta_1 s_1 + \beta_2 s_2$, where (s_1, s_2) are the spatial coordinates of the location $\mathbf{s}$; and fourthly that each $Y(\mathbf{s})$ has a normal distribution about this mean with the same constant variance σ^2. In short, our model says that $Y(\mathbf{s})$ are independent and the probability distribution of each is $N(\beta_0 + \beta_1 s_1 + \beta_2 s_2, \sigma^2)$.

Under such assumptions, we are relieved of the task of specifying a joint probability distribution for every subset of $Y(\mathbf{s})$, since independence implies this would just be the product of the distributions at each of the sites involved in any such subset. Furthermore, although each of these distributions differ, they do so only through a simple relationship involving the small number of parameters, β_0, β_1 and β_2. Hence, our observed data $(y_1, y_2, \ldots)$, at specific sites, $(\mathbf{s}_1, \mathbf{s}_2, \ldots)$ in $\mathcal{R}$, cease to be just rather unhelpful single observations from the joint probability distribution of the random variables $\{Y_1, Y_2, \ldots\}$, and become usable to estimate the unknown parameters of our postulated model—there is now the necessary 'replication' in observations to achieve this. We discuss how this parameter estimation is achieved in the next paragraph; the point here is simply that our assumptions provide a framework under which final model specification reduces to a problem of the estimation of unknown parameters. Furthermore, enough replication is present in the data to make this estimation feasible.

Given such a model framework exactly how are unknown parameters to be estimated from the data? In other words, how do we fit a proposed statistical model? The literature on the subject of parameter estimation is considerable and we certainly cannot go into all the various ramifications for particular kinds of models. However, in practical terms there is a single general approach that is most commonly used, that of *maximum likelihood*.

The idea of maximum likelihood is really quite simple. Our exploratory analyses, background knowledge, intuition and judgement have led us to a model framework, which, as discussed previously, will usually amount to the specification of a general mathematical form for the probability distribution appropriate to the phenomenon under study, but one which involves certain parameters whose values are left unspecified. In general let us denote these unknown parameters by the vector $\boldsymbol{\theta}$. In the above example of the linear regression model $\boldsymbol{\theta}$ would have four elements β_0, β_1, β_2 and σ^2; but in general it might have any number of elements. Now, if the model framework provides a general mathematical form for the probability distribution appropriate to the phenomenon under study, then from this we must be able to write down a general mathematical form for the joint probability distribution of the set of random variables $\{Y_1, Y_2, \ldots, Y_n\}$ for which our data $(y_1, y_2, \ldots, y_n)$ constitute a set of observed values. Suppose that this joint probability distribution is $f(y_1, y_2, \ldots, y_n; \boldsymbol{\theta})$. Here we have explicitly indicated that as well as being a function of the observed values $(y_1, y_2, \ldots, y_n)$, this will also in general depend on the values of the unknown parameters $\boldsymbol{\theta}$.

How is $f(y_1, y_2, \ldots, y_n; \boldsymbol{\theta})$ to be interpreted? It is a joint probability distribution so, given particular values for $\boldsymbol{\theta}$, it specifies the probability, or probability density, associated with Y_1 taking the specific value y_1, at the same time that Y_2 takes the specific value y_2 and so on. Hence, if $(y_1, y_2, \ldots, y_n)$ are our actual data observations, $f(y_1, y_2, \ldots, y_n; \boldsymbol{\theta})$ is effectively the probability or probability density associated with them occurring, given the general model framework proposed. We refer to this as the *likelihood* for the data, and would normally denote it as $\mathcal{L}(y_1, y_2, \ldots, y_n; \boldsymbol{\theta})$. Notice that given a known set of observed data $(y_1, y_2, \ldots, y_n)$, the likelihood depends only on the unknown parameter values $\boldsymbol{\theta}$. Recall that our objective is to estimate values for $\boldsymbol{\theta}$ and an obvious way of doing this is now apparent. We should choose values for $\boldsymbol{\theta}$ that maximise the likelihood for the observed data. Often, because it is easier and equivalent we maximise the logarithm of the likelihood or the *log likelihood*, which we denote $l(y_1, y_2, \ldots, y_n; \boldsymbol{\theta})$. Under the maximisation of either function, we are effectively 'tuning' our proposed model framework by choosing parameter values which give the greatest possible likelihood of observing the data that we actually observed, which would seem an entirely sensible way to proceed.

This, then, is the general approach to parameter estimation or model fitting, known as maximum likelihood. Of course, in any specific case the maximisation of $\mathcal{L}(y_1, y_2, \ldots, y_n; \boldsymbol{\theta})$, or its logarithm, with respect to $\boldsymbol{\theta}$, may not be particularly easy and may involve intensive computation and the development of algorithms to numerically approximate the solution. However, for some 'standard' and commonly used model frameworks it turns out to be relatively straightforward and mathematically equivalent to ways of fitting models with which you may be more familiar. For example, for the multiple linear regression model discussed earlier, where the model framework involves the assumption of independently distributed random variables, having normal distributions with the same variance, maximum likelihood parameter estimation reduces to using the method of *ordinary least squares*. That is, parameters are estimated by minimising the sum of squared 'residuals' or differences between the data values and those predicted under the model. If we relax the assumption of independence and equal variance, but maintain the other aspects of the model, then maximum likelihood leads to parameter estimation by *generalised least squares*, which is the minimisation of a weighted sum of squared 'residuals'. We will come across both of these particular approaches to model fitting in later chapters, as well as the use of maximum likelihood itself.

Maximum likelihood not only provides parameter estimates, but also general measures of how reliable these estimates are (their *standard errors*) and of how well alternative models fit a particular set of data. Standard errors are, in essence, derived by consideration of how 'peaked' the likelihood function is at its maximum. Informally, the idea is again simple. If the likelihood is highly 'peaked' at its maximum and falls sharply in value as you move away from this, then you can be pretty sure that the estimated parameter values are reasonably reliable (small standard errors). If, alternatively, the maximum occurs at some point on a slowly changing 'plateau', then the value of the likelihood is similar all over this 'plateau' and therefore for several different sets of parameter values; hence you should be less confident in the estimates derived (large standard errors). When it comes to comparing the overall fit of two different models we can consider the ratio of the likelihoods associated with each (duly maximised for the parameters involved). Informally, we ask whether one value is significantly better than the other.

We do not give details here, but these ideas are made more mathematically precise in the case of standard errors, by looking at the matrix of second derivatives of the log likelihood evaluated at the maximum, which, after some manipulation, yields approximate standard errors for the parameters. For reasonably large samples of data an approximate 95% confidence interval for a parameter is formed by taking two standard errors either side of its maximum likelihood estimate. Formal comparison of the fit of two models is effected through likelihood ratio tests. Again, as with parameter estimation, these general ideas are broadly equivalent to alternative and more commonly used ways of deriving such quantities in the case of 'standard' model frameworks. For example, the 't' and 'F' tests which you may have come across in relation to multiple regression are, for all practical purposes, equivalent to using results that would be derived from the general maximum likelihood method.

This is perhaps an appropriate point to indicate where statistical hypothesis testing fits into our picture of statistical modelling. Testing a hypothesis is a question of comparing the fit to the data of two models, one of which incorporates assumptions which reflect the hypothesis, the other incorporating a less specific set of assumptions. Usually a hypothesis will amount to the specification of values for certain of the parameters involved in the model. Testing hypotheses is therefore one facet of statistical modelling, we simply ask the question: 'does a model in which certain parameters have pre-specified hypothesised values, fit the data significantly less well than one where these parameters are allowed to be optimally estimated from the data?' No new theory is really involved, although of course we can wrap all this up in the language of 'p-values' and so on, if we so wish. Notice, however, that all modelling inevitably involves some assumptions about the phenomenon under study; hence hypothesis testing will always involve comparison of the fit of a hypothesised model with that of an alternative which also incorporates assumptions, albeit of a more general nature. The validity of a particular form of hypothesis test often relies critically on these alternative assumptions being, in turn, valid for the phenomenon in question. For example, in the ozone model discussed above, the standard multiple regression test of whether the parameters β_1 and β_2 are significantly different from zero depends on both the assumption of the independence of $Y(\boldsymbol{s})$ and of a normal distribution for these random variables.

Thus far, our very brief review of the general process of statistical modelling and hypothesis testing has not been oriented specifically towards models of spatial phenomena. However, hopefully we have now established sufficient general background to introduce some concepts of particular importance with regard to this latter area; themes which will recur in many of the subsequent chapters of the book.

As we have already emphasised earlier in this chapter, spatial phenomena often exhibit a degree of spatial correlation and spatial analysts need to incorporate the possibility of such spatial dependence into their models if the models are to provide realistic representations of such phenomena. For example, the independence assumption of the standard multiple regression model would be unlikely to be realistic in relation to our earlier ozone level example. In addition to the mean value of the ozone level varying in $\mathcal{R}$, the distribution of values about this mean at any site is likely to be related to that in neighbouring sites. In fact, a simple linear relation between mean value and location is also unlikely! But the point which we wish to emphasise here is that in general the behaviour of spatial phenomena is often the result of a mixture of both *first order* and *second order* effects. First order effects relate to variation in the mean value of the process in space—a global or large scale trend. Second order effects result from the spatial correlation structure, or the spatial dependence in the process; in other words, the tendency for deviations in values of the process from its mean to 'follow' each other in neighbouring sites—local or small scale effects.

A somewhat artificial illustration may help our reader to understand these ideas better. Suppose we imagine scattering, entirely at random, iron filings on to a sheet of paper marked with a fine regular grid. The numbers of iron filings landing in different grid squares can be thought of as the realisation of a spatial stochastic process. As long as the mechanism by which we scatter the filings is purely random, there should be an absence of both first and second order effects in the process—different numbers of filings will occur in each square, but these differences arise purely at random. Now suppose that a small number of weak magnets are placed under the paper at different points and we scatter the filings again. The result will be a process with spatial pattern arising from a first order effect—clustering in the numbers in grid squares will occur globally at and around the sites of the magnets. Now remove the magnets, weakly magnetise the iron filings instead, and scatter them again. The result is a process with spatial pattern arising from a second order effect—some degree of local clustering will occur because of the tendency for filings to attract each other. If the magnets are now replaced under the paper and the magnetised filings scattered again, we end up with a spatial pattern arising from both first and second order effects.

Because 'real life' spatial patterns frequently arise from this sort of mixture of both first and second order effects, independence in the random variables representing a spatial stochastic process is often too strong an assumption for the spatial modeller. By definition, such an assumption rules out second order effects and therefore needs to be replaced by some weaker alternative which allows for the possibility of a covariance structure. A common approach is to think of the variable of interest (such as the ozone level at a location) as comprising two components. The first order component represents large scale spatial variation in mean value. This is similar to the dependence proposed in the simple regression model used earlier, although the relationship between the mean and location need not be linear and 'covariates' might be included in the relationship, instead of, or together with, location. The second order component is concerned with the behaviour of stochastic deviations from this mean. Instead of assuming these to be spatially independent, they are allowed to have a covariance structure which may give rise to local effects.

The second order component is often modelled as a *stationary* spatial process. Informally, a spatial process $\{Y(\boldsymbol{s}), \boldsymbol{s} \in \mathcal{R}\}$ is stationary or *homogeneous*, if its statistical properties are independent of absolute location in $\mathcal{R}$. In particular, this would imply that the mean, $E(Y(\boldsymbol{s}))$, and variance, $VAR(Y(\boldsymbol{s}))$, are constant in $\mathcal{R}$ and therefore do not depend upon location, $\boldsymbol{s}$. Stationarity also implies that the covariance, $COV\big(Y(\boldsymbol{s}_i), Y(\boldsymbol{s}_j)\big)$, between values at any two sites, $\boldsymbol{s}_i$ and $\boldsymbol{s}_j$, depends only on the relative locations of these sites, the distance and direction between them, and not on their absolute location in $\mathcal{R}$ (Figure 1.9).

We say further that the spatial process is *isotropic* if, in addition to stationarity, the covariance depends only on the distance between $\boldsymbol{s}_1$ and $\boldsymbol{s}_2$ and not on the direction in which they are separated (Figure 1.9). If the mean, or variance, or the covariance structure 'drifts' over $\mathcal{R}$ then we say that the process exhibits *non-stationarity* or *heterogeneity*.

In terms of our earlier example, the case where weakly magnetised iron filings are scattered onto paper with no magnets underneath would roughly equate to an isotropic process. The case with unmagnetised filings and magnets underneath the paper, approximates a process with heterogeneity in the mean value and independence in deviations from mean value—a simple form of non-stationary model, similar in spirit to our simple regression model, although obviously not in respect of the relationship between mean and location, which would be non-linear in this case. The experiment with both magnetised filings

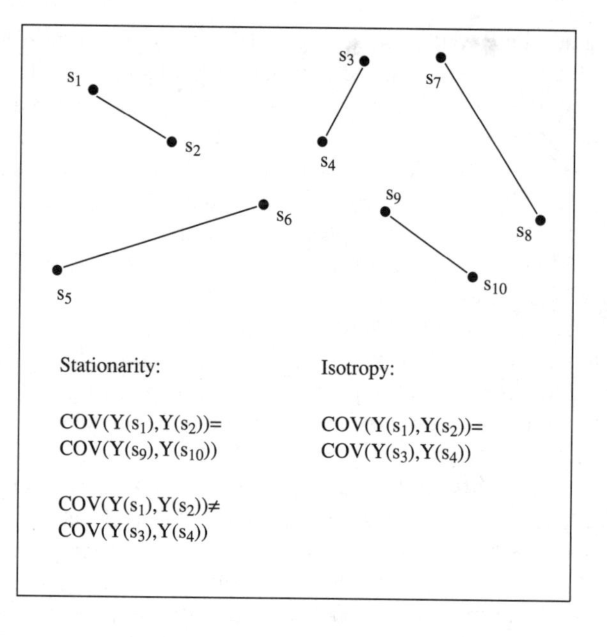

Fig. 1.9 Illustration of stationary and isotropic spatial processes

and magnets under the paper is a more complex non-stationary process that mixes the previous two cases and might be considered akin to the 'two component' model suggested earlier as useful in practice for spatial processes. It involves first order variation or heterogeneity in mean value combined with a stationary second order effect.

Heterogeneity in the mean, combined with stationarity in second order effects, is a useful spatial modelling assumption, where 'reasonable' and acceptable, since it implies that the covariance of the process has the same structure from area to area within the region studied. Without the assumption of some form of stationarity in spatial models, one begins to have great difficulty in fitting them to observed data, since the number of parameters involved becomes unmanageable. If all locations in space have potentially different covariance structures as well as means, and, as is usual, we have only a set of single observations at a particular subset of locations, then we stand little chance of estimating all parameters involved in the model.

So, in broad terms, the modelling of a spatial process often tends to proceed by first identifying any heterogeneous 'trend' in mean value and then modelling the 'residuals', or deviations from this 'trend', as a stationary process. This idea will crop up in various ways as we deal with specific types of spatial data in the various parts of the book. For example, in our various sections on exploratory methods, we will discuss both techniques designed to investigate first order variations and those more related to second order effects. In subsequent modelling sections we will often be concerned with estimating parameters relating to both first and second order effects. Sometimes stationarity assumptions are made explicit, at other times they are implicit, in the sense that the use of some methods would make little or no sense unless stationarity applied.

However, in concluding this section on spatial modelling, it must be acknowledged that concepts such as stationarity and first or second order effects are artefacts of the modeller and not reality. In practice, effects are confounded in observed data and the distinction between them is difficult and ultimately to some extent arbitrary. Both types of effect can, as we have said, give rise to similar spatial patterns. If high values of a process are found in one region and low values among a set of adjacent sites in another region, then how do we know whether the underlying process is non-stationary (heterogeneous) or if these are local effects resulting from a homogeneous spatial dependence in the data? In other words, how can we distinguish spatial dependence in a homogeneous environment from spatial independence in what is a heterogeneous environment?

These are difficult questions with no definitive answers. Certainly, we believe analytical methods can help to identify appropriate models and distinguish in some cases between effects that must clearly be global first order trends and those that are more likely to be the result of a second order covariance structure. However, the extent to which they can assist the modeller is limited. Ultimately, models are mathematical abstractions of reality and not reality itself. Statistically we may not be able to distinguish between two equally good but structurally different explanations for variations in a phenomenon of interest. We have already pointed out that the specification of a model will always involve using a combination of both data and assumptions about the nature of phenomena being modelled. Ultimately, judgement and intuition on the part of the analyst are always involved in statistical modelling. Statistical models are always at best 'not wrong', rather than 'right'.

1.6 Practical problems in spatial data analysis

Most of this introductory chapter has been concerned with laying the foundations for subsequent chapters and explaining their intended structure. Many of the points discussed will be returned to and reinforced in those chapters. We want in this final section briefly to turn our attention to a set of issues which we will not necessarily discuss explicitly again in the book, but which are always in the background of any spatial data analysis. We want to discuss some important practical problems that arise when we come to do applied spatial data analysis; problems of a general nature that confront the spatial analyst in particular and which we acknowledge will never be resolved by analytical techniques alone. We encourage the reader to bear these in mind in relation to any of the subsequent methods presented in the book.

The first of these that we would like to highlight is the rather obvious one of the geographical scale at which analysis is performed. Spatial data analysis is concerned with detecting and modelling spatial pattern, but pattern at one geographical scale may be simply random variations in another pattern at a different geographical scale. For example, local variations in disease rates may pale into insignificance against the national picture. The extent to which the spatial analyst should 'zoom' into particular areas looking for explanations of variations in data values is not something that can be determined in any absolute sense. Clearly, it depends on the phenomenon under study, the objectives of the analysis, and the scale at which data have been collected. However, largely it will be a matter of the judgement and experience of the analyst.

A second difficulty that the spatial analyst faces is concerned with the lack of any natural 'indexing' or ordering in space. Recall that mention was made at the start of this chapter of some similarities between time series analysis and spatial data analysis. However, most time series analysis deals with data that have a rather simple 'indexing'. The observations tend to be evenly spaced (for example, at monthly or hourly intervals). Further, we can usually assume that the dependence extends only in one direction; what happens in July 1994 may well depend on what happened in February 1994 but will not depend on what happens in February 1995 (unless we believe that expectations play a part). As we shall see in several places in the book, spatial data are not so simply 'indexed'. While some data, such as those from satellites, do come in the form of a regular grid or lattice, much spatial data is provided for a 'patchwork quilt' of areal units or an irregularly distributed set of sites. Moreover, the kind of spatial dependence, or correlation, to which we have alluded several times in this chapter, extends in general over several directions (though, for simplicity, we shall often have to assume isotropy, an absence of directional effects.)

An 'indexing' also implies that we have a natural notion of what is meant by the 'next' (or, conversely, the 'previous') observation in the sequence. In the spatial domain this is not at all obvious. On a regular grid or lattice there is a reasonably natural ordering of locations. We can, for example, speak of the neighbourhood of a zone as those areal units that share a common boundary (or, perhaps either a common boundary or vertex). Apart from sites on the boundary of our study area, all sites, or areal units, on a grid have an identical neighbourhood structure. They are, furthermore, equidistant from each other. On an irregular set of zones, or where we have a set of sites that are irregularly distributed in space, there is no obvious measure of 'proximity'. As we shall see in Part D, much of our analysis requires us to specify such a proximity measure. This might be taken to be simple contiguity (the sharing of a common border between areal units) or it might be based on distances between centroids of the areal units; other definitions are possible. Related measures are possible for a set of point locations.

As we have already hinted, a further set of problems arises when we deal with what happens on the edge or boundary of our study area. Consider, for example, what we have just said about proximity. In the 'middle' of our study area a site or zone is likely to be surrounded by others; it has a 'neighbourhood', however we choose to define it. At the edge of the map the 'neighbours' perhaps extend in one direction only. But, unless the boundary is a coastline there probably *will* be sites beyond the boundary; we have simply chosen not to collect data outside our study region. How serious is this problem? In time series analysis, the problem arises at the start and end of the series where, as a result of our data exploration or modelling we may 'lose' observations. In the spatial domain there is a potentially much greater set of observations around the edge of the map and the need to allow for the influence of 'edge effects' is therefore correspondingly greater. On a regular lattice the boundary is simple; in other instances the boundary may be quite convoluted. As we shall see, the irregularity of the boundary can be allowed for in the analysis but certainly makes for added complications.

An additional issue arises where our data are measurements on a set of zones. Often, and especially in social science applications of spatial analysis, the data in such zones are aggregated measurements for households or individuals living in such zones. For reasons of confidentiality data are not often released for these 'primary units', only for a set of rather arbitrary areal aggregations. These areas are 'arbitrary' in the sense that they are only one possible configuration that could have been drawn up. The important point to note is that any results we obtain from analyses of these area aggregations may be conditional upon the set of zones we are presented with, a problem which is often referred to as the *modifiable areal unit problem*. Sometimes it is possible to experiment with alternative configurations, and to assess the magnitude of the problem. More frequently we are presented, as a *fait accompli*, with a set of zones and accompanying data.

Another set of problems become apparent if we force ourselves to consider the real complexity of the topography which exists in many of the areas over which analyses may be performed. In spatial analysis we tend to think of space in very abstract terms, rather glibly talking of 'distances', 'boundaries' and 'edge effects'. Of course in reality the areas over which analyses are being conducted are vastly complex, criss-crossed with natural boundaries such as rivers or ranges of hills, or human constructions such as roads, industrial estates, recreational parks and so on. Any models that we may develop, or relationships that we may uncover, have to be viewed in the cold light of the relative simplicity of what we can cope with mathematically, as compared with what we know to exist in geographical reality. Humility is indeed wise for the spatial analyst!

In summary, we note that all sorts of practical difficulties may, and often do, arise in applied spatial data analysis. In this book we take a fairly phlegmatic view of this, feeling that all analysis has to begin somewhere. We recognise that in places we will be forced to over-simplify. In some instances this will be because the level of mathematical difficulty involved in a more rigorous treatment is simply not appropriate for this particular book. We will do our best to 'flag' these areas. In other cases it will simply be because there is no current way of dealing with the complexity involved and we are forced into ignoring certain issues, or making simplifying assumptions, in order to be able to proceed at all.

At all times we urge the reader to remain critical of what we say and, throughout the book, to bear in mind the kind of issues which we have discussed in this section.

1.7 Summary

This has been a rather long introductory chapter and in places somewhat involved. We have touched on several ideas, many of which may be new to our reader and some of which involve really quite difficult concepts, especially those relating to statistical modelling and spatial stochastic processes. We would not necessarily expect all this to be fully understood in a single reading. We hope that the reader will return to sections of this chapter in the light of the methods and examples we present later. Issues about which our reader may be unsure at present will, we trust, become clearer then.

However, we do hope that our reader has been able to grasp the main points of our discussion; in particular, that we have managed to convey successfully what our subject is about and exactly how we mean to structure our development of it in the subsequent chapters. To summarise, we are interested in methods for the statistical description and modelling of spatial data which give explicit consideration to the possible importance of spatial effects in drawing inferences from such data. We believe that such forms of analysis are important because 'space can make a difference'; in other words, that the recognition of a spatial dimension in analysis may sometimes yield different, and more meaningful, results, than an analysis which ignores it.

We have classified the spatial problems and types of data that we will be considering into four classes, namely: *point patterns, spatially continuous data, area data* and *spatial interaction data.* We intend to devote one of the subsequent Parts B–E of the book to each of these classes respectively. Within each of these parts we will try to adopt a consistent structure. In each (apart from Part E) we will have two chapters. The first of these will be concerned with a basic core of methods appropriate to the kind of data under discussion. In these chapters we shall introduce the kind of spatial problem we wish to consider and motivate this by considering a series of case studies. We shall identify the sorts of analytical questions that we wish to ask of the data. We then go on to consider ways of *visualising* the type of spatial data we are dealing with; in other words, the sorts of maps and other graphical displays we might produce. Next, we discuss methods for *exploring* the data and investigating relationships of interest. Finally, we consider more formal *modelling* and hypothesis-testing techniques relevant to that type of data.

The second chapter in each part of the book will be devoted to miscellaneous additions, extensions and modifications to the basic core methods. Here we relax our visualisation, exploration and modelling division. We do not attempt there to be entirely comprehensive in the methods we present, selecting methods for discussion on the basis of our particular interests and our experience of what has proved useful to us in practice. We hesitate to refer to these methods as more 'advanced' than those in the previous 'core' chapter; some are, but others simply address different questions. Instead, we refer to them as 'further' methods. We draw on the same data sets introduced in the 'core' chapter to illustrate their use.

In every chapter, at appropriate points in our discussion we invite the readers, if they wish, to tackle computer exercises that relate to the data sets we include and involve using the software we supply. We also end each chapter with a set of readings that will allow the reader to go further. These include references to the origins of any data sets that we have used.

This, then, is the general outline we will follow in Parts B–E. However, before doing so we would like to devote the next and final chapter in this introductory Part A of the book to discussing the relationship between computers and spatial data analysis; a relationship which we feel is increasingly important given the recent major advances in digital mapping, computer cartography and Geographical Information Systems. These potentially provide an environment within which to visualise spatial data and explore spatial patterns and relationships interactively—an aspect of spatial data analysis which we would particularly like to encourage in this book.

1.8 Further reading

The original Californian rainfall data come from:

Taylor, P.J. (1980) A pedagogic application of multiple regression analysis: precipitation in California, *Geography*, 65, 203–12.

The data were re-analysed in:

Jones, K. (1984) Graphical methods for exploring relationships, in Bahrenberg, G., Fischer, M. and Nijkamp, P. (eds) *Recent Developments in Spatial Data Analysis*, Gower, Aldershot, 215–27.

The study of space–time clustering of Legionnaire's Disease is taken from:

Bhopal, R. Diggle, P.J. and **Rowlingson, B.S.** (1992) Pinpointing clusters of apparently sporadic Legionnaire's disease, *British Medical Journal*, 304, 1022–27.

The example of rainfall modelling in England and Wales may be consulted in:

Oliver, M., Webster, R. and **Gerrard, J.** (1989) Geostatistics in physical geography. Part II: applications, *Transactions, Institute of British Geographers*, 14, 270–86.

For reference to the work on AIDS in Uganda see:

Cliff, A.D. and **Smallman-Raynor, M.** (1992) The AIDS pandemic: global geographical patterns and local spatial processes, *Geographical Journal*, 158, 182–98.

On the locations of swimming pools in London, Ontario you should see:

Goodchild, M. and **Booth, P.** (1980) Location and allocation of recreation facilities: public swimming pools in London, Ontario, *Ontario Geography*, 15, 35–51.

On the nature of spatial data see:

Laurini, R. and **Thompson, D.** (1992) *Fundamentals of Spatial Information Systems*, Academic Press, London, Chapter 3.

An important general reference on exploring spatial data, which also makes helpful comments on problems of spatial data analysis, is:

Haining, R. (1990) *Spatial Data Analysis in the Social and Environmental Sciences*, Cambridge University Press, Cambridge, especially Chapters 2 and 6.

and for a strong statement of the importance of data exploration see:

Openshaw, S. (1991) Developing appropriate spatial analysis methods for GIS, in Maguire, D.J., Goodchild, M.F. and Rhind, D.W. (eds) *Geographical Information Systems: Principles and Applications*, Longman, Harlow, 389–402.

On the Geographical Analysis Machine (GAM) see:

Openshaw, S., Charlton, M., Wymer, C. and **Craft, A.** (1987) A Mark 1 Geographical Analysis Machine for the automated analysis of point data sets, *International Journal of Geographical Information Systems*, 1, 335–58.

On spatial modelling, especially in an econometrics framework, a useful text is:

Anselin, L. (1988) *Spatial Econometrics: Methods and Models*, Kluwer, London, Chapter 2.

But see also:

Griffith, D. (1988) *Advanced Spatial Statistics*, Kluwer, London.

as well as the book by Haining cited above.

Other important general texts on spatial statistics and spatial data analysis which we would recommend, some of which are referred to in later chapters, include:

Ripley, B.D. (1981) *Spatial Statistics,* John Wiley, Chichester.

Ripley, B.D. (1988) *Statistical Inference for Spatial Processes,* Cambridge University Press, Cambridge.

Upton, G. and **Fingleton, B.** (1985) *Spatial Data Analysis by Example, Volume 1: Point Pattern and Quantitative Data*, John Wiley, Chichester.

Upton, G. and **Fingleton, B.** (1989) *Spatial Data Analysis by Example, Volume 2: Categorical and Directional Data*, John Wiley, Chichester.

Diggle, P.J. (1983) *Statistical Analysis of Spatial Point Patterns,* Academic Press, London.

Cressie, N.A.C. (1991) *Statistics for Spatial Data*, John Wiley, Chichester.

Webster, R. and **Oliver, M.A.** (1990) *Statistical Methods in Soil and Land Resource Survey*, Oxford University Press, Oxford.

Much of the literature on spatial data analysis requires familiarity with matrix algebra. There are numerous good introductions. One that is oriented towards spatial analysis is:

Tinkler, K. (1985) *Introductory Matrix Algebra*, Concepts and Techniques in Modern Geography, 48, Geo Abstracts, Norwich.

A good introduction to maximum likelihood methods, oriented towards a geographical audience is:

Pickles, A. (1985) *An Introduction to Likelihood Analysis*, Concepts and Techniques in Modern Geography, 42, Geo Abstracts, Norwich.

1.9 Computer exercises

Here are suggested exercises that you can try that relate to ideas discussed in this chapter, using INFO-MAP and our example data sets. These exercises have been referenced at appropriate points in the chapter by numbered symbols in the margin.

Exercise 1.1

Since this is the first of our sets of 'computer exercises', we need to make some introductory remarks about what we hope to achieve by including such exercises in the book; also about the software and the example data sets upon which they are based.

Because we wish our reader, at least to some degree, to be able to do some 'interactive' spatial data analysis, we have provided with the book a simple software package that allows small spatial data sets to be 'viewed' and some basic forms of analysis to be performed. We have supplied with it all of the data sets that we discuss in the various parts of the book. Each of these is described in detail in the 'case studies' section of one of the Parts B–E, depending on the type of data involved. In those descriptions we also discuss what issues may be of interest in the analysis of each of these data sets. References to the original sources of each of the data sets are given in the 'further reading' section of the relevant chapter.

Most of the illustrations of analyses that we give in each of our chapters draw on one or other of these data sets. However, we invite the reader, in the series of computer exercises associated with each chapter, to gain a better understanding of the ideas in the text by exploring these data sets further in conjunction with the software package that we have supplied. The reader who is familiar with and has access to other software for spatial analysis or mapping, may prefer to use our data sets and work through the general suggestions in our exercises, but in conjunction with another system. If so, the *User's Guide*, given as an appendix at the end of this book, and associated *User's Reference*, supplied with our software, gives details on how to use it to 'export' data from the file formats that we have supplied, so that the data sets can be 'imported' into another system. Alternatively, readers may like to use our software package, but with their own data sets, rather than those we provide. In this case the *User's Guide* and *User's Reference* should again be consulted to determine how to create new spatial data sets within our package.

We call this package INFO-MAP. In developing the package we have assumed that our reader has access to only the most basic computing facilities, so it is deliberately designed to be relatively unsophisticated and to run on any PC with standard colour graphics. Chapter 2 is devoted to the wider context of using computers to map, manipulate and analyse spatial data and there we discuss in detail the scope of recent and potential computer developments relevant to our subject. This discussion will give the reader some idea of the limitations of INFO-MAP in relation to what is currently possible. Nevertheless, we hope that INFO-MAP will allow the reader to 'view' our example data sets and experiment with some of the forms of analysis discussed in the text, without too steep a 'learning curve' and this is our main objective in including it with the book. By no means all of the analysis methods we discuss in the book are implemented as part of the package; however, we have included a range of the techniques, so that the reader can get a feel for practical interactive spatial data analysis.

An overview of the software is given in the *User's Guide* appendix to the book, and you may like to read that now to get a feeling for what the package is designed to do. If you wish to follow our exercises then you will obviously need to install the software and data sets onto your computer. Instructions on how to do this may be also be found in the *User's Guide* and you should carry these out now.

Assuming that you have succeeded in installing and running the software, we now turn to the first of our computer exercises. We start simply by using INFO-MAP to obtain various 'views' of the Californian rainfall data, discussed in the text, as well as getting a 'feel' for this data and some of the points we make concerning what spatial data analysis involves.

Run INFO-MAP and from the `File` menu select `Open`, and then 'Rainfall in California'. (Consult the section 'Getting started' in the *User's Guide* for an explanation of the general use of menus, mouse and keyboard within INFO-MAP.) A map should appear on the screen when you 'open' the file. This simply shows the outline of the state, which should look something like Figure 1.1. Next you should load the data associated with this file, by selecting `Open` from the `Data` menu. You are then in a position to draw a map of precipitation. Do this by selecting the `Map` menu, the `Dot map` option and the variable 'Precipitation'. This adds the locations of the stations and at each gives an indication of the corresponding precipitation value with a small circle, coloured according to the map legend displayed.

You could follow this by mapping some of the other variables in the file in a similar way in order to get a feel for their spatial variation. Do this by selecting the `Dot map` option again and then the variable in which you are interested. You might also try producing some scatter plots of precipitation against various of these other variables. You do this by selecting the `Scatter plot` option from the `Analysis` menu and then the appropriate variables for the vertical and horizontal axes.

Use these plots combined with maps of different variables to explore the relationship between the spatial pattern of precipitation and that of the other variables.

Exercise 1.2

As we note in the text, Taylor considers the 'rain shadow' effect to be of some significance in relation to this data. You may add an 'overlay' to the California map of the previous exercise, which shows the 'higher ground' in the 'study region'. Do this by selecting `Open` from the `Overlay` menu. We will now go on to create a new variable in INFO-MAP to indicate whether sites are in the 'rain shadow' or not.

Select `Profile locations` from the `Analysis` menu, with the option 'default order'. This gives a list of stations. Not all are immediately displayed, but the list will 'scroll' if required by using the cursor keys. As you scroll through the stations their corresponding location is indicated on the map with a 'cross-hair' cursor. Make a written list of the station names and put a '1' against those which seem to be in the lee of the mountains (in the rain shadow). Use of the <ESC> key will then exit from 'profiling locations' and you may select `Modify` from the `Data` menu. A spreadsheet appears. Select `Add` from the `Data` menu; this adds a blank column to the spreadsheet. For this new column add a '1' or a '0' to each row to indicate whether or not the station is or is not in the 'rain shadow'. After you have completed the column, select `Name` from the `Data` menu and <TAB> to your new column to give it a name, say 'Rain shadow'. Now choose `Save` from the `File` menu and save your modified data file (we suggest you use a different file name and description than that for the original file, so that the original data remains unchanged). Then choose `Exit` from the `File` menu, leaving you with the modified version of the original California rainfall data which you may now use in any subsequent analysis.

First try fitting Taylor's original regression. You may do this by choosing `Calculate` from the `Data` menu and then entering the following formula:

```
fits=regr({1},{2},{3},{4})
```

This calculates a new variable called 'fits' containing the fitted values of an ordinary least squares linear regression of 'precipitation' (variable 1 in the file) on three 'explanatory' variables, 'altitude' (variable 2), 'latitude' (variable 3) and 'distance from the coast' (variable 4). (By the way, instead of typing variable numbers in curly brackets into the formula, you can select them from a menu of data items, at any point in the formula, by using the up arrow key, which enters their 'column' numbers automatically in the appropriate brackets). To see the results of the regression analysis select `Regression parameters` from the `Analysis` menu. This gives the estimated regression coefficients, their standard errors (in brackets) and the 'R^2' value.

In order to create the regression residuals, use the `Calculate` option again, along with the formula

```
resids={1}-{6}
```

Note that variable 6 contains the new variable 'fits', the fitted values calculated in the previous step. New variables are signified on the left hand side of an INFO-MAP formula by use of a text name, but ones which exist in the file and appear on the right hand side must be referred to by their column numbers in curly brackets.

The calculated residuals can now be mapped using the `Dot map` option. Try to relate any spatial patterning of the residuals to the overlay showing higher ground, perhaps using the 'profile locations' option in the `Analysis` menu to help, or possibly a scatter plot of the fitted values against the 'rain shadow' variable.

Using the same kind of ideas as above, repeat the regression including the 'rain shadow' as an additional explanatory variable. This new variable is variable 5 in the file, so you would do this by using the regression formula:

```
newfits=regr({1},{2},{3},{4},{5})
```

You should be able to demonstrate an improved 'R^2' value and less obvious spatial patterning in the new residuals when you calculate these. By the way, you can 'turn off' display of the 'overlay' of the higher ground by using the `Clear` option in the `Overlay` menu. Note in passing that each time you create a new variable no check is made on whether the name you use duplicates one that already exists. Whenever a text string is used on the left hand side of a formula a new variable is always assumed. Should you wish to overwrite an existing variable with the result of a calculation, then its variable number in curly brackets must be used on the left hand side of the formula. Note also that the maximum limit on the number of variables you may have at any time is 32. For data files with more than 500 observations the limit may be as small as 8. (The chapter 'Calculating data and worksheet items' in the INFO-MAP *User's Reference* supplied on the disk gives full details of the syntax of INFO-MAP formulae.)

You may now like to save all your results temporarily, so that you can return to them at some convenient future time. You may do this by the `Save session` option on the `File` menu, supplying a file name of your choice and a description. At any time in the future you may retrieve this file by using the `Retrieve session` option in the `File` menu. You can also delete a 'saved session' by using the `Delete session` option in the `File` menu. Saving sessions in this way is equivalent to 'dumping' where you are—the 'time for a cup of coffee' option! It is not particularly advisable to litter your computer with such 'saved sessions'; so it is preferable to use the same file name repeatedly and overwrite any previous saved session each time. Note that if you wish to make permanent changes to an INFO-MAP data set, then it is better to do this by modifying and saving the *base file* or the *data file* separately; see the *User's Reference* for details of how to do this.

Exercise 1.3

Following on from the last exercise, you might try calculating a new interaction variable as:

```
interact={4}*{5}
```

where variables 4 and 5 are respectively 'distance from coast' and 'rain shadow'. You could now fit another regression including the interaction variable as one of the explanatory variables. (Note that you could avoid a step here, by including the term `{4}*{5}` directly in the regression function in place of a numbered variable). Does this improve the explained variation as claimed in the text?

Exercise 1.4

In order to explore the idea of spatial correlation in the residuals from the previous exercise, you might try plotting the value of each of the residuals against their corresponding value at the nearest site. First calculate the residuals with a formula which subtracts the fitted from the observed values (variable 1). Assuming these residuals are in variable number 10, you may then calculate a new variable:

```
neigh= {10}[near(1)]
```

Each row *i* of the new variable will contain the value of variable 10 at the site *j* which is closest to *i*. You can now plot the residuals against these neighbouring residuals, to convince yourself that there is little evidence of any spatial correlation in the residuals (unexplained variation) from your final regression model.

Exercise 1.5

To see an example of a 'point pattern', draw a `Dot map` of the age of cases for the Ugandan Burkitt's lymphoma data. Recall this involves first using `Open` from the `File` menu, selecting the appropriate file and then using `Open` from the `Data` menu to load the data, before using the `Dot map` option from the `Map` menu. This data set is described in detail in Chapter 3. For the moment simply explore the pattern of the ages of disease cases. What factors might be involved in deciding whether a significant pattern is present?

By using the `Map distribution` option from the `Analysis` menu you can get some idea of the age distribution of cases. But notice that the default class intervals employed in the map do not relate to 'conventional' age groups. You can alter these intervals by using the `Scaling` option in the `Map` menu. Choose the `User defined` option and select four intervals say 0–5, 5–10, 10–15 and over 15. Note the option to supply text names ('very young', 'young' etc.) for the intervals you have specified; you leave these blank and <TAB> over them if you do not wish to supply such names. Map the data using these new class intervals. The spatial distribution of the specified age bands is now more easily observed. Note that the default colour scheme doesn't provide much discrimination with only four intervals. If you wish you could use `Colours` from the `Map` menu to adjust the colours used for each interval. It is the last eight colour boxes in this option that relate to colours for class intervals. The first four involve the background screen colours, text of the map legend, and the colour for the boundary of the map. Using <ESC> from this option restores the original colours, whilst <CR> will retain any new colour settings.

We could also map just the subset of age five and under. First calculate a logical variable having the value 1 for those aged five or under and a missing value otherwise, by using the formula:

```
under5=if({2}<=5,1,miss())
```

Note that here variable number 2 is 'age'. If you now map the new 'under5' variable only the locations of those aged five or under will be displayed.

Exercise 1.6

The California rainfall data used previously were one example of 'spatially continuous data'. For another example, again involving climate data, choose the file 'England & Wales temperature'. This data set is described in more detail in Chapter 5. Load this data set (remember both the map *and* the data have to be loaded separately, before you can do any analysis). Make sure you reset the 'trimmed equal' scaling option with six intervals, which you changed in an earlier exercise. Do this by using the `Scaling` option within the `Map` menu. Then use the `Dot map` option to look at the temperature distribution in August 1981 and August 1991 successively. In each case use the `Summary statistics` option in the `Analysis` menu and make some informal comparisons of overall changes in temperature between 1981 and 1991.

To map the percentage change use the formula:

```
%change=100*(({2}-{3})/{3})
```

to calculate an appropriate variable and map it; by now it should be obvious what this formula is doing. What do you observe about the spatial pattern in temperature change?

To perform a standard 'paired t-test' of differences in mean temperature between the two time periods, use `Calculate/display worksheet` in the `Analysis` menu followed by the option 'calculate item' and then the formula:

```
work[1]=47**0.5*mean({2}-{3})/stdev({2}-{3})
```

Notice that with 47 degrees of freedom the answer indicates a significant difference in mean temperature.

The 'worksheet' in INFO-MAP allows the user to calculate and store up to 32 numeric values, which may be displayed at any time or used in INFO-MAP formulae by referring to them as `work[number]`, where number is an integer in the range 1 to 32. Initially all worksheet items have missing values. Any one of them can be assigned a value at any time regardless of whether or not items with smaller numbers have values. Assigning a new value to a worksheet item simply overwrites the old value. You could try another calculation, such as:

```
work[2]=(max({2})-min({2}))-(max({3})-min({3}))
```

to give the difference in the ranges of temperature, between 1981 and 1991.

Another idea might be to map the difference in the ranked temperature values, using a new calculated data item (a new data column not a worksheet item; remember that you use `Calculate` under the `Data` menu to calculate these). Try:

```
rdiff=rank({2})-rank({3})
```

How do the results correspond with your earlier observations about spatial pattern in temperature change?

Suppose we want to look at the west–east variation in temperature change. We already have percentage temperature change from an earlier calculation. The formula:

```
easting=east()
```

will generate a new data item containing the 'easting' or the horizontal spatial coordinate of each site. You can then use a scatter plot of the earlier percentage temperature change variable against 'easting' to explore any relationship. Assuming that the former is variable 4 and the latter variable 6, then any relationship could be summarised by calculating a new worksheet item:

```
work[3]=corr({4},{6})
```

to give the product moment correlation between them.

Exercise 1.7

To see an example of some 'area data', select the file 'US Presidential election (1992)'. This data set is described in detail in Chapter 7. Open both the map and the associated data file. Then use the `Choropleth map` option in the `Map` menu to obtain a shaded map of the vote for President Clinton.

Does the vote in any way depend upon distance from Arkansas, President Clinton's home state? We need to calculate a new variable to explore this. Try:

```
arkdist=dist([3])
```

This calculates the distance of all of the state locations (in this case a central point in each state) from that of the third state in the file, which is Arkansas. (Note that instead of typing [3] you could select Arkansas from the list of states by using a down arrow key when typing the formula—when entering formulae 'up arrow' menus the data items (columns), 'down arrow' menus the locations (rows)). This new variable can then be used to examine any possible relationship with the vote, using some of the methods discussed in earlier exercises.

An alternative measure of nearness to Arkansas would be states with a common boundary. Say we wished to compare the average vote in such states with the overall average vote in all states. The average vote in all states may be found by calculating the worksheet item:

```
work[1]=mean({1})
```

The mean vote in states bordering Arkansas (and Arkansas itself) is calculated by a second worksheet item:

```
work[2]=mean(if(adjac([3]),{1},miss()))
```

(You may like to consult the chapter 'Calculating Data & Worksheet Items' in the INFO-MAP *User's Reference* to understand the details of this formula.) How do the means differ, if at all?

A more direct way to obtain the mean vote in particular selected states, for example all those in New England, would be to use `Profile locations` in the `Analysis` menu with the 'default order' option. Then 'double click' with the mouse whilst holding the <CTRL> key down (or <CTRL> + <CR> on the keyboard) on those states you wish to include. (You 'de-select' in this process by simply selecting again). When you have 'lit up' the required selection use <SHIFT> + 'double click' (or <SHFT> + <CR> on the keyboard) to display either the total or average of all the data items in the file in those states selected.

Exercise 1.8

To see an example of some 'spatial interaction data', select the file 'Dutch migration'. This data set is described in detail in Chapter 9. For the present, simply open it and load the corresponding data, then use the `Choropleth map` option in the `Map` menu to obtain a shaded map of migration from Groningen. The map you see represents the flow of migrants from Groningen to each of the eleven provinces, including intra-provincial migration within Groningen. The other variables in the data file are mostly levels of out-migration for other provinces. Overall, the data comprise a square matrix of migration 'flows'.

By mapping out-migration from different provinces try to understand the pattern of Dutch migration and notice the clear distance effects. You might like to try calculating the proportion of the total out-migration from any province to other provinces, using a formula say for Groningen such as:

```
propout=100*{4}/sum({4})
```

If you then calculate the distances of all provinces from Groningen using:

```
distance=dist([4])
```

you could now plot the proportion against distance. What do you notice from this plot? How does it compare with similar plots for provinces other than Groningen?

Exercise 1.9

To get some further idea of the concept of first order variation in the mean and a second order covariance structure, select the file 'Mortality in English DHAs'. This data set is described in detail in Chapter 7. It contains a selection of standardised mortality ratios for various diseases within the 190 Health Districts of England, calculated over the five year period up to 1989. Such measures relate to the number of deaths observed in each of the Health Districts expressed as a ratio ($\times 100$) of the number expected during the same period, on the basis of the national rate, corrected for the age–sex structure of the District population.

Open both the map and the associated data file. Then use the `Choropleth map` option in the `Map` menu to obtain a shaded map of the SMR for myocardial infarction (heart attacks). These data relate to males aged 35–64. Use the 'trimmed equal' scaling option with six intervals; `Scaling` within the `Map` menu will enable you to change to this setting if the current setting is different.

You should observe a very marked spatial pattern in the map displayed. Most obvious is a strong north–south trend in the average value of the SMR—apparently a clear first order effect. However, is this the only non-random component in the spatial variation that we can identify?

To look at this in more detail, we might try to 'remove' the north–south trend and then examine what variations we have left. We can do this quite simply, by calculating the residuals of a regression of the SMR on the 'northings' of the Health Districts. Variable 1 in the file is the SMR for myocardial infarction, so we need a formula:

```
resids={1}-regr({1},north())
```

Now `Choropleth map` this new variable 'resids'. The pattern you see no longer contains the marked north–south effect; however, neither is it random. There is a distinct tendency for high residuals to occur together in 'clumps'. It would not be sensible to try to explain this in terms of any simple spatial trend in average value (we could try, but such an explanation would have to be complex and ultimately would not be very convincing); rather, we are seeing a covariance structure in the residuals—a second order effect.

Of course, we realise that here we are just examining the variations in SMR purely from a spatial perspective. We know that there are likely to be all kinds of explanations for variations in SMR in terms of non-spatial factors. It is possible that we may be able to find a set of such factors that account for the spatial patterns we are observing (what candidates can you suggest?). If so, all well and good. But it is also possible that even when all the factors that we can think of (and obtain data on) are included, there might still be non-random spatial patterns in the remaining variations arising from second order effects.

2

Computers and spatial data analysis

Here we talk briefly about recent computer developments that make it possible to visualise and manipulate spatial data. We start with a short discussion of computer mapping and digital cartography. We then consider the subject of Geographical Information Systems, systems for the computerised collection, integration, analysis and display of spatial data. We go on to discuss the potential that these systems offer for spatial analysis, and the challenges they present. Some systems for performing interactive spatial data analysis are then reviewed.

2.1 Introduction

Given that our emphasis in this book is on 'interactive' analysis an essential requirement is that we 'see' the results of our analyses quickly and easily. In practice this implies the use of computers and we therefore devote this chapter to a discussion of the relationship between computing and spatial data analysis. We begin with a brief look at developments in computer mapping. We then consider the way in which these display functions can be linked to database management systems; such a link gives rise to what are known as Geographical Information Systems (GIS). As we have said in the Preface, this is not a book about GIS, but our work has much in common with the needs of GIS and we feel it is useful to give a brief overview of what GIS involve. Finally, we review what work has been done in the area of interactive spatial data analysis and discuss what developments are taking place in this field.

We emphasise that our discussion is very much oriented towards computer developments related to spatial analysis. There are other important areas in the application of computers to mapping in general, which we do not consider here in any detail. For example, we only touch indirectly and briefly on the wide subject of using computers in the preparation, drawing, revision and production of detailed large-scale maps and plans, such as those used by utility companies; although this is in an area which has clearly been transformed by computer technology in recent years.

2.2 Computer mapping

Computer mapping (sometimes referred to as automated or digital cartography) involves using computers to collect, store, edit and display map information. All of these functions, and more, are also provided by a GIS, which we will discuss later in the chapter. However, in this section we pay particular attention to the display aspect, concentrating on the range of software available for the production of digital maps. We shall discuss the collection, storage and editing aspects in a little more detail in the following section on GIS.

Early work in computer mapping used rather crude line-printer graphics for output. Foremost among the packages used was SYMAP, produced by the Harvard Laboratory for Computers and Spatial Analysis. This is now little used and largely of historical significance and modern approaches generally use high-quality output devices, with much higher resolution.

Modern computer cartography involves either the provision of libraries of graphics functions that are called from a user's purpose-built program, or more often the use of specialist packages, which may be either 'command' or 'menu' driven. As computer mapping packages become increasingly sophisticated the demand for 'home-grown' software drawing on libraries of procedures is shrinking. Nonetheless, it still exists. The main advantages of the function library approach lie in the flexibility it gives to devising solutions to specialist problems; the main disadvantage is that it requires some proficiency in a high-level programming language (such as C, C++ or FORTRAN).

As far as 'software libraries' are concerned, there are various products available. One typical example is the GINO library where particular subroutines or procedures may be called from a user's program written in a high-level language. Such routines can, for example, draw lines and symbols and 'fill' selected areas with shading of chosen types. Other routines control selection of pen colour, line and text type, and so on. A more general solution is available via the Graphics Kernel System (GKS), which provides similar sorts of basic functions. A third example are the routines provided within UNIRAS, which offers a wide range of facilities for both two and three dimensional graphics. Such products are well established, fully documented and available for a wide range of operating systems and hardware platforms.

A good example of a subroutine library of more specifically cartographic functions has been developed by Wolf-Dieter Rase. These may be used to produce 'proportional' symbol maps, where the symbols may be circles, rectangles, spheres and 'pictograms' such as human figures, all scaled according to values of the variable of interest at the different locations. For example, a map of gross domestic product for the Federal States in Germany in 1991, produced from this system, is shown in Figure 2.1.

Turning to software packages for computer mapping, rather than libraries of functions, there is a considerable range available to users who do not wish to build their own solutions to mapping problems. One of the best known and most widely used of these within the UK is GIMMS, devised by Tom Waugh at Edinburgh University. A major use of GIMMS is to produce choropleth maps, in which a set of areal units or 'zones' are shaded according to their value on a chosen variable. However, GIMMS is very versatile and can be used to map a variety of types of spatial data, including lines representing transport routes, rivers, and so on. It has facilities for producing shaded contour maps and for shading zones that meet a particular 'spatial criterion' (for example, all those zones intersected by a circle centred on some location of interest). In this way it has embedded in it some of the functions we expect of a GIS. Other examples of such software packages include MAPICS and ATLAS GRAPHICS. These, like many others, run on standard IBM PCs. However, there is also a range of easy-to-use software for desktop mapping available for the Apple Macintosh hardware environment and we reference a recent review in the further reading at the end of this chapter. In addition to such packages, the UNIRAS library mentioned earlier can be used as a menu-driven package;

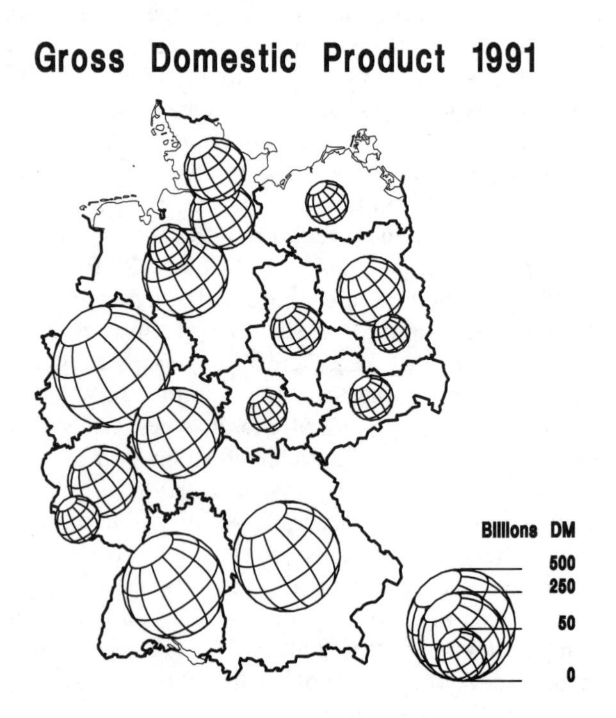

Fig. 2.1 Proportional symbol map generated from subroutine library

in other words, the user does not need to know a programming language in order to produce quite sophisticated plots.

Some specialist packages are also available to those wishing to experiment with map projections. The subject of map projections was formerly a key part of the undergraduate training of generations of geographers, with students being required to construct, by hand, the graticules of particular projections. Regrettably, little prominence is now given to the subject, but this may change as teachers discover easy-to-use software for viewing such projections. We draw the reader's attention to the excellent PC package called WORLD, developed by Philip Voxland at the University of Minnesota. This allows the user to view the graticules of over a hundred projections, as well as coastlines and country outlines. The distortion properties of such projections are readily displayed and analysed. One particular example (Figure 2.2) shows a view of Europe centred on Maastricht; note that Britain is very clearly on the periphery of Europe!

The advantages of computer mapping over manual techniques lie largely in the ability to edit the 'draft' product, by displaying the map on a screen and making desired changes before a 'hard copy' is generated. For example, packages such as GIMMS permit the user to experiment readily with text placement and styles, shading type and colour, viewing the map on a screen and satisfying oneself with the result before plotting a finished version. A second advantage lies in the ability to generate many different maps, typically of different variables, on the same digital base. For instance, if one is mapping Census data for a given urban area the same set of map boundaries may be used to map as many variables as one wishes. Indeed, it may be that the

Fig. 2.2 Projection produced by WORLD

only changes required to the 'command file' used to generate the maps will be the name of the variable and the title.

Virtually all of the software discussed above will run on micro-computers, though for speed of display a reasonably powerful PC is necessary. As the costs of both the computer and associated 'peripherals' (e.g. plotters and laser printers) continues to fall there is really no excuse for producing poor quality products. The aim of computer mapping, as with all cartography, should be to produce a clear and effective display. Certainly, the use of a computer is no excuse for ignoring the basic 'rules' and principles of cartography and we would advise anyone to arm themselves with a good cartography text before putting finger to keyboard or hand to mouse.

One drawback of packages that focus just on computer mapping is that in general one cannot 'interrogate' the map. One cannot point a cursor at a particular feature (an areal unit such as a county, or a linear feature such as a road) and ask questions about it. For instance, if we wanted to know the length of a road, or how many accidents had occurred on it over a given time period, we could not do this. Such 'querying' of the map is more a feature of a GIS. If we were interested, for example, in how many people live within 100 metres of a busy main road a GIS would be able to derive this for us. Moreover, if we wanted to overlay one or more maps and perhaps construct a new set of areal units from their 'intersection', then we would also turn to a GIS rather than a mapping package. This suggests that a GIS is rather more powerful than a computer mapping system for handling certain kinds of spatial problems. That this is indeed the case we shall discuss in the next section. But we shall also see that a GIS may itself fall short when it comes to allowing one to perform some of the sorts of statistical spatial analyses that are of interest in this book.

2.3 Geographical Information Systems (GIS)

A Geographical Information System (GIS) is a computer-based set of tools for capturing (collecting), editing, storing, integrating, analysing and displaying spatially referenced data. Embedded in this definition is a minimum set of 'functions' that a piece of software should include in order for it to qualify as a GIS. In our brief overview of computer mapping we have referred to the display function. We need here to say more about the other functions, beginning with data capture.

We should first distinguish *primary* data collection, such as that generated from field measurements or satellites, from *secondary* data collection, where we start with an existing hard copy map which we want to turn into digital information. We do not have much to say about the first of these, though we shall later make some reference to *remote sensing*. We should also point out that the collection of primary data through the use of Global Positioning Systems (GPS), where a constellation of satellites is used to obtain precise locational information, is a developing and important field, but not something we deal with here.

Turning to secondary data collection, there are essentially two ways in which this may be done within a GIS framework. The first of these is *vector digitising*; the second is *raster scanning*. Vector digitising captures a point object as a pair of (x, y) coordinates, while a line (such as a road, river or area boundary) is captured and represented in computer memory as an ordered string of such coordinates. Such coordinates are obtained by placing the hard copy map on a digitising table or tablet and tracing particular features using a cursor. How many points one uses to represent a complex line is a matter of judgement; clearly, the fewer the points the cruder the line representation, but the more points are digitised the greater the storage requirements. When digitising areal units such as soil polygons or administrative boundaries it is necessary to digitise a common boundary only once and then record with it the labels of the zones to the left and right. The software that is used in conjunction with the digitising uses such 'topological' information to reconstruct the map. Because of the resolution of the digitising table, lines that are supposed to meet at particular junctions will rarely do so exactly. Again, built in to the software are facilities for 'snapping together' points or 'nodes' that should indeed meet. Other features are usually also incorporated, such as those for the identification and removal of 'sliver' polygons—areas where two lines that should be coincident are very slightly displaced.

An alternative to doing one's own digitising is to make use of digital data captured by others. The obvious sources are the national mapping agencies, such as the Ordnance Survey (OS) in Britain and the US Geological Survey (USGS). The OS has in recent years undertaken a mammoth task of converting its large scale (1:1250 and 1:2500) maps into digital form. It sells products such as OSCAR (detailed road network data) and ADDRESSPOINT (digitised locations of all properties in Britain), together with digitised contour information at a scale of 1:50 000. The USGS provides digital data at three scales (1:24 000, 1:100 000 and 1:250 000) and collaborates with the US Census Bureau in providing boundary and network data (TIGER files) suitable for mapping results from the decennial Census.

In Britain, some digital data are freely available to government research establishments and universities, through centralised arrangements. For example, the boundaries of small areas (enumeration districts and electoral wards) used in the 1991 Population Census may be accessed via Regional Computer Centres, while data on roads, urban areas, hydrology, and other map 'layers', at a scale of 1:250 000 have been made available by the map publishers Bartholomew. This company has also made available for academic use digital data for Europe and the world, at smaller scales of course. In addition, the Ordnance Survey is also making available limited amounts of vector data for research and teaching.

Regardless of how we have collected our vector data, we might wish to 'generalise' the lines ('weed out' particular points that make up the line) in order to economise on storage space. We might want to have facilities for aggregating boundaries in order to construct larger regions. We might want to transform from one coordinate system, or one map projection, to another. All these are important functions within the GIS tool kit.

Raster data capture is much faster. Here, a scanner is used to sense the amount of light reflected from the surface of the hard copy map. Simplifying, we can imagine the map being converted into an array or fine grid of 'pixels'. If a section of linework intersects a pixel a value of 1 is recorded in that square; otherwise a zero is recorded (Figure 2.3).

Clearly, the value of raster scanning as a means of data capture depends upon the resolution of the scanning system; this is typically measured in dots (pixels) per inch. A density of about 200 dpi is rather crude but adequate for many purposes; a density of over 400 dpi resolves the linework very accurately. Now, since the ratio of linework to background on a typical map might be quite small we can imagine that the matrix of 1s and 0s contains a lot of redundancy. As a result, coding systems (such as 'run-length encoding' or 'quadtrees') are used to compact and efficiently store the data.

As with vector data, it is by no means always necessary to collect one's own raster data. Much of this is available for use in proprietary systems. For example, climate and vegetation data for the entire globe are available in raster format and these will come to play a very valuable role in research into, and teaching about, global environmental change.

The collection of data from satellites or *remote sensing* is essentially raster-based. The resolution of the data depends upon the satellite. For example, the LANDSAT 4 and 5 sensors yield data mostly at a resolution of 30 metres, while the French SPOT satellites have a resolution of 20 metres. These satellites are 'multi-spectral'; that is, they sense electromagnetic radiation at different wavelengths or in different 'bands' of the spectrum. In the case of the LANDSAT series seven bands are used by the Thematic Mapper (TM) scanner and these, singly or in combination, can be used to provide a rich picture of

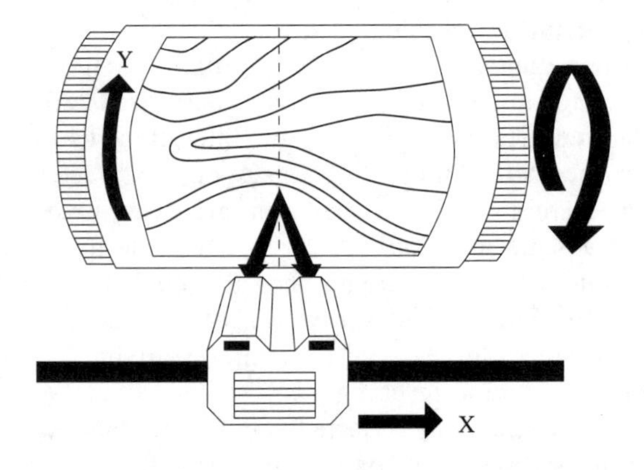

Fig. 2.3 Raster scanning using a drum scanner

land use, environmental quality, and so on. The digital data recorded for each pixel in each band vary from 0 (no reflectance) to 255. The analysis of satellite imagery involves converting this (literally) vast amount of data into meaningful information about the land surface. Although we include some discussion of satellite data in this book we do not deal in detail with the handling of such data; the reader interested in this area should consult a specialist text.

In remote sensing both locational and 'attribute' information are collected simultaneously. In some cases of raster data capture and in most cases of vector data capture it is usually the case that the locational information is captured separately from attribute information (though it may be feasible to 'tag' features such as roads with attribute data as they are being digitised). How, then, is the attribute data typically associated with the locational data?

This is usually done by using what are called relational database management systems (DBMS). We do not have space to go into detail about DBMS, which is a crucial aspect of GIS. Briefly, however, a DBMS is software that stores, manipulates, and allows the rapid retrieval of, data in a database. A relational DBMS can be thought of as a set of tables, where the rows are 'entities' and the columns are features or 'attributes' of these entities. For example, we might have a set of census zones (entities), the boundaries of which have been digitised. As part of the structuring of these vector data the GIS would be expected to calculate some simple geographical attributes such as the area of each zone, and to create a simple table with this information included. We might then want to add census data for each of these zones; for instance, percentages of the resident population owning one or more cars, owning their homes, aged over 75 years, and so on. This defines another table, where the entities are again zones but the attributes are the census variables. The two tables can then be linked together by what are called 'relational joins' so that, in a GIS context, particular features may be extracted for particular locations; for example, displaying all zones where the proportion of owner-occupation was less than 50 per cent.

The asking of questions, or 'querying' of the database is an important function of a GIS. 'What' and 'where' questions are typical. For example, we might want to retrieve and display all zones where owner-occupation is under 50 per cent and car ownership is less than 50 per cent. Or, we might want to point at a specific zone and enquire about certain of the attributes associated with it. In order to be able to answer more complicated questions like these it is likely that two or more tables, stored as part of the database, must be joined together. If two tables have different locational data as entities such joins become an 'overlay' operation. This can be visualised in terms of the physical overlay of two maps drawn on tracing film. For example, if we return to our census zones we might have one table consisting of these zones as rows and attributes comprising total population and the population aged under 5 years. Another table might be a set of road segments or 'arcs' in the same study area. Each of these might have attributes comprising road class, average daily traffic levels, and the proportion of heavy goods vehicles passing along each arc of the network. Suppose we wanted some assessment of the 'risk' (maybe in terms of

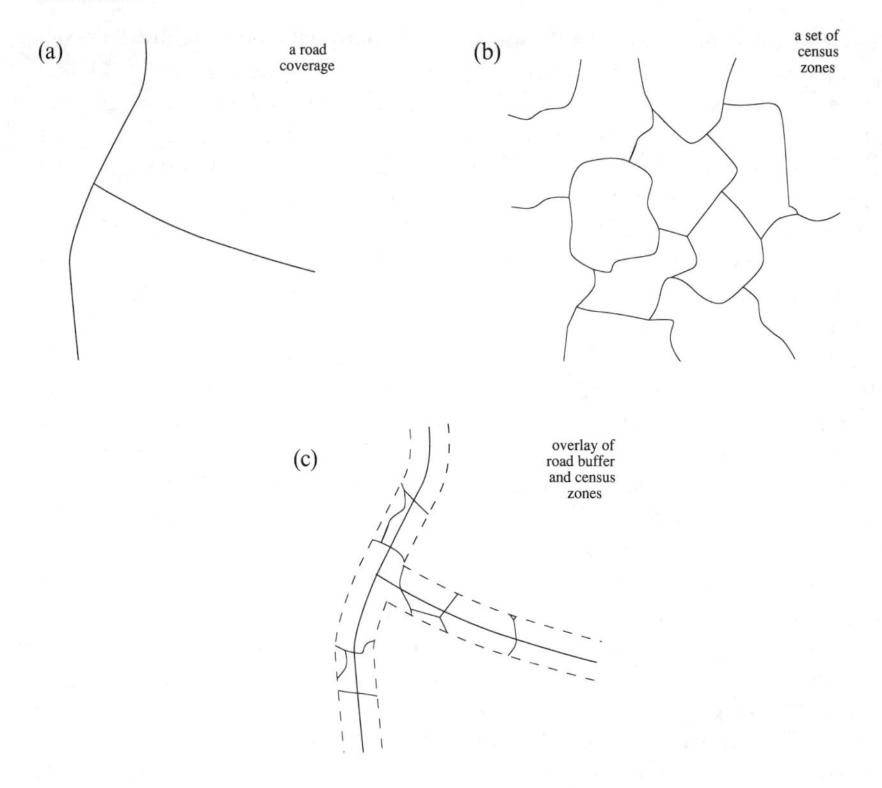

Fig. 2.4 Overlay and buffering of spatial features

traffic accidents or likely inhalation of noxious exhaust emissions) to residents living in the census zones. We could then perform an overlay of the two sets of spatial features, zones and roads, and then place a 'buffer' zone of specified width along each road arc (Figure 2.4). Such a buffer zone could be made proportional to the class of road. An estimate of the childhood population living near to such roads, and therefore 'at risk', could then be made.

Before leaving a consideration of GIS we ought to say something about the kinds of system that are available to the user community. Needless to say, there is a wealth of such systems. One way of getting a feel for this range, and the functionality each offers, is to attend one of the numerous conferences held regularly, in many different parts of the world, at which vendors come to set out their (soft)wares. Two particular systems have achieved prominence in the higher education market. These are ARC/INFO and IDRISI. The former is essentially vector-based, but has good functionality for handling raster data. The latter is fundamentally raster-based, but allows vector overlay. Both have facilities for converting from one data structure to another. ARC/INFO is available on a range of hardware platforms, including PCs; IDRISI is designed for the PC environment. Both offer all the GIS functionality we have dealt with above, as well as some of the analytical methods we consider in this book. For example, IDRISI contains modules that perform some simple quadrat analysis of the sort considered in Chapter 3; the kind of trend surface analysis outlined in Chapter 5; and the sorts of spatial correlation statistics we discuss in Chapter 7. The GRID module of ARC/INFO allows the user to perform many of the functions we consider in Chapter 5, especially interpolation methods such as 'kriging'. However, as with IDRISI much of this spatial statistical functionality requires that the data be in raster format. Turning to other sorts of spatial analysis, such as the spatial interaction models considered in Chapter 9, both ARC/INFO and IDRISI contain functions to perform some kinds of network and location-allocation modelling.

Of course, IDRISI and ARC/INFO are by no means the only proprietary systems that offer some spatial analysis functions. Others, such as GRASS, SPANS, MapInfo, INTERGRAPH and TransCAD, also contain valuable functionality. We have highlighted IDRISI and ARC/INFO simply because they are among the most widely used packages within the education and research communities.

2.4 GIS and spatial analysis

Given that some spatial analysis capabilities are available in widely used systems, is there a need for spatial analysis functions beyond those currently provided in GIS? At the present time there is; for example, a GIS will currently be able to overlay a set of points (representing childhood cancers, for example) onto a set of polygons (such as buffer zones constructed along high voltage power lines). The GIS will then be able to perform a 'point-in-polygon' operation to count how many points lie within particular polygons, or outside. But one is then hard put to find a system which evaluates statistically the nature of the association between the set of points and the set of polygons. If we want to know whether there is a statistically significant association between the incidence of childhood cancer and proximity to high voltage power lines, we cannot do this readily. In general, we need to go outside the GIS to test this sort of hypothesis.

There are several ways in which we might 'couple' these kinds of spatial analysis to GIS. The first of these is 'full integration' of spatial analytical functionality within the GIS; this is present to a limited degree in some existing systems. But it is unrealistic to expect the vendors of systems that are designed primarily for the commercial world to incorporate each and every method for exploratory spatial data analysis, or spatial modelling, into such systems. Fortunately, there are alternative ways of attaching this kind of functionality to GIS.

One of the approaches to the problem may be referred to as a 'loose coupling' between the GIS software and software used for spatial analysis. Here, data are exported from the GIS package for use within a statistical spatial analysis framework. This amounts to having the GIS and the separate analysis software 'talk to' each other simply by the interchange of files that each can read. While this offers some flexibility it is inefficient. A better strategy is 'close coupling', which involves calling a spatial analysis routine from within the GIS. This exploits the 'macro language' facilities available in many proprietary GIS. With this approach, one does not need to leave the GIS; one simply runs a command which, hidden from the user, calls a compiled program (or programs) and implements the required analytical function. This method has been explored by a number of researchers, working with particular systems. For example, a research team from Lancaster, Leeds and Newcastle Universities in the UK, has developed a suite of spatial analysis routines that may be called from within ARC/INFO. Research groups in North America have been pursuing similar goals. The package IDRISI also allows the user to link in compiled programs, while the system GRASS has the same kind of 'peg' on which to hang spatial analysis.

An alternative approach adopted by some researchers is to develop their own self-contained spatial analysis systems, adding in some of the features we might expect from a GIS. One example is the 'Geographical Analysis Machine' (GAM), developed by Stan Openshaw as an aid to detecting the existence of leukaemia 'clusters'. Another example is SpaceStat, developed by Luc Anselin. This offers the user a range of options for exploratory data analysis and model-fitting, of particular value to those with an interest in spatial econometrics.

There is yet another option. Some researchers have taken statistical analysis systems and added spatial analysis and limited GIS functions to these. For example, Dan Griffith has taken the widely used package MINITAB and developed a series of 'macros' that implement some exploratory and model-fitting methods for use with spatial data. We take up some related initiatives in more detail in the next section.

2.5 Interactive spatial data analysis

In this final section, we wish to draw the reader's attention to some of the recent developments taking place in the particular area of interactive spatial analysis. Such work builds upon dramatic changes that have taken place in recent years within the field of statistical computing. In particular, dynamic statistical graphics environments permit the user to see simultaneously on screen several different types of graphical display, such as histograms, scatter diagrams, numeric lists of data, and so on. These displays are known as 'views' of the data. Such views may be active, in that the user can, with a mouse, select and highlight subsets of the data. The views may also be linked, such that selecting and highlighting a subset of observations in one view causes those observations to be highlighted in other views or windows.

John Haslett and his colleagues at Trinity College, Dublin have exploited these ideas in relation to spatial data by developing a system called REGARD, within an Apple Macintosh environment. For example, one can display a map of locations at which stream sediments have been sampled, showing concentrations of, say, copper and then, in another window of the screen, display a scatter diagram of copper against molybdenum concentrations. Because of the dynamic linking of windows, if a small region of the map is highlighted using the mouse then the observations that have been defined spatially in this way are also automatically highlighted on the scatter diagram. Or, to give another example, we might wish to select high values from a histogram in one window; such selection would then automatically indicate the locations of such observations in the map view. More advanced uses of the system involve looking at correlation structures in the data, techniques which we discuss in more detail in Chapter 5.

A second example to which we draw attention relates to the statistical programming environment, S-Plus, which is widely used by professional statisticians for various kinds of statistical analysis and modelling. In addition to the facilities for exploratory data analysis and modelling that we would expect of any statistical package, S-Plus also includes high-level functions to perform various statistical and mathematical manipulations, and excellent dynamic graphics. Moreover, it embodies a programming language that can be used to link certain fundamental operations into more complex tasks; these then become new functions that can be used for particular purposes. For example, a spatial analysis system for handling point data has been written, based around S-Plus and operating within a UNIX environment. Called SPLANCS, this plots point maps, allows spatially smooth estimates of local intensity (density) of points to be made, and offers a number of options for summarising the spatial structure of a point pattern and for modelling such processes. These features form a 'library' of functions that may be called within the S-Plus environment. Some of these features we shall ourselves use in Chapters 3 and 4 where we discuss methods for the analysis of point patterns. One example of using SPLANCS might be in the investigation of space–time clustering; are events that are close together in geographical space also tending to occur at approximately the same points in time? Using the interactive graphics one can display the sequence of point events over time, as a kind of diffusion process, thus visualising local space–time interaction that subsequent analysis might confirm. Such methods have been put to real practical use in the study of Legionnaire's disease in Scotland, which we discussed in Chapter 1. Libraries of S-Plus functions, other than SPLANCS, also exist for the analyses of spatial data other than point patterns, although at the current time these are less well developed.

Recently, a 'close coupling', or interface, between S-Plus and ARC/INFO has also been developed. The virtue of this is that it allows the transfer of an ARC/INFO database to S-Plus for subsequent statistical analysis. Results from the analysis may then be returned to ARC/INFO to form additions to the database. Graphics, such as scatter plots and histograms, may be saved within S-Plus and then imported into ARC/INFO for use within the mapping module. For those interested in seeing further links between GIS and statistical analysis such developments have considerable promise. The price paid, however, is the need to get to grips with two quite demanding pieces of software!

A system that has much in common with S-Plus is known as XLISP-STAT. This is based on the computer language LISP (the 'X' comes from the use of the system within the so-called X-Windows environment) and is a further example of an interactive statistical analysis framework. Its main advantage over S-Plus is the ease with which dynamic linking of windows can be implemented, to reproduce the kind of facilities discussed earlier in relation to the 'stand-alone' system REGARD. As a result, using routines developed within XLISP-STAT, we could, for example, draw a map of crime incidence in one window, then view a histogram of the ages of victims in another window. We might then select graphically a subset of cases in the histogram window by taking all those aged over 65 years. The locations of these cases will then appear in the map view, by virtue of the link between the two windows.

In many of the kind of systems outlined above, the emphasis is not on cartographic excellence in the map displays. Although we should, as discussed earlier, recognise the importance of cartographic issues and high quality maps, we also need to remember that the aim of interactive data analysis is to view the data in a dynamic and imaginative way. Obviously it is important to be able to draw maps, but the objective of such maps will in general be to provoke questions and to gain immediate feedback from the analysis that has gone before. The purpose of interactive spatial analysis systems is not primarily to draw elegant maps; if this is the goal one must look to a high-quality mapping or GIS package.

Ex 2.4

Looking to the future, how might we see interactive spatial data analysis tools, such as those we have described, developing? Given the rapid changes in computer processing power presently occurring, we may well see enhancements becoming possible which are computationally infeasible at the current time. For example, one development might be to have the user define a window of interest on the map, compute some statistics or perform some exploratory data analysis on the data contained within that region, and then to 'drag' the window over the map in order to see, in real-time, how such statistics change over space. As we have seen in this chapter, there are a number of current research initiatives which involve the development of new tools for interactive spatial analysis. At the same time, proprietary GIS are being revised to incorporate enhanced functionality for spatial analysis. We hope that such developments will ensure a healthy and interesting future for the link between computing and spatial data analysis.

2.6 Summary

In this chapter we have only given a brief outline of the 'state of the art' concerning the relationship between computing, mapping and spatial data analysis. Developments in this field are occurring rapidly, and by the time this book is published we are only too well aware that some of the initiatives to which we have referred will have been overtaken by events and seem 'old hat'.

Undoubtedly some already have been and we have overlooked some recent developments that we should have mentioned! Nevertheless, we hope that the main structure of our discussion will continue to be of value in the future. In particular, we hope that we have made the point that computing has far more potential to offer to those interested in spatially referenced data than simply quick ways to draw maps. GIS involve much more than just display facilities and these other functions are ultimately far more important. However, when it comes to spatial analysis and computers, we have also made the point that analysis involves more than just fast and flexible spatial query and aggregation and we have argued the need for closer links between GIS or computer mapping systems and spatial analysis tools. We have discussed the ways in which various initiatives are beginning to achieve this, a trend which we hope will continue to develop until it reaches its full potential.

We have included this chapter in the book because we believe that an awareness of appropriate computing developments is important to the analyst dealing with spatial data. At the same time, we do not believe that GIS and related software, however sophisticated they become, will ever obviate the need for an awareness and understanding of basic spatial data analysis methods. Software developments will open up new possibilities and suggest new developments in spatial data analysis methodology, as they have already done; but the kinds of techniques which we now go on to discuss in the subsequent parts of this book, will, we are convinced, remain at the heart of extracting meaning and understanding from spatial data.

2.7 Further reading

A good introduction to computer mapping, including a brief introduction to GIMMS, may be found in:

Maguire, D. (1989) *Computers in Geography*, Longman, Harlow.

The proportional symbol mapping package we mention is described fully in:

Rase, W-D (1987) The evolution of a graduated symbol software package in a changing graphics environment, *International Journal of Geographical Information Systems*, 1, 51–65.

For those interested in computer mapping within an Apple Macintosh environment see:

Whitehead, D.C. and **Hershey, R.** (1990) Desktop mapping on the Apple Macintosh, *Cartographic Journal*, 27, 113–18.

An excellent discussion of good cartographic practice, with much of relevance to computer cartography, is:

Monmonier, M. (1991) *How to Lie with Maps*, University of Chicago Press, Chicago.

There is a substantial, and ever-growing, literature on GIS. The most comprehensive source is the collection of essays in:

Maguire, D.J., Goodchild, M.F. and **Rhind, D.W.** (1991) *Geographical Information Systems: Principles and Applications*, Longman, Harlow.

A fine introduction to remote sensing and the analysis of satellite data is:

Mather, P.M. (1987) *Computer Processing of Remotely-Sensed Images*, John Wiley, Chichester.

An excellent text on the use of interactive statistical graphics is:

Cleveland, W.S. and **McGill, M.E.** (1988). *Dynamic Graphics for Statistics*, Chapman Hall, London.

Research that explores links between GIS and spatial analysis is reported in the following edited collection:

Fotheringham, A.S. and **Rogerson, P.** (1993) (eds) *Spatial Analysis and GIS*, Taylor and Francis, London.

Luc Anselin's SpaceStat is described in:

Anselin, L. (1992) *SpaceStat: a program for the analysis of spatial data*, National Center for Geographic Information and Analysis, University of California, Santa Barbara, Ca. 93106–4060, USA.

while Dan Griffith's work on developing MINITAB macros for spatial statistics is outlined in:

Griffith, D. (1988) Estimating spatial autoregressive model parameters with commercial statistical packages, *Geographical Analysis*, 20, 176–86.

John Haslett's research is described in:

Haslett, J., Bradley, R., Craig, P., Unwin, A. and **Wills, G.** (1991) Dynamic graphics for exploring spatial data with application to locating global and local anomalies, *The American Statistician*, 45, 3, 234–42.

while the work on SPLANCS is reported in:

Rowlingson, B.S. and **Diggle, P.J.** (1993) SPLANCS: spatial point pattern analysis code in S-Plus, *Computers and Geosciences*, 19, 627–55.

For details on XLISP-STAT see:

Tierney, L. (1990) *LISP-STAT: An Object-Oriented Environment for Statistical Computing and Dynamic Graphics*, John Wiley, Chichester.

A good introduction to ARC/INFO is provided in:

Understanding GIS: the ARC/INFO Method, Longman, Harlow.

There are various user guides to, and instructional materials associated with, IDRISI. See:

Eastman, J.R. (1993) *IDRISI: A Geographical Information System*, Clark Labs for Cartographic Technology and Geographic Analysis, Clark University, Worcester, Massachusetts, 01610–1477, USA.

2.8 Computer exercises

Here are suggested exercises that you can try on ideas discussed in this chapter, using INFO-MAP and our example data sets. These exercises have been referenced at appropriate points in the chapter by numbered symbols in the margin.

Exercise 2.1

INFO-MAP is not designed as a computer mapping package and certainly not for high quality computer cartography. Nevertheless it does have a limited range of features which illustrate some of those that one might expect from such systems and it may be of interest for the reader to experiment critically with some of these. Most of the ideas in this exercise are conceptually very simple; however, their value in preliminary data analysis should not be underestimated.

Use `Open` from the `File` menu to open the file 'Chinese socio-economic measures' and then load the associated data using `Open` from the `Data` menu. This data set is described in more detail in Chapter 5; here we simply use it to illustrate a number of ideas concerned with computer mapping. First note the relative 'coarseness' of the digital display on the screen, both the map boundary and the text. If you look closely, the individual screen 'pixels' (dots) are quite clearly distinguishable. This demonstrates that INFO-MAP uses only a fairly low resolution graphics display. In general the 'resolution' (number of vertical by horizontal 'pixels') of any computer graphics is limited firstly by what any particular hardware will support and secondly by possible additional restrictions built into a specific software package. Most hardware is capable of operating at a number of different resolutions, with a maximum of about (1000×1000) (approximately eight times more 'dense' than that used by INFO-MAP) being fairly common on modern workstations. Digital mapping systems, particularly systems designed for computer cartography, are usually designed to operate at the highest possible resolution, allowing the maximum amount of detail to be displayed.

Resolution is also usually, although not necessarily, related to the range of colours able to be displayed. INFO-MAP uses a 'palette' of only 64 colours, only 16 of which can be displayed at any time. Select `Colours` from the `Map` menu, and simply experiment with using the left and right cursor keys to change the colour of just the first box displayed. In this way you can move through all 64 colours that INFO-MAP can display. Notice that none of these colours are particularly subtle. Most digital mapping systems would be able to display 256 or more colours, which enables a far greater degree of colour gradation to be employed. Use <ESC> to exit from the colour function and restore the original colours.

Next, note the range of different types of map available. These are described in more detail in later chapters, but at the moment simply experiment with using `Dot map`, `Choropleth map` and `Symbol map`, from the `Map` menu, to display different views of one or more of the variables in the data file on China. `Symbol map` allows the option to map information about two variables in the data file simultaneously, one which will relate to symbol size, the other to symbol colour. The same variable may be selected for both options if this flexibility is not required. Try using such a 'bivariate map' and note that it is possible to display a map legend showing the relationship of data values to size of symbol, in addition to that relating to colour classes. In order to do this use `Key` from the `Map` menu and set the option for 'symbol size key' to 'on'. Then redraw the map; legends for both colour and symbol size should then be displayed. In passing, note that by a similar process one could 'switch off' the display of both kinds of map legend. The default symbol type used is a circle, whose area is proportional to data value. However, by using `Symbol type` in the `Map` menu you can select an alternative, a rectangle, whose height is proportional to data value. Explore this idea.

The colours used for the map may be adjusted to better suit the requirements of a specific application. We referred to this earlier in one of our exercises concerned with mapping Burkitt's lymphoma cases in Chapter 1. Here try experimenting with the option again for the China data. Draw a `Choropleth map` of one of the variables and then use `Colours` from the `Map menu` to adjust the colour used for any of the class intervals, the text of the map legend, the boundary of the map, or the background screen colour. Using <ESC> from this option restores the original colours, whilst <CR> will retain the new colour settings. The ability to set particular colours in a mapping package may be particularly important when it comes to printing maps, since determining a suitable 'mapping' between screen and printer colours or grey scales is often largely a process of trial and error. Using `Colour mode` in the `Map` menu and selecting the 'monochrome' option will cause a subsequent `Choropleth map` to be shaded with various patterns of the same colour, rather than with solid blocks of different colour. By using `Colours` from the `Map` menu whilst in this 'monochrome' mode, the shading colour may be changed and the particular shading styles used for each class interval may be adjusted, to a limited extent. Experiment with these options, perhaps trying them with a different data set, such as the file 'US Presidential election (1992)', first seen in Chapter 1.

Turning to the map legend in more detail, note that by using `Scaling` from the `Map` menu you can vary the number of class intervals and the scaling method used to determine the range of data values in each. A number of different scaling methods are provided, including an option for the user to set their own class intervals. Text names can be assigned to class intervals, to be used in the map legend in place of the data range for the class, if required. An option is also provided to determine class intervals according to a fixed variable in the data file, rather than with respect to that being mapped at any time. This is useful for generating a series of maps of different variables on a common basis. Explore the effect of these options.

Also note that you can 'superimpose' one type of map on top of another within INFO-MAP. When the `Superimpose` option in the `Map` menu is set to 'on' the next map is drawn 'on top' of that already displayed, rather than clearing the map first. So, for example, you could produce a `Choropleth map` of 'Industry/All output', then set 'superimpose' to 'on' and draw a `Dot Map` of 'Coal output'. You should first use `Scaling` and set the 'scaling data item' to 'Industry/All output' so as to force both maps to be coloured in relation to the same class intervals. Remember to set 'superimpose' to 'off' again afterwards, otherwise all subsequent maps will continue to be superimposed.

Finally, note that most of the options discussed may be saved so that they become the 'defaults' used for subsequent runs of the package. This is done by using `Save settings` from the `Preferences` menu. This menu also contains a number of other optional settings not specifically related to mapping. These are described in more detail in the *User's Reference*.

Exercise 2.2

The objective of this exercise is to illustrate some of the general points made in the text concerning the capture of maps and spatial data in digital form for use on computer. At the same time this provides us with an opportunity to introduce the reader to the various ways in which maps and associated data can be prepared within INFO-MAP, so that if required the package may be used in conjunction with the reader's own data sets and maps rather than only the example data we provide. It is a long and detailed exercise and you might wish to skip over it if you are content to use just our example data sets and have no wish to create your own. If you decide you do want to create your own data sets at some future time, then you can always return to this exercise when you wish. On the other hand, if you are interested in finding out a bit more about how computers store maps and spatial data and in particular about how INFO-MAP does this, then you should carry on with the exercise.

INFO-MAP supports, to a limited extent, both vector digitising and raster scanning of spatial data. However, before discussing these in more detail it is useful to get a basic understanding of the way in which a spatial data set is structured within the package. Clearly, details of this are specific to INFO-MAP, but some of the ideas involved are more generally applicable and also relate to more sophisticated packages.

One general point to appreciate about storing spatial data on a computer, is a distinction between a spatial 'object' such as a 'point', 'area' or 'line', and particular data values which may be associated with that spatial 'object'. Since many different data values may be related to a single spatial 'object', it is typical to store data separately from their associated spatial 'objects' and maintain links between them. In GIS systems these links are sophisticated and, as described in the text, commonly involve a relational database management system. Within INFO-MAP, which has no pretensions to be a

GIS, a similar division of spatial 'objects' and data is present, but only a few very restricted types of spatial 'objects' are allowed (points or closed polygons) and the link between data and 'object' is implemented in a very basic way, which allows for little flexibility in how data and 'objects' may be manipulated and used.

The idea in INFO-MAP is as follows. Each data set consists of at least two components, firstly, a *base file* and secondly one or more *data files* associated with this. The *base file* actually fulfils two functions. Firstly, it contains an appropriately grid-referenced monochrome 'map' (screen image) showing a simple outline of the appropriate 'study region' and any relevant sub-divisions of this, such as administrative zones or districts. This is referred to as a *boundary* in INFO-MAP. Secondly, the *base file* contains the set of spatial coordinates within the 'study region' where data values have been observed. These are referred to as *locations* within the package. *Data files* associated with a particular *base file* then contain the observed data values for these *locations*. They are each best thought of as a 'spreadsheet' in which rows correspond to the *locations* specified in their associated *base file* and columns correspond to different data items, that is, attributes or variables recorded at each of these *locations*. Notice in particular that the spatial coordinates of data values are stored in the *base file* not the *data file*; everything 'spatial' about the data set resides in the *base file*. Most of our example data sets have only one *data file* associated with each *base file*, but the distinction between the two types of files allows for the possibility of several sets of attributes stored in separate *data files* each of which relates to the same set of *locations* and the same *boundary* as defined in their corresponding *base file*. These ideas are discussed in more detail in the *User's Reference*, and the reader may like to refer to this before embarking on the following exercise which is concerned with setting up a simple data set within INFO-MAP from scratch.

When using INFO-MAP to enter a new data set, the *base file* must always be created first, so establishing the *boundary* which defines the 'study region' and the set of *locations* for observed data values. There are a variety of ways to create or modify these two components of a *base file*. *Boundaries* can be drawn or edited on the screen, using a mouse or graphics tablet, or entered directly from hard copy using a scanner. Alternatively, *boundaries* may be 'imported' from ASCII files of polygon coordinates, or from raster 'bitmaps' of graphic images. *Locations* for data values can then be defined and named using the mouse and keyboard, or alternatively imported from ASCII files of grid coordinates. All these possibilities are described in detail in the relevant chapter of the *User's Reference*. In this exercise we consider just one option, that of importing a *boundary* from an ASCII file of digitised polygons, or line segments, and importing a set of *locations* from an ASCII file of grid referenced records.

Consider first the importing of the *boundary*. With the example data sets we have included a file '`iommap.txt`'. This comprises digitised coordinates for each of 25, 1991 census districts, on the Isle of Man (there are, in fact, only 24 administrative divisions, but one, Rushen, is split into two separate zones). You can, if you wish, inspect the file using any ordinary text editor on your computer. The format is very simple and is typical of that produced by various types of vector digitising software, although different packages will involve their own particular variations in format. Given the powerful facilities of modern text editors it is not difficult to transform most digitised coordinate data into the basic type of format used in '`iommap.txt`', which is the type required by INFO-MAP. Such files should contain coordinate pairs of digitised points for one or more contiguous line segments or closed polygons. Each of the line segments or closed polygons included in the file must start with a header record containing the number of points involved. This is then followed by a sequence of exactly that number of records, each containing the coordinate pair (easting, northing) for the successive points in turn. This may then be followed by the next header record, and so on. Coordinate pairs in a record are delimited by a comma, or one or more blank spaces. Any coordinate system may be used so long as it adheres to the convention that 'eastings' increase 'left to right' and 'northings' increase 'bottom to top' and that unit

Table 2.1 Format of ASCII file for digitised Isle of Man districts

127	
6914.	8175.
6893.	8201.
⋮	⋮
(127 pairs for zone 1)	
⋮	⋮
120	
3287.	2618.
3287.	2671.
⋮	⋮
(120 pairs for zone 2)	
⋮	⋮
(etc.)	

increases in 'eastings' or 'northings' imply equivalent straight line distance. In this particular case the file appears as in Table 2.1.

In order to import this file into INFO-MAP as a new *boundary* choose `New` from the `File` menu to indicate that you wish to set up a new *base file* and then select the 'import from ASCII file' option. (Notice the other possibilities in passing). Type in the file name '`iommap.txt`' followed by <CR> to start the import. INFO-MAP first reads the complete file to establish the coordinate range of the whole area involved, and then begins to draw each of the zones, scaling them so that the overall maximum and minimum 'eastings' and 'northings' fit onto the screen whilst at the same time preserving the 'aspect ratio' of the 'study area'. At the end of this process you are placed in the *base file* module of INFO-MAP (notice the menu bar at the top of the screen has changed), so that further modifications can be made to the imported *boundary* if required and then the *locations* for data observations added, before results are saved as a new INFO-MAP *base file*. Details of options available within this module of INFO-MAP are all described in the appropriate chapter of the *User's Reference*.

In this particular case, the imported *boundary* requires no editing and so it simply remains to set up the second component of the *base file*, namely the *locations* for data observations. During the import of the *boundary*, INFO-MAP automatically set up a grid referencing for the *boundary*, based upon the coordinate system which applied to the digitised polygons. Each point in the 'study region' is therefore identified in terms of this coordinate system. We may therefore import the *locations* of data observations from an ASCII file of similarly grid referenced points (other possibilities would be to add such *locations* manually with the mouse, or generate them to be on a regular grid within the 'study region'). The file '`iomdat.txt`', supplied with the example data sets, contains a suitable set of grid referenced records.

The file contains data records each comprising a set of ten variables (taken from the 1991 Census) for each of 25 districts on the Isle of Man (for the split district, Rushen, the original raw data values have been apportioned to each of the two zones). We will come to these data values later. At present our interest is in the first three fields of each record, which contain, respectively, the name of the district to which the record relates and then the 'easting' and the 'northing' of its district 'centre'. We can import just these three fields as the *locations* for data observations. Again you can inspect the file using any ordinary text editor. It appears as shown in Table 2.2.

Table 2.2 Format of ASCII file for Isle of Man census data

'Andreas'	6411.	8174.	1.64	76.04	...
'Arbory'	2970.	2594.	1.20	79.39	...
'Ballaugh'	5226.	6945.	1.50	74.86	...
⋮	⋮	⋮	⋮	⋮	...
		(etc.)			

To import *locations* from this file, use `Merge` from the `Locations` menu. This allows the position and labels of new *locations* to be imported from an ASCII file with an appropriate format and added to those existing in the current *base file*. The ASCII file is expected to contain one record for each *location* to be imported and each of these records must have a minimum of two fields, one for the easting and one for the northing of the corresponding *location*. In this case, confirm that the grid reference system of the current *base file* is appropriate to the coordinates in the file to be imported with a 'y' response. A window of user prompts is then displayed. Type in the file name '`iomdat.txt`' and <TAB> to enter the other prompts, entering 2 for the 'easting' field, 3 for the 'northing' field and 1 for the 'name' field. Leave the 'filter' prompt blank and enter <CR> to start the import.

You can then check the *locations* imported, by using `Display` in the `Locations` menu to display their position (use <ESC> to end this display). Alternatively, you could use `Modify` in the `Locations` menu, which allows you to <TAB> around their positions displaying the corresponding labels and grid references; again <ESC> ends this process. Now that both the *boundary* and *locations* have been imported, the *base file* may be saved by using `Save` in the `File` menu. Having set up and saved the *base file* for the new data set, you can now exit from the *base file* module by using `Exit` from the `File` menu; this returns you to the familiar 'top level' of INFO-MAP.

To complete the new data set we now need to set up a *data file* associated with the new *base file* (note that this file is still 'open'). The *data file* will contain values for data items of interest recorded at each of the *locations* specified in the *base file*. Again, several ways of doing this are possible; in this case we will simply import data values from the same ASCII file, '`iomdat.txt`', from which the *locations* were taken. In order to import this file into INFO-MAP as a new *data file* choose `New` from the `Data` menu to indicate that you wish to set up a new *data file*. Since we know that data records occur in the file in the same order as the *locations* already set up in the *base file* (because these were imported from the same file), we can use the 'import by location sequence' option. (Notice the other possibilities in passing). Then select 'free format'. A window of user prompts is then displayed. Type in the file name '`iomdat.txt`' and <TAB> to enter the other prompts. In this case the file contains ten data items. The variables are shown in Table 2.3 (each expressed as a percentage of either total population or numbers of private households).

These variables are in data fields 3–12 in the file (the district 'label' field is not counted as a 'data' field in INFO-MAP and fields 1 and 2 contain the 'easting' and 'northing' used earlier) so enter 3, 12 and 12 in response to the prompts concerning which 'columns' to import. There are no variable 'names' in the first record of the file, so specify 'n' to indicate this. We wish to import all records in the file so specify 1 in the 'rows from' prompt, and there are district 'labels' in the first field of the file, so specify 'y' in response to the 'row labels' prompt. Finally use <CR> to perform the import. When the import is complete, the *data file* module is entered (note the menu bar has altered) so that further additions or modifications can be made to the data values which are displayed, before they are saved to a *data file*. Details of options available within this module of INFO-MAP are all described in the appropriate chapter of the *User's*

Table 2.3 Description of Isle of Man census variables

1. Population speaking Gaelic Manx
2. Households that are owner-occupied
3. Households renting from Government or Local Authority
4. Households without a telephone
5. Households with central heating
6. Population aged 0–4 years
7. Population aged 5–9 years
8. Population aged 0–9 years
9. Population aged over 70 years
10. Population that is employed

Reference. Note the row labels of the data display are automatically derived from the *base file* that has already been set up; they are part of the *base file* not the *data file*.

In this case we only need to specify names for each of the data items imported. Use Name from the Data menu to do this. Then highlight a column name using the <TAB> or <SHFT> + <TAB> keys (these will scroll the display of columns if required) and enter an appropriate column name. Note that data item names are restricted to a maximum of 30 characters so use suitable mnemonics as appropriate ('gaelic', 'ownocc', etc). Entry of the name may be terminated by <TAB> or <SHFT> + <TAB>, allowing another column name to be added. <CR> will end the column naming process. Once all the column names have been entered, the *data file* may then be saved by using Save from the File menu. Having set up and saved the *data file*, the new data set is now complete. If you now exit from the *data file* module by using Exit from the File menu, this returns you to the familiar 'top level' of INFO-MAP and you can then proceed to map and analyse the data. For example, you might now try using Choropleth map in the Map menu, to map one of the variables you have in the data file.

This has been a long and in some ways quite involved exercise, but hopefully it has conveyed to the reader some of the general ideas behind the capture of spatial data on computer, albeit in the specific context of one particular package. As has been mentioned, even within INFO-MAP, which is unsophisticated compared with some of the systems described in the text, there are several possibilities for setting up maps and data values other than import from ASCII files. All these are fully described in the *User's Reference*.

Before leaving the subject of capturing spatial data, let us finally see how to add a further variable to an existing *data file* within INFO-MAP. Consider the data in Table 2.4, which relate to the percentage of the Isle of Man population with a long-term illness, again taken from the 1991 Census. To add these data to the existing *data file* that you have already set up, use Modify from the Data menu. This places you back into the *data file* module and allows you to change an existing *data file*. Next use Add from the Data menu which adds a blank column to the end of the data file. Use the cursor keys to scroll the screen display until this column is visible and then add the data listed above into the blank column. Next use Name from the Data menu and <TAB> to the column to enter a name for it. Finally use Save from the File menu to save the modified file. You can now exit from the *data file* module by using Exit from the File menu and return to the 'top level' of INFO-MAP. The data set is now modified and you can then proceed to map and analyse the data again, including the new variable that you have just added. Produce a choropleth map of this new variable, with appropriate interval classes and speculate on why the district of Braddan has such a high proportion of its population with long-term illness.

Table 2.4 Isle of Man % 1991 population with long-term illness

Andreas	9.0	Marown	7.1
Arbory	9.5	Maughold	8.5
Ballaugh	9.9	Michael	5.3
Braddan	16.7	Onchan	8.4
Bride	7.4	Patrick	5.3
Castletown	8.2	Peel	9.5
Douglas	9.6	Port Erin	10.3
German	5.8	Port St Mary	12.4
Jurby	6.7	Ramsey	11.8
Laxey	10.4	Rushen	8.6
Lezayre	7.1	Rushen (det)	8.6
Lonan	6.7	Santon	5.9
Malew	8.3		

Exercise 2.3

INFO-MAP is not a GIS and has few of the functions that one would expect of such a system. Nevertheless we can use it to illustrate some simple ideas concerned with spatial query. Open the file 'Barnet census data' and load the data file entitled 'census counts'. This data set is described in detail in Chapter 7; it contains a number of Census counts (note: not percentages) derived from the 1981 UK Census, for enumeration districts in the area of Barnet in North West London.

Suppose we wished to select those enumeration districts where owner-occupation is under 50 per cent and car ownership is less than 50 per cent. We could do this by creating a new variable with the value 1 if this is true and 0 otherwise. Use `Calculate` from the `Data` menu and enter the formula:

```
interesting=if({2}/{1}<0.5 and {4}/{1}>0.5, 1, 0)
```

Here, data item 1 is the number of private households, 2 is the number which are owner occupied and 4 is the number with no car. A choropleth map of the new variable 'interesting' (with discrete class intervals) clearly highlights the districts meeting the criteria specified. If we now wanted to calculate the overall percentage of households with pensioners in just these districts, we could use `Calculate/display worksheet` from the `Analysis` menu, choose the 'calculate' option and enter the formula:

```
work[1]=sum(if({11}=1,{8},miss()))/sum(if({11}=1,{1},miss()))
```

where data item 11 is 'interesting', the variable just created, and variable 8 is the number of households with one or more pensioners. In fact we didn't need two steps to do this since the worksheet formula:

```
work[1]=sum(if({2}/{1}<0.5 and {4}/{1}>0.5, {8}, miss()))
        /sum(if({2}/{1}<0.5 and {4}/{1}>0.5, {1}, miss()))
```

would have given the same result in one step.

Nevertheless our new variable 'interesting' becomes more useful if we want to carry out several analyses on just the subset of districts we have selected, rather than calculate a single value. Instead of repeatedly using formulae that involve a logical condition relating to the selection of these areas, we could temporarily 'restrict' the data file by removing all but the selected districts. To do this choose `Select locations` from the `Analysis` menu with the option 'Select using data item'. A menu of data items is then presented. Choose the item 'interesting'. You have now created a 'select list' containing the districts of interest. Such a list may be used in various ways in INFO-MAP. In particular it may be used to remove *locations* from a data set, so creating a new temporary data set. To do this choose `Zoom/Restrict` from the `File` menu and then the option 'Restrict to locations already selected'. The result is to remove from the data set all but the selected districts, so that all subsequent analysis involves only those districts. For example, draw a choropleth map of the percentage of 'Local Authority' housing (you should now know how to calculate this) and only the selected districts will be mapped. Similarly the worksheet formula:

```
work[1]=sum({8})/sum({1})
```

now gives the same result as the more complex expression used earlier. Plots of data items will now also involve only the selected districts.

If we wished to point at one of these selected districts and enquire about all of the attributes associated with it, then `Profile locations` from the `Analysis` menu may be used. Select the 'default order' option and a list of the selected districts is displayed. Use of <SHFT> + <CR> will cause a window of attributes to be displayed for the district highlighted in the list. <SHFT> + 'double click' whilst pointing to a district in the map will cause the same display for the district pointed to. A number of districts may be selected by successively highlighting each one in the displayed list and using <CTRL> + <CR> together to select it (or <CTRL> + 'double click' whilst pointing at the same district in the map). Notice that the corresponding areas in the map are highlighted as they are selected. To 'deselect' an area already selected, simply select it again; to start the selection again from scratch, use <CR> rather than <CTRL> + <CR> for the first area selected. Once you have a subset of districts highlighted then use <SHFT> + <CR>, which allows a window of either overall total, or average, data values to be displayed for the subset so selected.

Experiment with these various ideas to see if you can identify any distinct subgroups of districts within those originally selected on the basis of owner-occupation being under 50 per cent and car ownership less than 50 per cent.

Exercise 2.4

At the end of this chapter on 'computers and spatial analysis', it is appropriate to place INFO-MAP into the wider context discussed in the chapter. You should appreciate that INFO-MAP falls short of the kind of sophistication that we have described as being deliverable by some recent computing developments related to the handling of spatial data and, having read this chapter, you should understand in what ways this is so. In particular, within INFO-MAP the digital 'map' files associated with data have a simple 'run length encoded' raster format at a fixed and fairly low level of resolution. They are simply graphic images with no structure. Data are stored in a simple 'flat' file structure in which rows are geographical locations and columns are attributes or variables. 'Maps' are linked to data only through a crude spatial indexing involving the grid references of a pre-defined set of point locations. The package does not recognise different types of spatial object such as lines or areas explicitly. All data in INFO-MAP are, in effect, related to point locations. It does have some functions related to areas, but areas are only recognised by virtue of their being 'closed polygons' in the map surrounding a point location and not as explicit spatial objects in their own right. The 'overlay' files in INFO-MAP that can be created to contain contextual information, such as topographic data or road networks, are simply 'pictures' to aid in the interpretation of data. No facilities are provided to logically intersect this information with the data, to create new zones or buffers.

Hence we can now see clearly that INFO-MAP is certainly not a GIS, nor is it a system designed to produce high quality digital maps. It also has few functions which allow truly dynamic spatial analysis of the kind described in the later sections of the chapter. Nevertheless, many of the issues that we have discussed in this chapter, concerning digital mapping, manipulation of spatial data and the idea of interactive spatial data analysis are highlighted by the use of INFO-MAP and this is our main purpose in including the software with the book. In later chapters we will pick up on many of the points concerned with spatial analysis. In the final exercise in this chapter we look at some further ideas concerned with the manipulation and summarising of spatial data that have been touched on in this chapter.

Open the file 'Lancashire lung and larynx cases' and load the corresponding data file. This data set is described in detail in Chapter 3. For the present, simply notice that it contains three data items. The first is the total population in each of the areas in the map. You could display the variation in population density, by first calculating a new data item as:

```
popden={1}/area()
```

and then drawing a `Choropleth map` of the new data item.

The second data item relates to the (approximate) point locations of cases of lung cancer in the region collected over a 10 year period; these can be displayed by using `Dot map`. The third data item relates to point locations of larynx cancer over the same period; again `Dot map` will display the locations.

Firstly, it may be useful to get a better general idea of the geography of the area involved. Choose the `Grid references` option with the `Map` menu. This displays a 'cross-hair' cursor on the screen and gives the grid reference of its location in the bottom right hand corner. Use <SHFT> + cursor key (those on the numeric key pad), to move the cross-hair cursor, or 'drag' it by holding down the left button and moving the mouse. The grid reference updates as the 'cross hair' moves. <CR> or <ESC> will end the process. Use this facility to determine the extent of the area and then locate the same area on a similarly grid referenced map of England and establish the positions of major towns and communication links in the area. If you wish, you might try creating an INFO-MAP *overlay file* which contains this information. You would do this by using `New` from the `Overlay` menu and then placing symbols and text 'on top of' the underlying map, finally saving this to an appropriately named file. You will need to consult the relevant chapter in the *User's Reference* for details of the process if you wish to do this.

Next, suppose we wished to know how many lung cancer cases had occurred in the district 'Z31MEAK' (one of the electoral wards) in the north of the region. We use `Calculate/display worksheet` from the `Analysis` menu, choose the 'calculate' option and enter the formula:

```
work[1]=sum(if(inside([31]) and {2}=1,1,0))
```

Here 'Z31MEAK' is location 31 in the file and data item 2 is a variable which has the value 1 for a location if a cancer case occurred there and a 'missing value' otherwise. Notice that to get this result, INFO-MAP has to apply a 'point in polygon' algorithm to determine whether any of the lung cancer cases lie in the irregular polygon which constitutes 'Z31MEAK'; the calculation may take some time. What is the corresponding result for the area 'Z31MEAF'?

Finding out how many lung cancer cases are within 2 km of the 'centre' of the district 'Z31MEAK' is an easier operation. The worksheet formula:

```
work[2]=sum(if(dist([31])<2000 and {2}=1,1,0))
```

gives the result (note that the map is grid referenced in units of 1 m). Here we have applied a very simple kind of 'buffering'.

Finally, let us illustrate the idea of 'zooming' into a spatially defined subset of data. Draw a Dot map of the lung cancer cases. They are particularly dense around the urban area centred approximately on grid reference (354000, 422000). Suppose we wished to 'zoom' into this area to look at the distribution in more detail. To do this choose Zoom/Restrict from the File menu. You are now required to define a rectangular 'source area' to be rescaled; any observations outside this area will be removed from the data set. Define a 'source area' which roughly encloses the dense cluster of lung cancer cases discussed above. Do this by locating the cursor at the position of the first corner with the mouse or cursor keys. Fix this by double clicking the mouse or using <CR>. Then 'pull' the opposite corner to the required position by 'dragging' with the mouse or using <SHFT> + cursor key (those on the numeric key pad). Double click or use <CR> to finalise. This source area is then delineated with a dashed line. A 'target area' for the rescaling must now be defined. The process is the same as for the definition of the 'source area'. Mark out a moderate sized area in the middle of the screen. Note that INFO-MAP will 'fit' the 'rescaled' source area into your 'target' area as best as possible whilst preserving aspect ratio during a rescale; hence the target area need not have the same relative proportions as the source area. A final option is now provided between 'polygon' or 'raster' zoom. Choose the 'polygon' option. You should then be left with a temporary 'zoomed' map and the data set will have been restricted to the 'source area' that you defined. Now try a Dot map of the lung cancer cases; the distribution within the area of interest can now be seen in much greater detail. Can you think of some deficiencies in this 'temporary' data set, when it comes to carrying out further analyses on just this subset of data? What, for example has INFO-MAP done with the 'population' data values? What perhaps should it have done with them?

Part B

The Analysis of Point Patterns

3

Introductory methods for point patterns

Here, and in the following chapter, we consider methods for the analysis of a set of 'event' locations, often referred to as a 'point pattern'. In general, our objective is to ascertain whether there is a tendency for 'events' to exhibit a systematic pattern; in particular, some form of regularity or alternatively clustering. We wish to estimate how the 'intensity' of 'events' varies over the study region, and possibly seek models which account for this. Such methods are relevant to the study of the patterns of occurrence of various diseases, types of crime, epicentres of earthquakes, distribution of plants, and many similar problems.

3.1 Introduction

We start our discussion of spatial data analysis methods by considering perhaps the simplest example of spatial data, a *point pattern*. By this we mean a data set consisting of a series of point locations, $(\mathbf{s}_1, \mathbf{s}_2, \ldots)$, in some 'study region', $\mathcal{R}$, at which 'events' of interest have occurred. Here, we use the general notation introduced in Chapter 1, where the (2×1) vector, $\mathbf{s}_i = (s_{i1}, s_{i2})^T$ is a shorthand way of referring to the 'x' (s_{i1}) and 'y' (s_{i2}) coordinates of the ith 'event'. We use the term 'event' in a very general sense since the events in question could relate to a wide variety of spatial phenomena that can be considered to occur at point locations, at least at the scale of our investigation. Examples include: the locations of cell nuclei in a microscopic tissue section, certain types of tree in a forest, cases of a disease or incidents of a type of crime in a geographical region, or locations of a specific type of star on a photographic plate of an area of the cosmos. One advantage of referring to observations on a spatial point pattern as 'events' is that often we will need to distinguish between these observed occurrences and other arbitrary locations in the study region; for these other locations we will reserve the term 'points'.

A spatial point pattern is a 'simple' example of spatial data because the data comprise only the coordinates of events, at least at the most basic level. This does not necessarily mean that analysis is any easier than for the other types of spatial data which we will meet in later chapters. In fact, from a statistical perspective, point patterns can in some ways be mathematically more complex to handle. In any case, often the data do not comprise solely the locations of events. We may also have additional 'attributes' relating to the events which we might wish to include in some types of analyses. Examples of such attributes include a labelling of type of event (tree species, nature of crime, for example) or the time of occurrence of the event (date of disease notification, for instance).

Our basic interest in analysing spatial point patterns is in whether the observed events exhibit any systematic pattern, as opposed to being distributed at random in $\mathcal{R}$. The general possibilities are either some form of *clustering* or alternatively *regularity*. If there does appear to be some systematic pattern present, then we might be interested in over what spatial scale this occurs and in whether particular spatial aggregations, or clusters, are associated with proximity to particular sources of some other factor. In cases where events are of differing types or occur at different points in time, or in a region with a heterogeneous background 'population at risk', then various other hypotheses become possible. For example, is the pattern in one type of event similar to that of another, are events which are clustered in space also close in time, and so on.

The important point to appreciate in any of these analyses is that it is patterns in the event locations themselves which are of interest. The stochastic process we are studying relates to where events are likely to occur. This should be distinguished from the situation in later parts of the book, where the analysis is concerned with patterns in an attribute which pertains to the whole region studied, values of which have been recorded at a set of sampled spatial locations. There, it is the spatial behaviour of values of the attribute which is the stochastic process studied, not whether and where they occur.

Throughout our discussion of point pattern analysis we shall assume that the data we are studying represent a complete map of events in the study region $\mathcal{R}$. This is sometimes referred to as a *mapped point pattern*; all relevant events which occurred in $\mathcal{R}$ have been recorded. Some point pattern analysis is directed towards extracting limited information about a point process, by recording events in a sample of different areas of the whole study region—a *sampled point pattern*. This requirement often arises in field studies in forestry, ecology or biology, where complete enumeration is not feasible. We shall not consider the sampling of point patterns explicitly, but where methods we discuss could be used in conjunction with sampling schemes we will indicate this.

In general, our study region $\mathcal{R}$ might be of any arbitrary shape. Some of the methods we discuss can be applied only to regions which are square or rectangular, in which case we assume that a suitable sub-area of the original study region has been selected. In addition, we may have to work with a sub-area in order to avoid *edge effects* by leaving a suitable *guard area* between the perimeter of the original study region and the sub-region within which analysis is performed. In all cases we must make the assumption that the final area selected for study is in some sense representative of any larger region from which it has been selected.

From a statistical point of view a spatial point pattern can be thought of in various ways. One possibility is in terms of the number of events occurring in arbitrary sub-regions or areas, $\mathcal{A}$, of the whole study region, $\mathcal{R}$. As a result, the process is represented by the set of random variables $\{Y(\mathcal{A}), \mathcal{A} \subseteq \mathcal{R}\}$, where $Y(\mathcal{A})$ is the number of events occurring in the area $\mathcal{A}$. In Chapter 1 we discussed the behaviour of a general spatial stochastic process in terms of first order and second order properties. We related these to the behaviour of the expected value (mean) and the covariance of the process respectively. With the representation of a spatial point process we have just introduced, both the mean of the random variables, $E(Y(\mathcal{A}))$, and their covariance, $COV\big(Y(\mathcal{A}_i), Y(\mathcal{A}_j)\big)$, relate to numbers of events in arbitrary areas of $\mathcal{R}$, and clearly depend on the size of the particular areas involved. They are not therefore particularly useful as they stand and it is usual instead to characterise first and second order properties on the basis of the limiting behaviour of these quantities 'per unit area'.

In particular, first order properties are described in terms of the *intensity*, $\lambda(\boldsymbol{s})$, of the process, which is the mean number of events per unit area at the point $\boldsymbol{s}$. Formally, this is defined as the mathematical limit:

$$\lambda(\boldsymbol{s}) = \lim_{\mathrm{d}s \to 0} \left\{ \frac{E(Y(\mathbf{d}\boldsymbol{s}))}{\mathrm{d}s} \right\}$$

where $\mathbf{d}\boldsymbol{s}$ is a small region around the point $\boldsymbol{s}$, and $\mathrm{d}s$ is the area of this region. For a stationary point process (see Chapter 1) $\lambda(\boldsymbol{s})$ is a constant over $\mathcal{R}$, say λ, and then $E(Y(\mathcal{A})) = \lambda A$, where A is the area of $\mathcal{A}$.

The second order properties, or spatial dependence, of a spatial point process involve the relationship between numbers of events in pairs of areas in $\mathcal{R}$. This can be formally described by the *second order intensity*, $\gamma(\boldsymbol{s}_i, \boldsymbol{s}_j)$, of the process, which again involves events per unit area and is formally defined as the mathematical limit:

$$\gamma(\boldsymbol{s}_i, \boldsymbol{s}_j) = \lim_{\mathrm{d}s_i, \mathrm{d}s_j \to 0} \left\{ \frac{E\big(Y(\mathbf{d}\boldsymbol{s}_i)\, Y(\mathbf{d}\boldsymbol{s}_j)\big)}{\mathrm{d}s_i \mathrm{d}s_j} \right\}$$

with notation similar to that used previously.

For a stationary process $\gamma(\boldsymbol{s}_i, \boldsymbol{s}_j) = \gamma(\boldsymbol{s}_i - \boldsymbol{s}_j) = \gamma(\boldsymbol{h})$. That is, the second order intensity depends on only the vector difference, $\boldsymbol{h}$ (direction and distance), between $\boldsymbol{s}_i$ and $\boldsymbol{s}_j$ and not on their absolute locations. The process is isotropic (see Chapter 1) if the dependence is purely a function of the length, h, of this vector $\boldsymbol{h}$ and not its orientation; that is, purely a function of the distance between $\boldsymbol{s}_i$ and $\boldsymbol{s}_j$ and not direction. Then $\gamma(\boldsymbol{s}_i, \boldsymbol{s}_j) = \gamma(h)$.

An alternative, but closely related characterisation of the second order properties of an isotropic point process, which we shall find to be of more practical use than the second order intensity, is the *reduced second moment measure*, or *K function*. We leave a formal discussion of this function until later in this chapter. There is also a close connection between second order properties and the distribution of the distances between pairs of events. Accordingly, we shall find the distributions of various inter-event distances to be useful tools to describe the spatial interaction or dependence, in point patterns.

Our approach to the analysis of point patterns will involve a mixture of methods, some designed to examine variations in the intensity, $\lambda(s)$, of the process over $\mathcal{R}$, others where we shall look for spatial interactions, through the use of inter-event distances, or the K function. Essentially, this corresponds, in the case of spatial point processes, to the general idea of spatial phenomena exhibiting first and second order effects, as outlined in Chapter 1. When it comes to proposing possible statistical models in order to try to 'explain' any effects we may detect, we may choose to model some effects as first order, others in terms of spatial dependence. Sometimes our analyses will be able to distinguish clearly between which effects are present. However, as we shall see towards the end of this chapter, some clustering models, which assume independent events with heterogeneous intensity (purely first order effects), cannot be distinguished on the basis of any observed data, from others which assume dependent events with homogeneous intensity (purely second order effects). Therefore, the reader should bear in mind that there will inevitably be cases where clustering, explained in terms of spatial dependence with the assumption of homogeneity, could equally well arise from heterogeneity.

In line with the general structure we outlined in Chapter 1, we commence by describing some case studies we shall use throughout this part of the book. We then move on to consider analysis methods, under the general headings of visualisation, exploration and modelling.

3.2 Case studies

Throughout this and the following chapter, we shall use a selection of point pattern data sets to illustrate various forms of analyses. We have provided copies of these data sets on disk. They include:

- The locations of craters in a volcanic field in Uganda
- The locations of granite tors on Bodmin Moor
- The locations of redwood seedlings in a forest
- The locations of the centres of biological cells in a section of tissue
- The locations of the homes of juvenile offenders on a Cardiff estate
- Locations of 'theft from property' offences in Oklahoma City
- Locations of cases of cancer of larynx and lung in part of Lancashire
- Locations of Burkitt's lymphoma in an area of Uganda

We begin by describing all of these data sets in more detail. Full references to the sources of data are given at the end of this chapter.

The first data set involves the locations of the centres of the craters of 120 volcanoes in the Bunyaruguru volcanic field in west Uganda. A map of the distribution (Figure 3.1) shows a broad regional trend in a north-easterly direction, this representing elongation along a major fault. We might wish to obtain a smooth map of such broad regional variation and we will see early in this chapter how we might accomplish this. At a smaller scale we may want to use point pattern methods to explore and model the distribution of craters. Is this distribution essentially random within the study region, or is there evidence for clustering or regularity? We would expect that rift faults would guide volcanic activity to the surface, along fractures or lines of weakness. We might expect some evidence of regularity in spacing at very short scales, simply because of the large size of some of the craters and the representation of each by a single point; but we shall want to see whether this holds true at other scales.

Another data set relating to geomorphology is that for the locations of granite tors on Bodmin Moor, first analysed in one of the references we give at the end of this chapter. There are 35 such locations (Figure 3.2) and on a large scale there is clear spatial patterning. However, we may wish to detect any evidence for departures from randomness at smaller scales. As with the volcanoes, the wisdom of treating these features as points may be doubted and we can expect some 'inhibition' or regularity of spacing at very short distances; but we shall be interested to see if the spatial distribution shows other patterning at slightly longer distances.

Many of the methods we shall encounter in this chapter have their origins in plant ecology, where researchers wish to describe and analyse the spatial distribution of plants, frequently within small areas of perhaps only a few square metres. As an example, we include a small set of data that many others have also analysed, comprising the locations of 62 redwood seedlings, distributed in a square region of $23\,m^2$ (Figure 3.9). From an ecological point of view we might expect some evidence of clustering around existing parent trees.

At a very different scale we also include data on the centres of 42 biological cells in a section of tissue (Figure 3.9). Microbiologists might wish to know whether there is evidence for departures from randomness in such data. Are such cells clustered or regular? As with the volcanic craters, such analysis requires some careful thought; the 'centres' of cells recorded here are simply convenient but perhaps slightly arbitrary locations for what might be regarded as 'area' objects (the cells themselves). Consequently, we might anticipate evidence for regularity, at least at very small scales.

We now consider data at a very different spatial scale, and turn attention to the human, rather than physical or biological, domain. We have included two data sets, both concerned with the distribution of crime in urban areas. One relates to the locations of the homes of juvenile offenders on a housing estate in Cardiff, Wales, UK (Figure 3.8). The data were recorded in 1971 and are therefore of historical and certainly not contemporary interest. The second data set is on the locations of offences (theft from property) rather than offenders, and is taken from research done on crime in Oklahoma City, USA, in the late 1970s (Figure 4.1). These data are rather different from the Cardiff data as they comprise two distinct categories of events. One set refers to offences committed by whites, the other by blacks. We shall want to see if the spatial patterns differ. Do the two sub-groups have different 'activity spaces', based perhaps on different cognitive or mental maps of the city? Do the crimes committed by different groups display different spatial patterns? Are those for one group clustered or aggregated in some way, while those for the other group are more random? We shall see how to examine such issues.

As we shall be at pains to point out in this chapter, departures from randomness in patterns of events, such as a tendency for crimes to cluster, are not necessarily a function of any underlying 'interaction' between offences committed in one small area and those in another. Rather, any such apparent aggregations might result from the fact that the spatial distribution of crime 'opportunity' (such as the distribution of residential property) is itself spatially varying. Nowhere is this kind of problem more acute than in the study of data concerned with disease incidence. What is the purpose of trying to assess whether rare diseases come in 'clusters' unless we know something about the underlying distribution of the population at risk?

The next data set allows us to explore these kinds of issues and includes data on the occurrence of two sorts of cancer, that of the larynx and that of the lung. The data are for part of Lancashire in Britain and have been collected over a 10 year period 1974–83 (Figure 4.5). Unfortunately, lung cancer is quite a common disease and there are 917 cases in the study area. Larynx cancer is rare and there were only 57 cases notified during the study period. We also include the total population in 42 census wards which comprise the study region. This is not, of course, an entirely sensible measure of disease 'risk', since, in general, only older adults are at risk from these diseases. Nonetheless, this simple total does give some idea of population distribution.

The study from which these data are taken was motivated by the following problem. There had been concerns expressed by residents living near the site of an old industrial waste incinerator that their health had been affected by exposure to the by-products of the incineration process. These health complaints ranged from fairly non-specific complaints about breathing difficulties (for which no data had been collected), to concerns about raised levels of some cancers. Larynx cancer was a possible candidate, though the risk factors for this disease are mostly exposure to tobacco smoke and to excess alcohol consumption. Now, although the cancer is rare, it will tend to 'cluster' simply because population distribution is itself clustered. This natural background variation in population needs to be allowed for in any analysis. The population in the 42 wards covering the area represents one measure of such background variation, but a very crude one. As an alternative we may also consider using the occurrences of lung cancer over the same time period as a surrogate for the specific population at risk, again imperfect, but which may at least serve to better 'mimic' the appropriate background population distribution.

The data take the form of Ordnance Survey grid references, and it is of general interest to say something about how they were obtained. Their form (they have a resolution of 1 metre on an Ordnance Survey grid) implies a level of detail which is quite spurious! No names or addresses for these kinds of data are usually available to the researcher. However, in Britain, much epidemiological data is tagged with a postcode, an alphanumeric code that contains valuable spatial information. This is because it can be matched against a large machine-readable file that contains a grid reference for each of the approximately 1.7 million postcodes in Britain. These grid references have a resolution of 100 metres. However, we have randomly shifted the locations in order to protect any patient confidentiality (note that the data are at least ten, and in some cases, twenty, years old) and in the process have added some (slightly spurious) detail. However, the broad patterns that we shall see later are genuine.

Our final data set is again concerned with cancer epidemiology, this time in the West Nile District of Uganda, bordering southern Sudan and eastern Zaire. The data comprise information on 188 cases of Burkitt's lymphoma, a cancer affecting usually the jaw or abdomen, primarily in children (Figure 4.3). It is quite common in much of east Africa and has been the subject of research by epidemiologists for many years in efforts to understand its aetiology (causes). These data, which relate to the time period 1961–75, differ from the larynx and lung cancer data by virtue of the fact that we have other attribute information attached to each case—both the age of the child and also the date of onset of the disease (measured in days elapsed since January 1st 1960). We shall want to use these data, not to look at evidence for spatial clustering (again, we would expect this *a priori*), but rather in order to assess evidence for space–time clustering. Are cases that are 'near' each other in geographic space also 'near' each other in time? If so, this might be evidence in support of the hypothesis that suggests an infective aetiology for the disease.

3.3 Visualising spatial point patterns

The obvious way to visualise a spatial point pattern is to plot the data in the form of a *dot map*. This will give an initial impression of the shape of the study region and any obvious patterns present in the distribution of events. However, intuitive ideas about what constitutes a 'random pattern' can be misleading. Generally it will be hard to come to any conclusions purely on the basis of a visual analysis, particularly when the data set is large and there are multiple occurrences of events at the same or nearly the same location. If attributes other than just the location of events are involved, then this compounds the problem. Labelling of events or the time of their occurrence can be included in a dot map by using different colours or symbols to display different types of events or occurrence in different time periods. However, such displays get 'cluttered' very quickly and are very difficult to interpret by eye. What information can we glean from a simple dot map of the distribution of, for example, volcanic craters in part of Uganda (Figure 3.1) or tors on Bodmin Moor (Figure 3.2)?

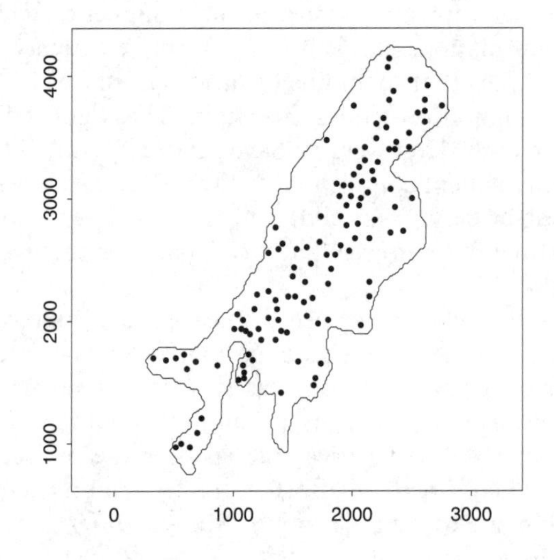

Fig. 3.1 Volcanic craters in west Uganda

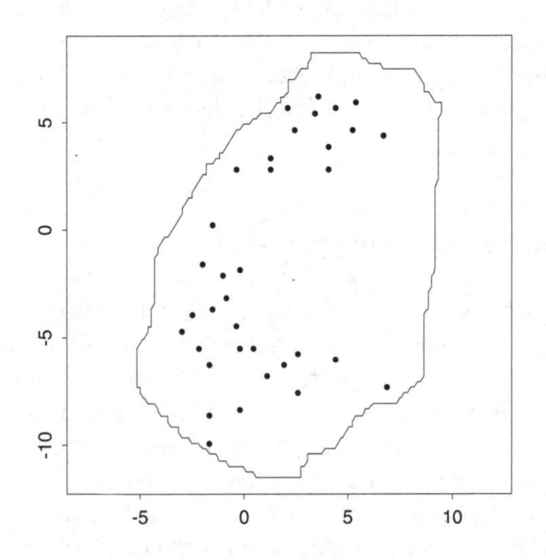

Fig. 3.2 Tors on Bodmin Moor

In the craters example we get an intuitive impression of groups of events aligned along the 'spine' of the study area. We might think we can detect small aggregations of events in the south of the study region. If we look instead at the tors, we might argue for evidence of regularity of spacing, though in two sub-regions. But with both these examples we need statistical tools to confirm (or refute!) our subjective impressions. This is one of the tasks we set ourselves in this chapter.

If the 'population' from which events arise is varying over the study region we would automatically expect events to be clustered in regions of high population, as would be the case in the example of cancer cases in Lancashire. We need other tools for helping us to visualise such point event data. Our interest is now in whether the events are more or less clustered than we would expect on the basis of the population alone. A possible way to deal with this on a visual display might be to use symbols of different sizes to represent events, for example by letting the area of an event symbol be inversely proportional to the population density at that location. Events in areas of low population would then show up 'larger' than those in high population areas and the impression gained from the display is whether the map is evenly covered with symbols, or there are significant 'gaps'.

In general, population information relevant to event occurrences is more likely to be available in the form of population counts in some coverage of administrative districts such as census tracts over the study region. One way of displaying such data is to geometrically transform each of the population zones in such a way as to make its area proportional to the corresponding population value, whilst at the same time maintaining the spatial contiguity of the zones. The resulting transformed map is known as a *density equalised* map or *cartogram*. We consider cartograms in a little more detail in Chapter 7. Suffice it to say here that the possibility exists of transforming event locations by the same algorithm that is used to transform the zonal boundaries. The result is then a set of event locations in a transformed space where the population density is constant, which makes it easier to identify clustering in a study region with a heterogeneous 'at risk' population. In effect, a spatial point pattern that one expects to be heterogeneous because of population variation, has been transformed to one that we would expect to be homogeneous. This may also be useful in a more general context than simply as a visualisation tool.

3.4 Exploring spatial point patterns

We move on now to consider methods of analysis where we derive various summary statistics or plots from the observed distribution of events and attempt to use these informally to investigate hypotheses of interest or to suggest possible models. Some of these methods are more concerned with investigating first order effects in the process; others address the possibility of spatial dependence or second order effects.

3.4.1 Quadrat methods

One simple way of summarising the pattern in the locations of events in some region $\mathcal{R}$ is to partition $\mathcal{R}$ into sub-regions of equal area, or *quadrats* (so called because of historical use of square sampling frames in field sampling) and to use the counts of the number of events in each of the quadrats to summarise the spatial pattern. Essentially we are simply creating a two-dimensional histogram or frequency distribution of the observed event occurrences. We impose a regular grid over $\mathcal{R}$, count the number of events falling into each of the grid squares and convert this to an intensity measure by dividing by the area of each of the squares. The result will give some indication of whether and how the intensity of the process $\lambda(\boldsymbol{s})$ is changing over $\mathcal{R}$. Clearly, the same kind of idea can also be used in conjunction with the sampling of a point pattern. Quadrats may be randomly scattered over $\mathcal{R}$ and all events within each quadrat counted, to give a crude estimate of how intensity varies over $\mathcal{R}$.

Once we have obtained a set of quadrat counts, we have effectively transformed our original point pattern into a set of counts within areas, in other words a set of area data. We can therefore bring to bear many of the powerful analysis tools that we discuss later in Chapters 7 and 8. However, the problem with quadrat counts is that although they may give a global idea of sub-regions with high or low intensity they throw away much of the spatial detail in the observed pattern. As quadrats are made smaller to retain more spatial information we get a very high variability in quadrat counts; this finally degenerates into a mosaic with many empty quadrats, making meaningful interpretation impossible.

A way round this problem is through the use of counts per unit area in a 'moving window'. One defines a suitable size 'window' which is then moved over a fine grid of locations in $\mathcal{R}$ and the intensity at each grid point is estimated from the event count per unit area within the 'window' centred on that point. This produces a more spatially 'smooth' estimate of the way in which $\lambda(\boldsymbol{s})$ is varying. The problem, of course, is that no account is taken of the relative location of events within the particular 'window' and it is difficult to decide what size of 'window' to use.

A natural extension to this kind of idea is a technique known as *kernel estimation.*

3.4.2 Kernel estimation

Kernel estimation was originally developed to obtain a smooth estimate of a univariate or multivariate probability density from an observed sample of observations; in other words, a smooth histogram. Estimating the intensity of a spatial point pattern is very like estimating a bivariate probability density and bivariate kernel estimation can be easily adapted to give an estimate of intensity. If $\boldsymbol{s}$ represents a general location in $\mathcal{R}$ and $\boldsymbol{s}_1, \ldots, \boldsymbol{s}_n$ are the locations of the n observed events then the intensity, $\lambda(\boldsymbol{s})$, at $\boldsymbol{s}$ is estimated by

$$\hat{\lambda}_\tau(\boldsymbol{s}) = \frac{1}{\delta_\tau(\boldsymbol{s})} \sum_{i=1}^{n} \frac{1}{\tau^2} k\left(\frac{(\boldsymbol{s} - \boldsymbol{s}_i)}{\tau}\right)$$

Here $k()$ is a suitably chosen bivariate probability density function, known as the *kernel*, which is symmetric about the origin. The parameter $\tau > 0$ is known as the *bandwidth* and determines the amount of smoothing—essentially it is the radius of a disc centred on $\boldsymbol{s}$ within which points $\boldsymbol{s}_i$ will contribute 'significantly' to $\hat{\lambda}_\tau(\boldsymbol{s})$. The factor

$$\delta_\tau(\boldsymbol{s}) = \int_{\mathcal{R}} \frac{1}{\tau^2} k\left(\frac{(\boldsymbol{s} - \boldsymbol{u})}{\tau}\right) \mathrm{d}\boldsymbol{u}$$

is an edge correction—the volume under the scaled kernel centred on $\boldsymbol{s}$ which lies 'inside' $\mathcal{R}$. For any chosen kernel and bandwidth, values of $\hat{\lambda}_\tau(\boldsymbol{s})$ can be examined at locations on a suitably chosen fine grid over $\mathcal{R}$ to provide a useful visual indication of the variation in the intensity, $\lambda(\boldsymbol{s})$, over the study region.

Choice of the specific functional form of the kernel, $k()$, presents little practical difficulty. For most 'reasonable' choices of possible probability distributions for $k()$ the kernel estimate $\hat{\lambda}_\tau(\boldsymbol{s})$ will be very similar, for a given bandwidth τ. A typical choice for $k()$ might be the quartic kernel

$$k(\boldsymbol{u}) = \begin{cases} \frac{3}{\pi}(1 - \boldsymbol{u}^T\boldsymbol{u})^2 & \textit{for } \boldsymbol{u}^T\boldsymbol{u} \leq 1 \\ 0 & \textit{otherwise} \end{cases}$$

When this is used the above expression for $\hat{\lambda}_\tau(\boldsymbol{s})$, ignoring the edge correction factor, becomes:

$$\hat{\lambda}_\tau(\boldsymbol{s}) = \sum_{h_i \leq \tau} \frac{3}{\pi\tau^2}\left(1 - \frac{h_i^2}{\tau^2}\right)^2$$

where h_i is the distance between the point $\boldsymbol{s}$ and the observed event location $\boldsymbol{s}_i$, and the summation is only over values of h_i which do not exceed τ. The region of influence within which observed events contribute to $\hat{\lambda}_\tau(\boldsymbol{s})$ is therefore a circle of radius τ centred on $\boldsymbol{s}$. In Figure 3.3 we picture a 'slice' through the radially symmetric function corresponding to this choice of $k()$. The function we show is simply shifted to be centred on the location $\boldsymbol{s}$ and then 'scaled up' by a factor of τ to provide the weighting applied to observed events around $\boldsymbol{s}$. At the site $\boldsymbol{s}$ (a distance of zero) the weight is $3/\pi\tau^2$ and it drops smoothly to a value of zero at distance τ.

From a visual point of view, when we perform such kernel estimation over a fine grid of locations in $\mathcal{R}$ we may think of a three-dimensional floating function that 'visits' each point $\boldsymbol{s}$ on the fine grid (Figure 3.4). Distances to each observed event $\boldsymbol{s}_i$ that lies within the region of influence, i.e. a distance τ, are

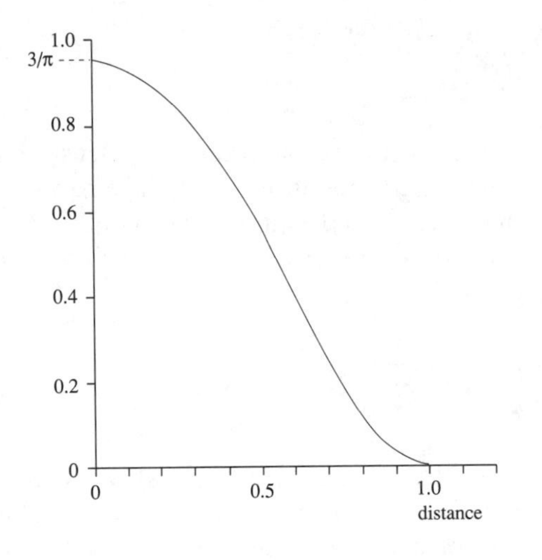

Fig. 3.3 Slice through a quartic kernel

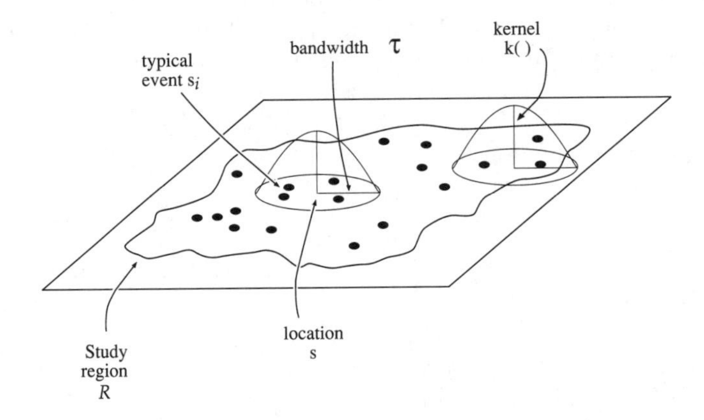

Fig. 3.4 Kernel estimation of a point pattern

measured and contribute to the intensity estimate at $\boldsymbol{s}$ according to how close they are to $\boldsymbol{s}$.

Whatever the choice of $k()$, the effect of increasing the bandwidth, τ, is to 'stretch' the region around $\boldsymbol{s}$ within which observed events influence the intensity estimate at $\boldsymbol{s}$. For very large τ, $\hat{\lambda}_\tau(\boldsymbol{s})$ will appear flat and local features will be obscured; if τ is small then $\hat{\lambda}_\tau(\boldsymbol{s})$ tends to a collection of spikes centred on the $\boldsymbol{s}_i$.

A 'rough' choice for τ has been suggested as $\tau = 0.68n^{-0.2}$ for estimating the intensity when $\mathcal{R}$ is the unit square and n is the number of observed events in $\mathcal{R}$. This recommendation can be scaled up to fit the particular dimensions of other study regions. In practice of course the value of kernel estimation is that one can experiment with different values of τ, exploring the surface $\hat{\lambda}_\tau(\boldsymbol{s})$ using different degrees of smoothing in order to look at the variation in $\lambda(\boldsymbol{s})$ at different scales.

There are methods which attempt to optimise the value of τ given the observed pattern of event locations, but we do not go into details here. It is worth commenting however on adjusting the value of τ locally at different points in $\mathcal{R}$, in order to improve the kernel estimate. In areas where observed events are more closely packed than others, more detailed information on the variation in intensity is available. It follows that, locally to these regions, a smaller value of the bandwidth τ would be more appropriate to avoid smoothing out too much detail. In other words, setting a bandwidth which would be appropriate for sparse areas, unfortunately obscures detail in dense areas. Local adjustment of bandwidth may be achieved by a technique known as *adaptive kernel estimation*. Here, τ is replaced by $\tau(\boldsymbol{s}_i)$ (some function of the presence of 'events' in the neighbourhood of $\boldsymbol{s}_i$); so that if for simplicity we ignore the edge correction factor:

$$\hat{\lambda}_\tau(\boldsymbol{s}) = \sum_{i=1}^{n} \frac{1}{\tau^2(\boldsymbol{s}_i)} k\left(\frac{(\boldsymbol{s} - \boldsymbol{s}_i)}{\tau(\boldsymbol{s}_i)}\right)$$

Of course this requires us to specify $\tau(\boldsymbol{s}_i)$ in some way. One suggestion that seems to work reasonably well in practice is to first perform non-adaptive kernel estimation with some reasonable fixed bandwidth τ_0, so achieving a *pilot estimate* $\tilde{\lambda}(\boldsymbol{s})$. The geometric mean $\tilde{\lambda}_g$ of the pilot estimates $\tilde{\lambda}(\boldsymbol{s}_i)$ at each $\boldsymbol{s}_i$ is then computed, that is the *n*th root of their product. Adaptive bandwidths for subsequent estimation are then taken as:

$$\tau(\boldsymbol{s}_i) = \tau_0 \left(\frac{\tilde{\lambda}_g}{\tilde{\lambda}(\boldsymbol{s}_i)}\right)^{\alpha}$$

where $0 \leq \alpha \leq 1$ is known as the *sensitivity parameter*. Clearly, a choice of $\alpha = 0$ corresponds to no local adjustment of bandwidth and the subsequent adaptive estimation reproduces the pilot estimate; $\alpha = 1$ corresponds to maximum local adjustment. A choice of $\alpha = 0.5$ has been found to produce reasonable results in practice.

We may explore this idea of kernel estimation with reference to the data set on volcanic craters, a dot map of which we saw earlier. In effect, any number of kernel estimate maps could be produced, corresponding to different values of τ. Three examples are presented (Figure 3.5) where values of $\tau = 100$, $\tau = 220$ and $\tau = 500$ reveal that the higher value 'smooths' the distribution rather too much, while the lower value of 100 gives a rather too 'spiky' impression of the spatial distribution. A value of $\tau = 220$ gives an adequate representation of the regional distribution. While there is indeed a broad band of craters running down the 'spine' of the study region, there are local peaks in intensity. This is masked in the two maps where inappropriately low or high values of τ are chosen.

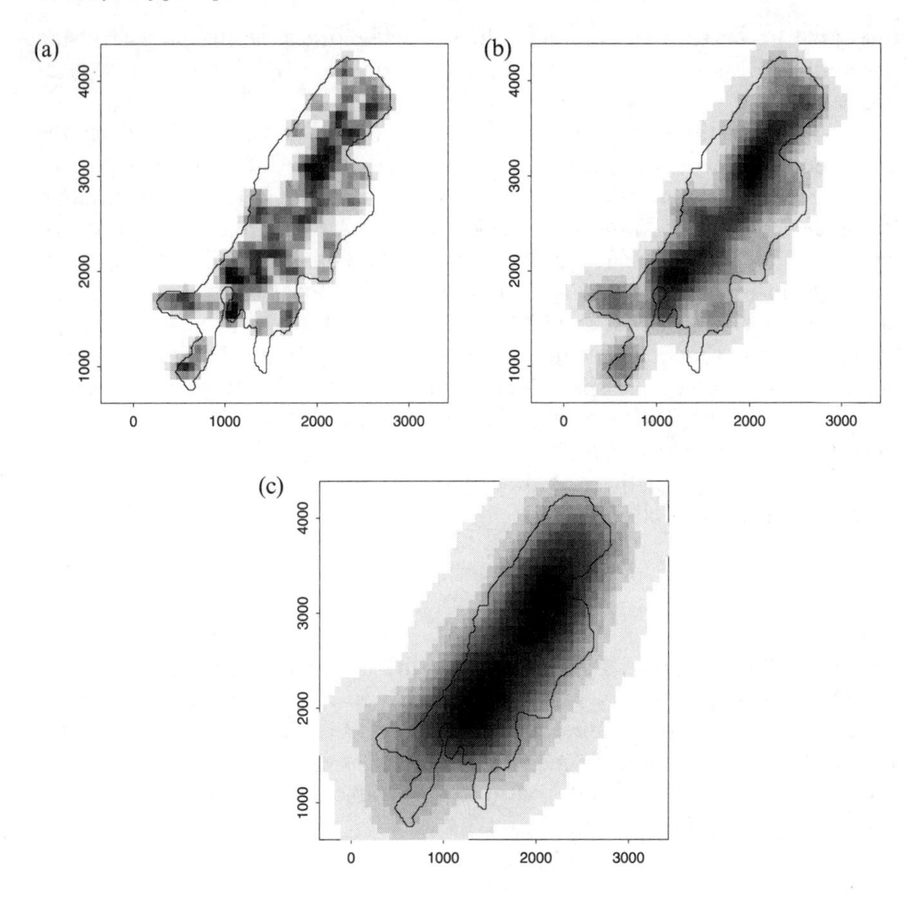

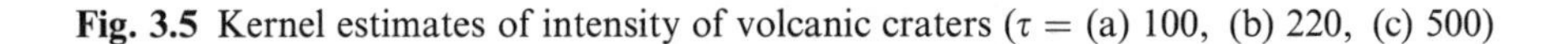

Fig. 3.5 Kernel estimates of intensity of volcanic craters (τ = (a) 100, (b) 220, (c) 500)

3.4.3 Nearest neighbour distances

Quadrat counting and kernel estimation are essentially concerned with exploring the first order properties of a spatial point pattern, in other words with estimating the way in which the intensity varies in the study region. In our introductory remarks in this chapter, we also discussed ways of characterising the second order properties of such processes. In particular, we commented that for isotropic processes there was a close relationship between the distribution of inter-event distances and second order properties. We now turn to exploratory methods which are designed more to investigate second order properties using distances between observed events in the study area.

In general we could define several different kinds of 'inter-event' distances that could throw light on second order properties. However, the methods that we discuss will be based on *nearest neighbour* distances. In particular, we will be interested in the nearest neighbour *event–event* distance W, which we take as being the distance between a randomly chosen event and the nearest neighbouring event, and also the nearest neighbour *point–event* distance X, the distance between a randomly selected point in the study region and the nearest neighbouring observed event. Both of these measures may be used when all events in $\mathcal{R}$ have been enumerated, as is the general assumption with our data sets in this chapter. It should be noted that W is strictly undefined when sampling from point patterns since we cannot randomly choose an event unless all the events in the study area have been enumerated. However X would be potentially useful in sampling—in a field study we could choose a random point in $\mathcal{R}$ and then find the nearest event and measure the distance to it.

One way of investigating the degree of spatial dependence in a point pattern is to examine the observed distribution of one or both of these nearest neighbour distances. Note that nearest neighbour distances only provide information about inter-event interactions at a small physical scale, since, by definition, they use only 'small' inter-event distances. However this may be a very sensible approach if there is also the possibility of large scale variation in the intensity of the given point pattern over the whole of $\mathcal{R}$, since in that case it is only sensible to study second order effects at a scale small enough to avoid variation in intensity becoming confounded with interaction.

A simple approach to summarising pattern using either event–event nearest neighbour distances, W, or those for point–event distances, X, is to estimate the empirical cumulative probability distribution or distribution function, $\hat{G}(w)$ of W, or $\hat{F}(x)$ of X from the data as:

$$\hat{G}(w) = \frac{\#(w_i \leq w)}{n}$$

and

$$\hat{F}(x) = \frac{\#(x_i \leq x)}{m}$$

where # means 'the number of' and n is the number of events in the study area, or m is the number of random points sampled.

The resulting empirical distribution function, either $\hat{G}(w)$ or $\hat{F}(x)$, can then be plotted against suitable values of w or x and examined in a purely exploratory way to throw light on possible evidence of inter-event interaction. For example, if the distribution function climbs very steeply in the early part of its range before flattening out, then the indication would be an observed high probability of short as opposed to long nearest neighbour distances, which would suggest clustering due to inter-event attraction. Alternatively, if it climbs very steeply in the later part of its range, then the suggestion might be one of inter-event repulsion or regularity. Another idea would be to plot $\hat{G}(w)$ against $\hat{F}(x)$. If there is no interaction then these two distributions should be very similar and we would expect to obtain roughly a straight line in the plot. In cases of positive interaction, or clustered patterns, the point–event distances x_i will tend be large relative to the event–event distances w_i and so we might expect $\hat{G}(w)$ to exceed $\hat{F}(x)$. The opposite would hold in a regular pattern.

At this stage examination of these distributions is purely subjective. Ultimately, of course, we would like to be able to provide a more objective comparison of either $\hat{G}(w)$ or $\hat{F}(x)$, against that which might be theoretically expected according to various models for the point pattern (random, clustered, or regular). We defer such formal comparisons until our later section on modelling spatial point patterns.

Note also that the estimation of $G(w)$ or $F(x)$ discussed above, makes no correction for *edge effects*. The nearest neighbour distance for an event near the boundary of $\mathcal{R}$ will be biased, tending to be greater than that for one well inside the region, because an event near the boundary is denied the possibility of neighbours outside the boundary. We really ought to allow for this in our estimates $\hat{G}(w)$ or $\hat{F}(x)$. One way to avoid the problem is to construct a *guard area* inside the perimeter of $\mathcal{R}$. Nearest neighbour distances are not used for events (or points) within the guard area, but events in the guard area are allowed as neighbours of any events (or points) from the rest of $\mathcal{R}$. Another approach which can be employed when the study region is rectangular is to use a *toroidal edge correction*. The top of the study region is assumed to be joined to the bottom and the left to the right. This implies that the study region is regarded as the central region of a 3×3 grid of rectangular regions, each identical to the study region; events in the copies are allowed to be neighbours of any events (or points) which are selected in the study region.

As an alternative to both these approaches it has been suggested that $\hat{G}(w)$ can be approximately estimated as

$$\hat{G}(w) = \frac{\#(b_i > w \geq w_i)}{\#(b_i > w)}$$

where b_i is the distance from event i to the nearest point on the boundary of $\mathcal{R}$. What this effectively does is to ignore w_i values for events 'close' to the boundary. The same sort of idea can be used for $\hat{F}(x)$.

We may illustrate these ideas by estimating the $\hat{G}(w)$ function for the volcanic crater data (Figure 3.6). Note that, from a distance of about 50 up to 150 (approximately 1 to 3 kilometres) the function climbs quite rapidly. This implies that there are, cumulatively, rather a lot of short event–event distances; in other words, we are getting an impression of local clustering in the data. This confirms the earlier visual impression we obtained from simply plotting the locations of the events.

3.4.4 The *K* function

One of the problems mentioned in respect of the nearest neighbour distance methods discussed above is that they use distances only to the closest events

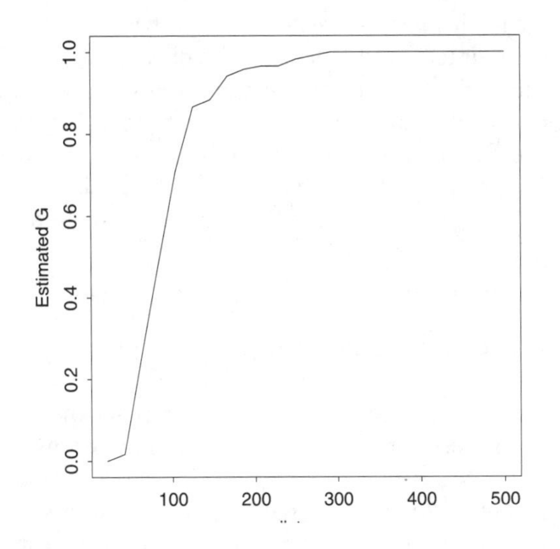

Fig. 3.6 Nearest neighbour distribution function for volcanic craters

and therefore only consider the smallest scales of pattern. Information on larger scales of pattern is ignored. An alternative approach is to use an estimate of the *reduced second moment measure* or *K function* of the observed process, which provides a more effective summary of spatial dependence over a wider range of scales. This is closely related to, but more practically useful than, the second order intensity of the process which we formally defined earlier, since the latter cannot be estimated directly from an observed event distribution.

We need to remind the reader at this point that once we start to examine for spatial dependence over anything but very small scales in $\mathcal{R}$, we are effectively making the implicit assumption that the process can considered to be homogeneous or isotropic over such scales. If this is not the case, then any attempt to estimate such effects does not really make any sense. Second order properties are not then necessarily constant over the scale considered and may also be confounded with first order variation in intensity. One therefore has no repetition available to estimate them. If, for example, it is clear that there is large scale variation in intensity of the given point pattern over the whole of $\mathcal{R}$, and this is truly a first order effect and not a result of spatial dependence, then it would only be sensible to study second order effects over scales within $\mathcal{R}$ small enough for the assumption of isotropy to hold. On the other hand, if variation in intensity does not seem to be present, or that which exists may reasonably be due to spatial dependence and not a *bona fide* first order effect, then we might be justified in examining second order effects over larger scales in the study region.

The K function which we describe below essentially relates to the second order properties of an isotropic process. However, the above comments imply that if it is used in a situation where there are large scale first order effects, then any spatial dependence it may indicate could well be due to these first order effects rather than to interaction effects. In such a case we might wish to examine smaller sub-regions of $\mathcal{R}$ for interaction effects, where isotropy can reasonably be assumed to hold, rather than the whole region at once.

With this proviso in mind we now go on to define and describe the K function on the assumption that we are dealing with an isotropic process over the whole of $\mathcal{R}$. Adjustments for application in sub-regions are straightforward.

The definition of the K function we shall use is:

$$\lambda K(h) = E(\#(\text{events within distance h of an arbitrary event}))$$

where again # means 'the number of', $E()$ is our normal expectation operator, and λ is the intensity or mean number of events per unit area, assumed constant throughout $\mathcal{R}$.

The practical value of $K(h)$ as a summary measure of second order effects is that it is feasible to obtain a direct estimate of it, say $\hat{K}(h)$, from an observed point pattern, whereas it would not be possible to estimate the second order intensity directly. We may do this in the following way. If R is the area of $\mathcal{R}$ then the expected number of events in $\mathcal{R}$ is λR. It follows from the definition of the K function that the expected number of *ordered* pairs of events a distance at most h apart, with the first event in $\mathcal{R}$ is $\lambda^2 RK(h)$. If d_{ij} is the distance between the ith and jth observed events in $\mathcal{R}$ and $I_h(d_{ij})$ is an indicator function which is 1 if $d_{ij} \leq h$ and 0 otherwise, then the observed number of such ordered pairs is $\sum\sum_{i \neq j} I_h(d_{ij})$. Therefore a suitable estimate of $K(h)$ is given by:

$$\hat{K}(h) = \frac{1}{\lambda^2 R} \sum\sum_{i \neq j} I_h(d_{ij})$$

Again, as with our estimation of nearest neighbour distribution functions earlier, we need to consider the effect of the edge of $\mathcal{R}$ on our estimate. The summation above necessarily excludes pairs of events for which the second event is outside $\mathcal{R}$ and therefore unobservable and so we need to correct for this. Consider a circle centred on event i, passing through the point j, and let w_{ij} be the proportion of the circumference of this circle which lies within $\mathcal{R}$. Then w_{ij} is effectively the conditional probability that an event is observed in $\mathcal{R}$, given that it is a distance d_{ij} from the ith event. Thus a suitable edge corrected estimator for $K(h)$ is

$$\hat{K}(h) = \frac{1}{\lambda^2 R} \sum\sum_{i \neq j} \frac{I_h(d_{ij})}{w_{ij}}$$

To complete our estimate we need to replace the unknown intensity λ with an estimate of its value. An obvious choice would be $\hat{\lambda} = n/R$. Thus finally we obtain:

$$\hat{K}(h) = \frac{R}{n^2} \sum_{i \neq j} \sum \frac{I_h(d_{ij})}{w_{ij}}$$

Again, it would be useful to have an intuitive, graphical idea of what is embodied in this notion of a *K* function (Figure 3.7). We may imagine that an event is 'visited' and that around this event is constructed a set of concentric circles at a fine spacing. The cumulative number of events within each of these distance 'bands' is counted. Every other event is similarly 'visited' and the cumulative number of events within distance bands up to a radius h around all the events becomes the estimate of $K(h)$ when scaled by R/n^2. Notice that this simple explanation ignores the edge correction factor.

The 'edge corrected' estimate of the *K* function for an observed point pattern that we have given will be reasonable as long as h is not taken too large relative to the size of $\mathcal{R}$. This restriction on h is necessary because the weights w_{ij} can become unbounded as h increases. In practice this is not a serious problem because we would usually be interested in relatively small values of h—it is not realistic to attempt to explore second order effects which operate on the same physical scale as the dimensions of $\mathcal{R}$.

Clearly, in practice the calculation of $\hat{K}(h)$ is not easy, since for arbitrary shaped $\mathcal{R}$, the weights w_{ij} are hard to derive, and there could be many of them. Explicit formulae for w_{ij} can be written down for simple shapes such as rectangular or circular $\mathcal{R}$; in other cases the derivation of w_{ij} will require quite complex algorithms and can be computationally intensive.

Once obtained, $\hat{K}(h)$ can be plotted and examined for suggestions of spatial dependence in the point process at different values of h. However, unlike the

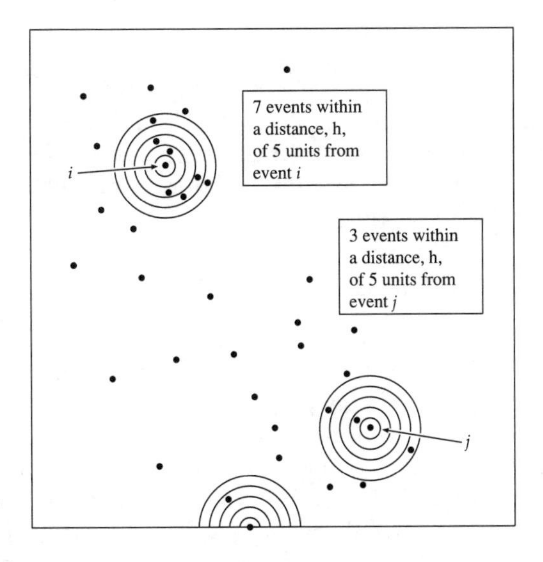

Fig. 3.7 Estimation of *K* function

case of the estimated distribution of nearest neighbour distances, where we had an intuitive idea of how to interpret the plot involved, with the K function we really do not know what we would expect the plot to look like if there was no spatial dependence. In our later modelling section we shall see that certain types of models give rise to various theoretical K functions which can be compared with that estimated from the observed point pattern. This can then be used to assess whether they might form reasonable models for the process generating that pattern. Here we are simply concerned with the K function as an exploratory device and so simply require some basic comparison which allows us to detect whether clustering or regularity is present. We can achieve this quite simply by asking what $K(h)$ would look like for a 'random' spatial point process. At this point we have not formally said what we mean by such a process, but intuitively it is clear that 'random' will imply that the probability of the occurrence of an event at any point in $\mathcal{R}$ is independent of what other events have occurred and equally likely over the whole of $\mathcal{R}$. Thus for a 'random' process the expected number of events within a distance h of a randomly chosen event would be just $\lambda\pi h^2$. Hence, looking back to our definition of the K function, we would expect $K(h) = \pi h^2$ for a homogeneous process with no spatial dependence.

Under regularity $K(h)$ would be less than πh^2, whereas under clustering $K(h)$ would be greater than πh^2. This suggests an obvious approach to exploratory use of the K function—compare $\hat{K}(h)$ estimated from the observed data with πh^2. One way of doing this is through a plot $\hat{L}(h)$ against h where:

$$\hat{L}(h) = \sqrt{\frac{\hat{K}(h)}{\pi}} - h$$

In this plot peaks in positive values tend to indicate spatial attraction of events or clustering and troughs of negative values indicate spatial repulsion or regularity, at corresponding scales of distance h in each case.

An alternative to the square root transformation would be to use a logarithmic transformation, plotting $\hat{l}(h)$ against h where:

$$\hat{l}(h) = \frac{1}{2}\log\left(\frac{\hat{K}(h)}{\pi}\right) - \log h$$

Again, peaks tend to indicate clustering and troughs of regularity, at different distances h. One could also simply plot $\hat{K}(h) - \pi h^2$ against h.

Let us now explore the use of these ideas with reference to one of the case studies outlined earlier. The data on residential locations of juvenile offenders on a Cardiff estate (Figure 3.8) reveals some visual evidence of clustering, notably in the northern part of the estate. We do not have any information on the distribution of properties on the estate, and so our assumption of underlying homogeneity may be a little optimistic. Nonetheless, there does not

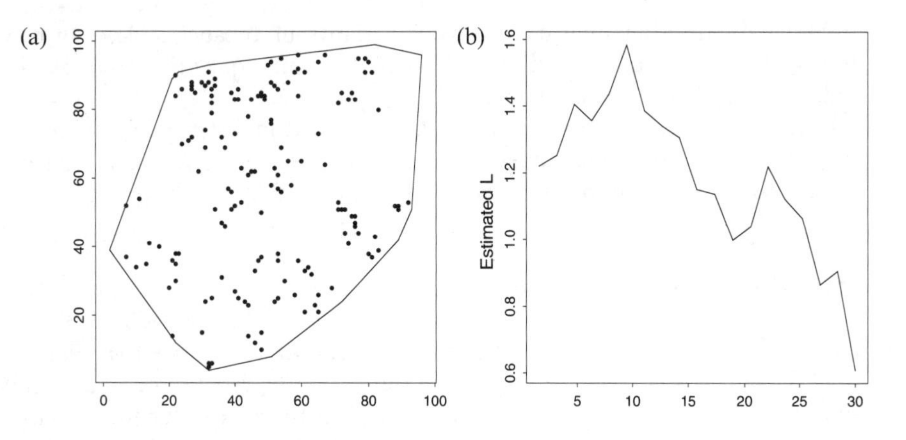

Fig. 3.8 (a) Juvenile offenders in Cardiff and (b) associated $\hat{L}$ function

appear to be any gross variation in intensity across the study region. The estimated $L()$ function gives some indication of clustering in the data, since the values are positive for all distances examined here. There is a peak at $h = 10$, suggesting clustering at this scale; there is a further peak at $h = 20$, suggesting possible 'clustering' of the smaller 'clusters' at this distance. Of course at this stage we do not know whether the magnitude of these peaks is at all significant; we shall examine such issues later in the next section, and see then whether our initial impressions can be confirmed.

As a method of data summary and exploration the K function has several attractive features—it presents information at various scales of pattern, involves use of the precise location of events and includes all event–event distances, not just nearest neighbour distances. Another attraction, as mentioned earlier, is that the theoretical form of $K(h)$ is known for various possible spatial point process models. $\hat{K}(h)$ can therefore not only be used to explore spatial dependence, but also to suggest specific models to represent it and to estimate the parameters of such models. We discuss this point further in the subsequent section.

3.5 Modelling spatial point patterns

So far we have been concerned with exploring spatial patterns in a fairly informal way. Ultimately these sorts of exploratory analyses may not be sufficient, and we may want to go further to consider explicit tests of various hypotheses or construct specific models to 'explain' an observed point pattern. The more formal methods of analysis that we will discuss in this section consist of statistically comparing various summary measures calculated from the observed distribution of events, mostly variants of those we have already discussed in the preceding section, with what we would expect these to look like under various hypothesised models.

In the first instance we shall mostly be interested in comparisons with one particular model, that of complete spatial randomness or *CSR* and so we need to start by saying more precisely what we mean by this.

3.5.1 Complete spatial randomness (CSR)

The standard model for CSR is that events follow a *homogeneous Poisson process* over the study region. In terms of our earlier representation of a spatial point process as the set of random variables $\{Y(\mathcal{A}), \mathcal{A} \in \mathcal{R}\}$, this would imply that $Y(\mathcal{A}_i)$ and $Y(\mathcal{A}_j)$ are independent for any choices of $\mathcal{A}_i$ and $\mathcal{A}_j$ and further, that the probability distribution of $Y(\mathcal{A})$ is a Poisson distribution with mean value λA, where A is the area of $\mathcal{A}$. That is, the probability distribution of $Y(\mathcal{A})$ is:

$$f_{Y(\mathcal{A})}(y) = \frac{(\lambda A)^y}{y!} \mathrm{e}^{-\lambda A}$$

where λ is a constant—the *intensity* or mean number of events per unit area. A further implication is that, conditional on n (the total number of events in $\mathcal{R}$), events are independently and uniformly distributed over $\mathcal{R}$.

Intuitively, this amounts to saying firstly, that any event has an equal probability of occurring at any position in $\mathcal{R}$; and secondly, that the position of any event is independent of the position of any other—events do not interact with one another.

One could simulate n events from such a process by enclosing $\mathcal{R}$ in a rectangle $\{(x, y) : x_1 \leq x \leq x_2,\ y_1 \leq y \leq y_2\}$ then generating events with x coordinates from a uniform distribution on (x_1, x_2) and y coordinates from a uniform distribution on (y_1, y_2) and then 'rejecting' any that do not lie in $\mathcal{R}$, continuing until the required n events have been obtained in $\mathcal{R}$.

Our interest in CSR is that it represents a baseline hypothesis against which to assess whether observed patterns are *regular*, *clustered* or *random*. Of course, having established a departure from randomness we might then be interested in the particular nature of the departure and might need to compare the observed process with other specific models for patterns that exhibit either clustering or regularity. We leave a discussion of formal models with these properties until later, and concentrate in the first instance purely on comparison with CSR.

3.5.2 Simple quadrat tests for CSR

One crude way of testing for CSR is to use the kind of quadrat counts we have discussed previously. Probably the simplest way to do this is the so called *index of dispersion test*. Let $(x_1, \ldots, x_m)$ be the counts of the number of events in m quadrats, either randomly scattered in $\mathcal{R}$, or alternatively forming a regular grid covering the whole of $\mathcal{R}$. Then a simple test statistic for randomness can be based on the idea that if these counts follow a Poisson distribution we would expect the mean and variance of counts to be equal. We therefore use as a test statistic:

$$X^2 = \frac{(m-1)s^2}{\bar{x}} = \frac{\sum_{i=1}^{m}(x_i - \bar{x})^2}{\bar{x}}$$

where $\bar{x}$ is the mean of the observed counts and s^2 is their observed variance. Under CSR the theoretical probability distribution of X^2 is χ^2_{m-1} to a good approximation provided that $m > 6$ and $\bar{x} > 1$. This provides the basis for a significance test, that is comparing the observed value with the percentage points of the χ^2_{m-1} distribution. Significantly large or small values indicate clustered or regular departures respectively from CSR.

$s^2/\bar{x}$ is often called the *index of dispersion* and $(s^2/\bar{x}) - 1$ is often called the *index of cluster size* (ICS). Note that for CSR, $E(ICS) = 0$, and if $ICS > 0$ then clustering is implied ('number of extra events'), whilst if $ICS < 0$ regularity is implied ('deficiency in events').

One advantage of the above index of dispersion test is that it can be used in conjunction with the sampling of point patterns. In this case m quadrats will be randomly scattered in $\mathcal{R}$, and events exhaustively counted in each quadrat. Of course we can also use such a sampling method to estimate the intensity, λ, of events in $\mathcal{R}$ if we are prepared to believe that this is a constant and CSR holds. An obvious estimate based on random quadrat counts, $(x_1, \ldots, x_m)$, where the area of each quadrat is Q, would be:

Ex 3.4

$$\hat{\lambda} = \frac{\bar{x}}{Q}$$

Confidence intervals for this estimate may be derived from the result that $VAR(\hat{\lambda})$ is approximately $\hat{\lambda}/mQ$. A 95% confidence interval would therefore be approximately $\hat{\lambda} \pm 2\sqrt{\frac{\hat{\lambda}}{mQ}}$.

Randomly scattered quadrats may well overlap and this may produce a problem if it is a frequent occurrence, since the x_i counts will not be independent. To avoid this we might need to use a sampling scheme that guarantees disjoint quadrats. Also, overlaps with the edge of $\mathcal{R}$ can be a problem. We might need to introduce a guard area inside the perimeter of $\mathcal{R}$, and only randomly scatter quadrats throughout that part of $\mathcal{R}$ which is not in the guard area, allowing events in the guard area to be counted in any quadrats which overlap into this area. Another problem is how big the quadrats ought to be. An empirical suggestion is to aim for a mean quadrat count of about 1.6.

One general problem with the above approach is that no account is taken of the relative position of quadrats or the relative position of events within a quadrat. For the basic situation considered in this chapter where all events in the study area have been enumerated and counts in a large grid of contiguous quadrats can be obtained, it becomes feasible to consider using information about the relative position of each of the quadrats in addition to the quadrat counts. One common method is known as the *Greig–Smith procedure*, which calculates the variance of quadrat counts for the original grid and then for further sub-grids each formed by successive combination of adjacent quadrats in the original grid into 'blocks' of increasing size. Variance estimates at each block size are then plotted against block size and peaks or troughs in the graph are interpreted as evidence of 'scales of pattern' (clustered or regular respectively). Several variations of this kind of quadrat method have also been suggested. We do not give details of any of these methods here, considering them to be of less practical use than the methods based on nearest neighbour distances or K functions which we take up in the next section. Neither do we discuss adaptations of these sort of ideas which are suitable for use in sampling point patterns. There are, for example, several suggested methods to test for pattern based on 'transects' (a row of contiguous quadrat counts). Such techniques can form the basis of useful sampling schemes in field studies.

3.5.3 Nearest neighbour tests for CSR

In the quadrat tests described above the locational information relating to observed events that is used is necessarily limited and imprecise. We now turn attention to some simple methods which are based on the nearest neighbour distances we described earlier, and which utilise more precise locational information.

In our section on exploratory methods we defined the nearest neighbour event–event distance W and point–event distance X, and discussed how to estimate their cumulative probability distributions, $G(w)$ and $F(x)$ respectively. Unfortunately the theoretical distribution functions $G(w)$ of W or $F(x)$ of X under CSR are not known when dealing with any specific study region $\mathcal{R}$. Because of edge effects they would in general depend on the particular shape of $\mathcal{R}$ in a possibly complex way. However we can derive some theoretical distribution results for W and X if edge effects are ignored.

Let the mean intensity of events per unit area be λ. Then, under CSR, events are independent and the number of events in any area is Poisson distributed, so the probability that no events fall within a circle of radius x around any randomly chosen point is given by $e^{-\lambda \pi x^2}$. Hence the distribution function $F(x)$ of nearest neighbour point–event distances X for CSR is given by:

$$F(x) = Pr(X \leq x) = 1 - e^{-\lambda \pi x^2} \quad x \geq 0$$

This implies that πX^2 follows an exponential distribution with parameter λ, or equivalently, that $2\pi\lambda X^2$ is distributed as χ^2_2. It may therefore be deduced that:

$$E(X) = \frac{1}{2\sqrt{\lambda}}$$

$$VAR(X) = \frac{(4-\pi)}{4\lambda\pi}$$

It also follows that if $X_1, \ldots, X_n$ are independent nearest neighbour point–event distances then $2\pi\lambda \sum X_i^2$ is distributed as χ^2_{2n}.

Exactly similar arguments would apply to the nearest neighbour event–event distance W, for a CSR process. That is, under CSR, the distribution function $G(w)$ is given by

$$G(w) = Pr(W \leq w) = 1 - e^{-\lambda\pi w^2} \quad w \geq 0$$

and $E(W)$, $VAR(W)$ are as given above for X.

Knowledge of the theoretical distribution of W and X under CSR allows one to derive (at least approximately) the sampling distributions under CSR of various summary statistics of observed nearest neighbour distances. These can then be used as a basis for a test of CSR. It should be noted that the distribution theory for many such tests assumes firstly that the nearest neighbour distances used to compute the summary statistic are independently sampled from the study region $\mathcal{R}$ and secondly that they have not been biased by edge effects.

The independence assumption is unlikely to be valid if the total number of events in the study area is small and the proportion of them that are included in the sample of nearest neighbour distances used is large; think of the case where there are only two events, the two nearest neighbour distances are identical! One suggestion is that the number, m, of nearest neighbour measurements sampled should be such that $m \leq 0.1n$ where n is the total number of events in the study region. The general effect of lack of independence will be that the test statistics will have a larger variance than their theoretical values under independence, which implies that the standard test may flag as significant a departure from CSR which would not be so if the dependence had been taken into account. In particular, the reader should be cautioned against applying nearest neighbour tests, without appropriate corrections, to the set of all nearest neighbour event–event distances, when a complete map of events in $\mathcal{R}$ is available. That such tests require a random sample of such distances, is of course an advantage when point patterns are sampled.

Correcting for edge effects also provides a difficulty. As discussed previously, nearest neighbour distances for events near the boundary of $\mathcal{R}$ will be biased, tending to be greater than those for events well inside the region. Some tests have explicit edge corrections built in, but generally these only apply to a restricted set of regularly shaped study regions. In the general case we have already commented on ways to avoid the problem, either by constructing a guard area inside the perimeter of $\mathcal{R}$, or employing a toroidal edge correction when the study region is rectangular.

Among the various tests suggested to detect departures from CSR based on summary statistics of m randomly sampled nearest neighbour event–event distances $(w_1, \ldots, w_m)$, or point–event distances $(x_1, \ldots, x_m)$, the following three are amongst the most commonly used:

Clark–Evans—compare $\bar{w} = \sum w_i/m$ with percentage points of the distribution:

$$N\left(\frac{1}{2\sqrt{\lambda}}, \frac{(4-\pi)}{4\lambda\pi m}\right)$$

Ex 3.5

The test is based on event–event distances and so requires a completely enumerated point pattern to be available, from which events can be randomly sampled and their nearest neighbour distances determined. This is reinforced by the fact that λ is unknown and needs to be replaced by an appropriate estimate, the obvious one being $\hat{\lambda} = n/R$, where n is the number of events in $\mathcal{R}$. In such a case it would then seem desirable to use all n event–event distances if possible, rather than a random sample of m of them. An approximate correction to $E(\bar{W})$ and $VAR(\bar{W})$ has been suggested which allows all nearest neighbour distances to be used ($m = n$), as opposed to a sample. That is:

$$E(\bar{W}) = 0.5\sqrt{\frac{R}{n}} + 0.051\frac{P}{n} + 0.041\frac{P}{n^{\frac{3}{2}}}$$

$$VAR(\bar{W}) = 0.070\frac{R}{n^2} + 0.037P\sqrt{\frac{R}{n^5}}$$

where P is the perimeter of the study region which has area R. These approximations break down for very convoluted regions $\mathcal{R}$.

Hopkins—compare $\sum x_i^2/\sum w_i^2$ with percentage points of the distribution $F_{2m,2m}$. The rationale behind this test is that in clustered patterns the point–event distances x_i will be large relative to the event–event distances w_i and vice versa in a regular pattern. Because the test uses w_i it requires again a complete enumeration of all the n events in the study region so that event–event distances can then be randomly sampled. However it has been suggested that it may be applied in conjunction with the sampling of point patterns, if a 'semi-systematic' sampling scheme is employed whereby a regular grid of study points is imposed over the study region. Alternate study points are used as points for calculating point–event distances x_i. Around each of the remaining study points a small circular area is prescribed within which all events are enumerated (big enough to contain about five events); the selection of random events for calculation of event–event distances w_i is based upon the population of events within the collection of these fully enumerated areas.

Byth & Ripley—compare $\frac{1}{m}\sum\frac{x_i^2}{(x_i^2+w_i^2)}$ with percentage points: $N\left(\frac{1}{2},\frac{1}{12m}\right)$, where the x_i values are randomly paired with the w_i values. The considerations for this test are very much the same as those for the Hopkins test.

As an illustration of one of these simple tests, we can apply the Clark–Evans test to the data set on the distribution of the centres of biological cells, and to the distribution of redwood seedlings, both of which were described in our case study section earlier. The two distributions are pictured in Figure 3.9 and the visual impression is very much one of regularity in the biological cells and clustering in the redwood seedlings. These impressions are confirmed by the application of the Clark–Evans test. In the case of the redwood seedlings a highly significant value of the test statistic is obtained in the lower tail of the normal distribution which would apply under the assumption of CSR; so indicating strong departure from CSR in the direction of clustering. For the biological cells the value of the test statistic is highly significant in the upper tail, indicating strong departure from CSR in the direction of regularity.

The reduction of complex point patterns to a single nearest neighbour summary statistic in order to test for CSR using the above methods obviously results in a considerable loss of information. The tests only indicate departure from CSR; little is known about the behaviour of these statistics when CSR does not hold and no information is provided as to the form of any alternative model when CSR is rejected. In general, then, one would not recommend this sort of approach when a completely mapped pattern is available, except perhaps as a preliminary procedure in analysis.

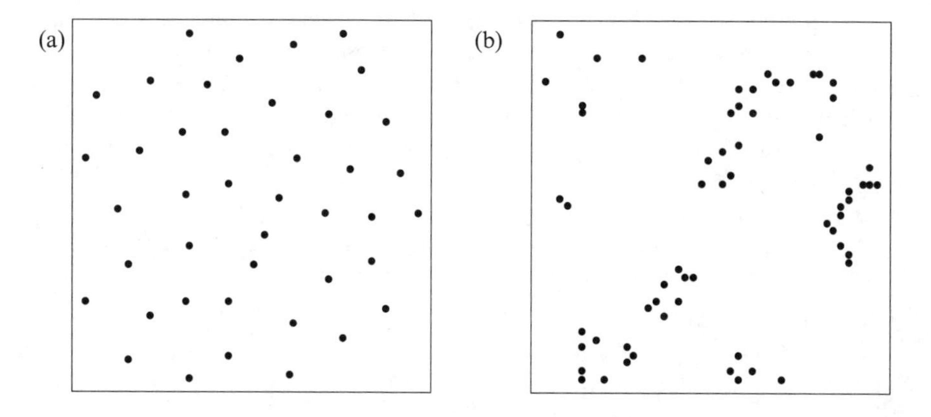

Fig. 3.9 Distribution of (a) biological cells and (b) redwood seedlings

For completely enumerated point patterns the possibility exists of looking at the complete estimated distribution function of W or X rather than just a single summary statistic. In our earlier exploratory section we discussed estimation of the empirical distribution function $\hat{G}(w)$ of W or $\hat{F}(x)$ of X from an observed point pattern. Can we construct a formal method for comparing the whole of this distribution with its theoretical form under CSR?

Note that the theoretical distribution functions $G(w)$ of W or $F(x)$ of X under CSR, which we gave earlier, assume no edge effects are present. Therefore if we wish to compare $\hat{G}(w)$ or $\hat{F}(x)$ directly with these, we will have to correct for edge effects in $\hat{G}(w)$ or $\hat{F}(x)$ by one of the methods discussed earlier, that is either a guard area, a toroidal correction, or alternatively use of:

$$\hat{G}(w) = \frac{\#(b_i > w \geq w_i)}{\#(b_i > w)}$$

where b_i is the distance from event i to the nearest point on the boundary of $\mathcal{R}$. (Equivalently for $\hat{F}(x)$). Having done this we can plot either $\hat{G}(w)$ or $\hat{F}(x)$ directly against their theoretical equivalents under CSR; that is, $G(w) = 1 - e^{-\lambda\pi w^2}$, or $F(x) = 1 - e^{-\lambda\pi x^2}$.

Examination of this plot is certainly less subjective than that suggested earlier in our exploratory section, where plots of $\hat{G}(w)$ or $\hat{F}(x)$ against w or x respectively were examined simply in respect of their general shape. At least here we are comparing with a known theoretical form under CSR. However we still do not have any formal way of assessing the significance of differences in the plot. Neither is one available, because of the complexities introduced by the approximate edge corrections used in estimating $\hat{G}(w)$ or $\hat{F}(x)$.

Perhaps a more satisfactory (although computationally intensive) approach is to compare the estimated distribution functions (without edge correction) with a simulation estimate of their theoretical distribution functions under CSR in the presence of the particular edge effects arising from the given study region $\mathcal{R}$.

We present the method here in the case of W; an exactly analogous approach would be applicable for X. The simulation estimate for $G(w)$ under CSR is calculated as $\bar{G}(w) = \sum \hat{G}_i(w)/m$, where $\hat{G}_i(w)$, $i = 1, \ldots, m$ are empirical distribution functions each of which is estimated, without edge correction, from one of m independent simulations of n events under CSR in $\mathcal{R}$; that is, n events independently and uniformly distributed in $\mathcal{R}$.

For the purposes of assessing the significance of departures between the simulated CSR distribution, $\bar{G}(w)$, and that which we actually observed, $\hat{G}(w)$, we also define upper and lower simulation envelopes:

$$\mathcal{U}(w) = \max_{i=1,\ldots,m} \{\hat{G}_i(w)\}$$

$$\mathcal{L}(w) = \min_{i=1,\ldots,m} \{\hat{G}_i(w)\}$$

We now plot $\hat{G}(w)$ (estimated without edge correction) against $\bar{G}(w)$ and add to the plot $\mathcal{U}(w)$ and $\mathcal{L}(w)$. If the data are compatible with CSR then the plot of $\hat{G}(w)$ against $\bar{G}(w)$ should be roughly linear and at 45 degrees. If clustering is present then the plot will lie above the line and vice versa for regularity. $\mathcal{U}(w)$, $\mathcal{L}(w)$ help to assess the significance of departures from the 45 degree line in the plot since they have the property that:

$$Pr(\hat{G}(w) > \mathcal{U}(w)) = Pr(\hat{G}(w) < \mathcal{L}(w)) = \frac{1}{(m+1)}$$

This also illustrates what value we should use for m, i.e. how many simulations we need to perform, in order to able to detect departure at a specified significance level.

Figure 3.10 gives the results of applying this procedure to the data on the residential locations of juvenile offenders on a Cardiff estate discussed earlier. The results confirm as significant our earlier impression of clustering gained from an exploratory examination of the K function.

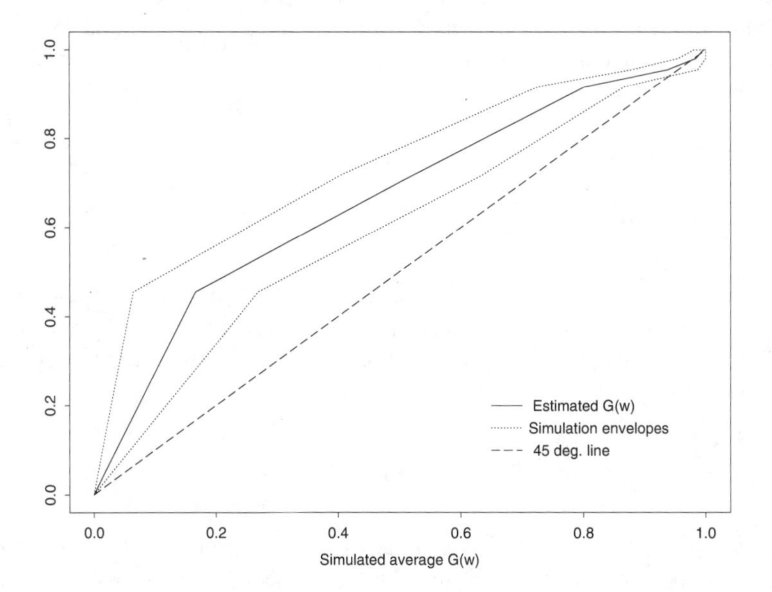

Fig. 3.10 $\hat{G}(w)$ v. simulated $\bar{G}(w)$ under CSR for Cardiff offenders

3.5.4 *K* function tests for CSR

As we saw earlier, the K function provides a more effective summary of spatial dependence over a wide range of scales than measures based just on nearest neighbour distances. In our exploratory section we have already given an informal argument as to the theoretical form of the K function for a 'random' point pattern. Now we have a more formal model for a 'random pattern' in the form of CSR we can confirm the result suggested there. Under CSR the expected number of events within a distance h of a randomly chosen event would indeed be $\lambda \pi h^2$ and hence theoretically under CSR, $K(h) = \pi h^2$.

We now seek formal ways to compare the empirical K function, $\hat{K}(h)$, estimated from the observed data with this theoretical value. Earlier, $\hat{K}(h)$ was compared with its theoretical value via a plot of $\hat{L}(h)$ against h where:

$$\hat{L}(h) = \sqrt{\frac{\hat{K}(h)}{\pi}} - h$$

In this plot positive peaks tend to indicate clustering and negative troughs regularity at different scales of distance h. Formal assessment of the significance of observed peaks or troughs requires knowledge of the sampling distribution of $\hat{L}(h)$ and hence $\hat{K}(h)$, under CSR. This is unknown and complex because of the edge corrections built in to $\hat{K}(h)$. However, it is possible to use an analogous approach to that used in the previous section on nearest neighbour distributions, to obtain a simulation estimate of this sampling distribution. The usual approach is to construct upper and lower simulation envelopes:

$$\mathcal{U}(h) = \max_{i=1,\ldots,m} \{\hat{L}_i(h)\}$$

$$\mathcal{L}(h) = \min_{i=1,\ldots,m} \{\hat{L}_i(h)\}$$

from m independent simulations of n events in $\mathcal{R}$ under CSR and their associated estimates $\hat{L}_i(h)$. These envelopes are then included on the plot of the actual observed $\hat{L}(h)$ against h. The significance of peaks or troughs is assessed on the basis that:

$$Pr(\hat{L}(h) > \mathcal{U}(h)) = Pr(\hat{L}(h) < \mathcal{L}(h)) = \frac{1}{(m+1)}$$

As before, this also illustrates what value we should use for m, i.e. how many simulations we need to perform in order to able to detect departure at a specified significance level.

Returning to the distribution of juvenile offenders, earlier exploration of the K function gave evidence for clustering at all scales. This is confirmed by the formal test, (Figure 3.11), since the estimated function, $\hat{L}(h)$, lies well outside the upper simulation envelope.

The envelopes were generated on the basis of 99 simulations under the hypothesis of CSR. We have, then, some confidence in concluding that the locations of offenders are clustered on this estate. Further research would be required to explain such clustering. Does it arise because of local social networks, for example, with offences committed by individuals who are

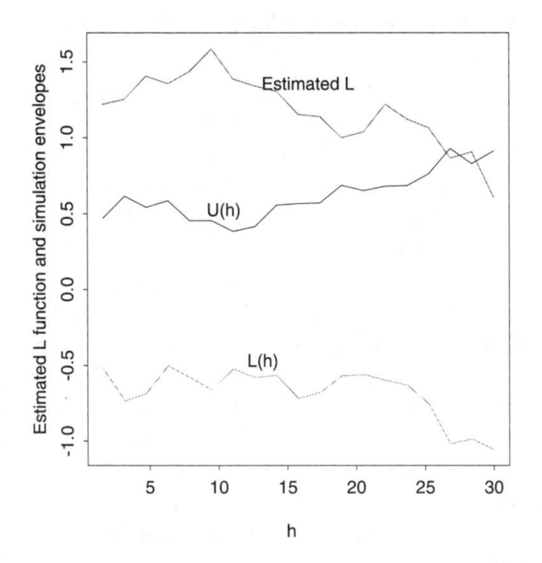

Fig. 3.11 $\hat{L}$ and simulation envelopes for Cardiff offenders

similarly 'disaffected'? Without having further data or local knowledge it is difficult to say.

3.5.5 Other models for spatial point patterns

So far the focus in this section on modelling spatial point patterns has been on comparing an observed point pattern with CSR. CSR is often a useful preliminary model to consider and that usefulness may even be applicable where, because of variations in 'population at risk', it is not tenable for the original data. This may be done by transforming the map using the kind of cartogram discussed in our visualisation section. However, there will often be situations where CSR is simply not a feasible model for the pattern in question. The exploratory analyses we have discussed may still be valuable in suggesting the kind of structure that may be present in the data, but in general will only indicate broad regularity or clustering, and this may be obvious *a priori*, given the nature of the data. We have not considered any formal statistical tests related to models other than CSR. If we wish to 'explain' the particular nature of regularity or clustering, then we need formal models to allow us to do this.

Our model for CSR was a homogeneous Poisson process over the study region $\mathcal{R}$. There are really two important aspects of this model. Firstly, homogeneous means that λ is constant over $\mathcal{R}$; there are no first order effects. Secondly, Poisson implies spatial independence in the occurrence of events. This means that the number of events occurring in any two neighbouring regions is uncorrelated; there are no second order effects. For a general spatial process we know that it is quite possible that λ varies over $\mathcal{R}$ (heterogeneity) and/or occurrences of events in neighbouring regions are correlated (spatial dependence).

In this section we consider briefly some basic alternative models to CSR. We do not develop these in any detail, but simply give an outline of the ideas behind each. In the subsequent section we outline how one might go about fitting one of the simpler examples of these to an observed point pattern. Further coverage of the analysis of spatial point patterns using such models is beyond the scope of this book.

Some simple possible models for clustering in a spatial point process are the *heterogeneous Poisson process*, the *Cox process* or the *Poisson cluster process*. *Simple inhibition processes* can be used to model regular point patterns. *Markov point processes* can model both clustering and regularity, through small scale regularity and large scale clustering. We discuss here only very brief details of these models to give a general flavour of the basic idea behind each model.

The heterogeneous Poisson process is perhaps the simplest alternative to CSR. The constant intensity λ of CSR is replaced by a variable intensity function $\lambda(\boldsymbol{s})$, but the occurrence of any event remains independent of that of any other. The resulting process is obviously a simple kind of non-stationary point process, with only first order effects. The simplest way to simulate such a process is to simulate CSR on $\mathcal{R}$ with intensity $\lambda_{max} = \max_{\mathcal{R}} \lambda(\boldsymbol{s})$ and then to independently retain an event at $\boldsymbol{s}$ with probability $\lambda(\boldsymbol{s})/\lambda_{max}$.

The Cox process is a natural extension to the heterogeneous Poisson process where the intensity $\lambda(\boldsymbol{s})$ varies randomly, rather than deterministically. Such processes are often described as *doubly stochastic*—$\lambda(\boldsymbol{s})$ is drawn from a probability function over $\mathcal{R}$ and then conditional on the value of $\lambda(\boldsymbol{s})$, events form a heterogeneous Poisson process with intensity $\lambda(\boldsymbol{s})$. The resulting process can either be stationary or non-stationary. It will be stationary if and only if the probability distribution from which the intensity is generated is stationary. In principle, such a process may be simulated by first simulating the probability distribution for $\lambda(\boldsymbol{s})$ on $\mathcal{R}$ then using the rejection sampling technique described above for the heterogeneous Poisson process.

The Poisson cluster process arises from the explicit incorporation of a spatial clustering mechanism into the model. Parent events form a CSR process and each parent then produces a random number of offspring realised independently and identically for each parent according to a probability distribution $f()$. Finally the positions of the offspring relative to their parents are independently and identically distributed according to some bivariate probability density $g()$. The final process consists of the offspring only. Such processes are stationary. They are also isotropic if $g()$ is radially symmetric. The method for simulating such a process follows directly from its definition. First, simulate the parent process to obtain the location of parent events; second, for each parent event simulate independently a number of offspring according to $f()$; and, third, independently locate each one of these around its parent according to $g()$; the offspring then form a realisation of the process. To avoid edge effects the parents must be simulated over a region which is larger than $\mathcal{R}$, so that the contribution of offspring falling in $\mathcal{R}$ from parents falling outside of $\mathcal{R}$ is not lost.

Certain Poisson cluster processes and certain Cox processes can be statistically indistinguishable. In particular, when the offspring distribution, $f()$, in the Poisson cluster process is another Poisson distribution, it can be shown that it is always possible to find a probability distribution for $\lambda(\boldsymbol{s})$ in the Cox process that will produce exactly the same effect as the cluster process for any given $g()$. This is an interesting and in some ways unfortunate result. We referred to it indirectly in the introduction to this chapter. It implies that in such cases no method of statistical analysis will be able to distinguish between observed data from the two processes, although the interpretation of the two models is substantially different, one relating to a heterogeneous rate of occurrence (first order heterogeneity and spatial independence), the other to spatial interaction or dependence (stationary second order effects). So, for example, in a study of the incidence of a rare disease, we may be unable ultimately to discriminate between an explanation in terms of varying exposure to risk, due perhaps to environmental factors, or that in terms of infection due to contacts between individual cases.

Simple inhibition processes arise from various different ways of incorporating into the model a minimum permissible distance between any two events. In the simplest form of such models, the so called *hard core* processes, a CSR process is 'thinned' by deletion of all pairs of events a distance less than δ apart. A different 'dynamic' version of such a model would be to generate a CSR process one point at a time, and discard points if they lie within a distance δ of any previously generated point, or in a slightly different version again, within δ of any previously retained point. Such processes are stationary. They can be simulated directly by mimicking the process by which they are defined.

Markov point processes represent a more general family of probability models for point processes. In particular, some members of this family provide a more flexible framework for the modelling of regularity than the simple inhibition process. For example, such models allow for the fact that it might be unlikely but not impossible that two events occur in close proximity. In general, Markov processes are theoretically somewhat complex models. Amongst the simpler examples are the class of *pairwise interaction processes*, for example the *Strauss* process. Here points in $\mathcal{R}$ are considered 'neighbours' if they are less than some distance δ apart and then the joint density function for n point locations $(\boldsymbol{s}_1, \ldots, \boldsymbol{s}_n)$ in $\mathcal{R}$ which contains m distinct pairs of neighbours is specified as

$$f(\boldsymbol{s}_1, \ldots, \boldsymbol{s}_n) = \alpha\beta^n\gamma^m \quad \beta > 0, \quad 0 \leq \gamma \leq 1$$

where α is a normalising constant, β reflects the intensity of the process, and γ describes the interactions between neighbours. The case $\gamma = 1$ gives a CSR process whilst $\gamma = 0$ gives a hard core inhibition process with inter-point distance δ intermediate values of γ represent a form of non-strict inhibition.

The general simulation of Markov processes can be somewhat complex and computationally intensive and we do not give any details here.

3.5.6 Fitting models other than CSR to spatial point patterns

In the previous section we have outlined a number of alternative models to CSR for a spatial point pattern. Here we consider briefly how one might go about identifying from these a possible candidate for a given observed point pattern and then evaluating how well this 'fits' the observed data.

In order to identify possible models we can use the general exploratory methods which we have described earlier. The main issues are the degree to which the observed intensity of events varies over $\mathcal{R}$ and how much of this can be considered to be due to heterogeneity in the process as opposed to second order spatial dependence. An obvious way to estimate intensity over $\mathcal{R}$ is using the kind of kernel methods we discussed earlier, whereas the reduced second moment measure or K function, captures spatial dependence between different regions, on the assumption that the process is isotropic. Information gleaned from these exploratory analyses may then be used to suggest a model for the process—for example, a heterogeneous Poisson process or a Cox process.

We are then faced with the question of how to estimate any unknown parameters in such a model and how to formally assess the 'fit' of the model to the data. This is a difficult area and one in which the approach used will depend very much on the particular model involved. For example, in a case where a purely first order effect, as in the heterogeneous Poisson process, was thought likely, we might attempt to use kernel methods to estimate the parameters of simple 'trend' models for the first order variations in intensity. These may perhaps include covariates thought to influence the intensity. In other cases where a model involving isotropic spatial dependence was suspected, we might try to compare the theoretical K functions of various isotropic point process models to that estimated for the observed point pattern. This approach might also be used to estimate the parameters of such a model. Suppose, for example, that our proposed model incorporates a vector of parameters $\boldsymbol{\theta}$. Then the theoretical K function for this model will also involve $\boldsymbol{\theta}$ and we can write it as $K(h; \boldsymbol{\theta})$. A reasonable way to estimate $\boldsymbol{\theta}$ might then be to choose the value $\hat{\boldsymbol{\theta}}$ which minimises:

$$\int_0^{h_0} \left([\hat{K}(h)]^c - [K(h; \boldsymbol{\theta})]^c\right) \mathrm{d}h$$

where $\hat{K}(h)$ is the estimated K function for the observed point pattern and h_0 and c are 'tuning constants' chosen to provide desirable estimation properties.

As an illustration of this idea consider fitting a particular type of Poisson cluster process to an observed point pattern. Suppose we take a specific example of a Poisson cluster process with Poisson numbers of offspring per parent and where $g()$ is a radially symmetric normal distribution with variance σ^2. Then it may be shown that for such a process

$$K(h) = \pi h^2 + \frac{(1 - e^{-h^2/4\sigma^2})}{\rho}$$

where ρ is the intensity of the parent CSR process. Here $\boldsymbol{\theta} = (\rho, \sigma)^T$ and these may then be estimated as described above.

It is clear that these approaches are far from simple, and may involve both complex theoretical considerations and intensive computation. Often, techniques will need to rely on simulation approaches. We may not know, for example, what the theoretical K function, or other relevant property, should actually look like for a realisation of a particular proposed model in our study region $\mathcal{R}$. Frequently this is because of the difficulties introduced by edge effects. If this is the case then we could repeatedly simulate the proposed model over $\mathcal{R}$ and estimate the property of interest from these simulations. This could then be compared with the corresponding measure derived from the observed point pattern, and the significance of departures assessed through ideas similar to the 'simulation envelopes' that we have described earlier, in relation to comparisons with CSR. This explains why we indicated in the last section how one might simulate the various models discussed. We might for example, using this approach, compare the distribution function of nearest neighbour distances for a Poisson cluster process with various choices of parameters, with that of the observed data. The same kind of approach could be used in comparison of the estimated K function.

We have only been able to give a brief outline here of approaches to the general modelling of spatial point patterns. Further details are discussed in some of the references contained at the end of this chapter.

3.6 Summary

In this chapter we have introduced the reader to spatial point patterns, given some 'real life' examples of this kind of spatial data and discussed the objectives of analysis. We have described how such patterns can be represented statistically and how their first and second order properties may be characterised through the concept of intensity and second order intensity; pointing out that for isotropic point processes, second order properties may also be investigated either through the distributions of various inter-point distances, or the K function.

We have then gone on to discuss methods of analysis in detail, under the classification of visualisation, exploration and modelling. The nature of spatial point patterns does not provide much scope for visualisation techniques, and so our discussion here was necessarily brief, focusing on variations of the simple 'dot map'. We did however comment on cartogram transformations as providing a possible way to 'remove' *a priori* heterogeneity from a point pattern prior to subsequent analysis. In our exploratory section, the most important techniques introduced were those of kernel estimation of the intensity of a point pattern, and those concerned with estimation of the distributions of nearest neighbour distances and of K functions, both of the latter being particularly useful to explore spatial dependence. The main focus in the subsequent modelling section was on formal comparisons of an observed point pattern with one particular model, that for complete spatial randomness. We discussed a number of simple tests based on quadrat counts and summary measures of nearest neighbour distances and then moved on to consider more powerful techniques which involved the full distributions of nearest neighbour distances and the K function. Finally, we outlined some alternative models to that of CSR and discussed ways of evaluating how suitable they might be as representations of a specific observed pattern.

Throughout the chapter we have rarely referred explicitly to the use of additional information related to an observed pattern, although we acknowledged in discussion of case studies that this is often present and of interest in analysis. We refer here to, say, a 'labelling' of different types of events, a time of their occurrence or measurements of variation in the background 'population' from which events arise. It is not hard to see how we might incorporate some of these considerations into the forms of analysis that we have discussed, for example by comparing kernel estimates, or K functions between different types of events, or events in different time periods. In our next chapter we discuss details of some of these approaches, and others, more explicitly.

3.7 Further reading

We first give the sources of the data sets used in this and the following chapter. Readers are referred to these for further substantive discussion and, in some cases, interpretation, though by no means all of these references have analysed the data using point pattern methods.

The data on volcanic craters come from:

Tinkler, K.J. (1971) Statistical analysis of tectonic patterns in areal volcanism: the Bunyaruguru volcanic field in west Uganda, *Mathematical Geology*, 3, 335–55.

The original source of the Bodmin tors data is:

Pinder, D.A. and **Witherick, M.E.** (1977) The principles, practice and pitfalls of nearest neighbour analysis, *Geography*, 57, 277–88.

but the data are re-analysed in:

Upton, G. and **Fingleton, B.** (1985) *Spatial Data Analysis by Example: Volume 1: Point Pattern and Quantitative Data*, John Wiley, Chichester.

The data on redwood seedlings and biological cells come from:

Diggle, P.J. (1983) *The Statistical Analysis of Spatial Point Patterns*, Academic Press, London.

The locations of juvenile offenders on a Cardiff estate are taken from:

Herbert, D.T. (1980) The British experience, in Georges-Abeyie, D.E. and Harries, K.D. (eds.) *Crime: a Spatial Perspective*, Columbia University Press.

The locations of offences committed in Oklahoma City were digitised from maps in:

Carter, R.L. and **Hill, K.Q.** (1979) *The Criminals' Image of the City*, Pergamon Press, Oxford.

The original source of the Lancashire cancer data was the North West Regional Cancer Registry, though, as explained in the text, an element of randomisation has been applied to the coordinates. The data have been analysed in:

Diggle, P.J., Gatrell, A.C. and **Lovett, A.A.** (1990) Modelling the prevalence of cancer of the larynx in part of Lancashire: a new methodology for spatial epidemiology, in Thomas, R.W. (ed.) *Spatial Epidemiology*, Pion, London, 35–47.

However, we apply methods described in the next chapter to these data and further references are supplied there. Similarly, the source of the Burkitt's lymphoma data is:

Williams, E.H., Smith, P.G., Day, N.E., Geser, A., Ellice, J. and **Tukei, P.** (1978) Space–time clustering of Burkitt's lymphoma in the West Nile District of Uganda: 1961–1975, *British Journal of Cancer*, 27, 109–22.

but we shall analyse these data and give references on space–time methods in Chapter 4.

Turning to methods for analysing point patterns, the idea of mapping point events onto a transformed map or cartogram is discussed in:

Selvin, S., Merrill, D., Schulman, J., Sacks, S., Bedell, L. and **Wong, L.** (1988) Transformations of maps to investigate clusters of disease, *Social Science and Medicine*, 26, 215–21.

A full discussion of quadrat analysis is given in:

Thomas, R.W. (1977) *Introduction to Quadrat Analysis*, Concepts and Techniques in Modern Geography, No. 12, Geo Abstracts, Norwich.

The classic reference on kernel estimation is:

Silverman, B.W. (1986) *Density Estimation*, Chapman and Hall, London.

Other examples of kernel estimation are given in:

Gatrell, A.C. (1994) Density estimation of spatial point patterns, in Hearnshaw, H.J. and Unwin, D.J. (eds.) *Visualisation and Geographical Information Systems*, John Wiley, Chichester.

Good general statistical accounts of the analysis of spatial point patterns can be found in:

Ripley, B.D. (1981) *Spatial Statistics*, John Wiley, Chichester, Chapters 6–8.

Upton, G. and **Fingleton, B.** (1985) *Spatial Data Analysis by Example: Volume 1: Point Pattern and Quantitative Data*, John Wiley, Chichester, Chapters 1–2.

Diggle, P.J. (1983) *Statistical Analysis of Spatial Point Patterns*, Academic Press, London, Chapters 1–5; on which we have based much of our discussion.

Cressie, N.A.C. (1991) *Statistics for Spatial Data*, John Wiley, Chichester, Chapter 8.

Getis, A. and **Boots, B.N.** (1978) *Models of Spatial Processes*, Cambridge University Press, Cambridge.

Good geographical introductions to some of the methods discussed here are:

Haggett, P., Cliff, A.D. and **Frey, A.E.** (1977) *Locational Methods in Human Geography*, Edward Arnold, London.

Boots, B.N. and **Getis, A.** (1988) *Point Pattern Analysis*, Sage Scientific Geography Series, Volume 8, Sage Publications, London.

Unwin, D.J. (1981) *Introductory Spatial Analysis*, Methuen, London, Chapter 3.

3.8 Computer exercises

Here are suggested exercises that you can try on ideas discussed in this chapter using INFO-MAP and our example data sets. These exercises have been referenced at appropriate points in the chapter by numbered symbols in the margin.

Exercise 3.1

You can explore the idea of kernel estimation using various data sets. For example, select and open the 'Bodmin tors' file and load the data. Draw a `Dot map` from the `Map` menu and then select `Kernel map` from the `Analysis` menu and choose the 'intensity' option. You then need to specify a bandwidth, by marking the start and end points of a line on the map, the length of which will be used as the bandwidth. You mark the start point by a 'double click' at some convenient point on the map and then 'drag' the line (hold down the left button and move the mouse) to an end point which is fixed with another 'double click'. This will produce a 'raster' image of the kernel estimate of intensity. Experiment with different bandwidths, bearing in mind that the procedure is computationally demanding and will take some time to produce estimates; also, large bandwidths drastically over smooth the data and produce little of interest.

It may be useful to see the original events superimposed on the kernel estimate. In order to do this, use `Superimpose` from the `Map` menu, choosing the option 'on'. A `Dot map` produced after a kernel map will then superimpose the events onto the kernel estimate.

You could try similar ideas with the data on 'Cell centres', or 'Redwood seedlings', or try to reproduce the results in the text on 'Volcanoes in Uganda'.

Exercise 3.2

Load the data on 'Redwood seedlings' and create a dot map of these data. Calculate the distances of events to their nearest neighbours as follows:

```
nearneigh=dist([near(1)])
```

This new variable may itself be mapped using `Dot map` if you wish. Use the `Cumulative distribution` option within the `Analysis` menu to derive the estimated cumulative nearest neighbour distribution function, $\hat{G}(w)$. What information does this provide about the spatial distribution of redwood seedlings?

Repeat this analysis using the data on 'Cell centres' and check that the results confirm your intuitive impressions about the spatial arrangement of these events.

Exercise 3.3

Load the data on 'Bodmin tors' and create a `Dot map`. Now estimate a K function for these data, by selecting the `K function` option within the `Analysis` menu. You are then prompted for the maximum distance at which the K function is to be estimated. This is specified in exactly the same way as for bandwidth selection in kernel estimation, discussed in an earlier exercise. Mark the start point by a 'double click' somewhere on the map and 'drag' a line to the end point that defines your distance limit; fix this with a 'double click'. Specify zero for the number of simulations required. The calculations may take a while, especially if the distance limit is large. In general, it makes sense only to estimate the function over very short distances relative to the size of the study region. Note that what is plotted in INFO-MAP is the square root transformation $\hat{L}()$ described in the text.

What information does this plot convey about the spacing of tors?

Repeat this exercise using the 'Cell centres' and the 'Redwood seedlings' data. Make observations on the form of the functions and relate the estimated functions to the map distributions. Try also with the 'Cardiff offenders' data comparing your results with those in Figure 3.8.

Exercise 3.4

INFO-MAP has no direct facility for performing quadrat analysis, but it is possible to form quadrat counts for a point pattern, by using the general functions for 'importing' data. This is a little 'clumsy' but relatively straightforward and it illustrates some general points about aggregating spatial data from one frame of reference to another.

Open the file 'Cardiff offenders'. Then choose `Grid references` from the `Map` menu and locate the point on the estate with grid reference (20,90) and that with reference (70,25). Note roughly the position of these two points; the idea is to create a regular grid of quadrats with these two points as top left and bottom right corners. To do this select `Modify` from the `File` menu and you are placed in the *base file* module of INFO-MAP as described in one of the exercises in Chapter 2 (note that the menu bar at the top of the screen has changed).

Now choose `Grid` from the `Annotate` menu, which allows you to add a grid to the displayed map. You do this by marking the top left and bottom right corners and then specifying the number of horizontal and vertical divisions you require. Mark the top left corner roughly at the point identified earlier as having grid reference (20,90). Do this by moving the cursor to the point and 'double clicking' (if you don't have a mouse, see the *User's Guide* for how to do this from the keyboard). Then 'drag' the box you have initiated (hold down the left button and move the mouse) so that its bottom right corner is roughly at the second point identified earlier as having grid reference (70,25). Fix the box with another 'double click'. Finally, specify that you require an (8×8) grid. The grid should then be drawn on the screen.

Next, choose `Delete` from the `Locations` menu with the option 'Delete all locations'. You have now deleted all the original locations in the file (that is, the residences of the offenders) and you now replace them with a set of locations which correspond to the centres of the quadrats you have formed. Do this by choosing `Generate` from the `Locations` menu, marking a box (using the same technique as before) which corresponds exactly to that which you delineated when you set up the quadrats; again specify that you want an (8×8) set of locations. Finally, choose `Save` from the `File` menu to save the results of your labours to file. Then use `Exit` from the `File` menu to return to the familiar 'top level' of INFO-MAP.

The net result of this process is that you now have a new file consisting of an (8×8) array of quadrats. You have no data for these quadrats at present; but the new file is grid referenced on the same basis as the original 'Cardiff offenders' file; hence you can 'import' data from the 'Cardiff offenders' file into the quadrats. To do this, select `New` from the `Data` menu with the option 'Import file by grid references'. A window of prompts is then displayed. The file name from which you wish to 'import' is that for the 'Cardiff offenders' so enter 'cardest.dta' into the file name box. Do not terminate your entry with <CR>, since you wish to alter some of the other prompts; use <TAB> to move to these prompts. INFO-MAP knows where the eastings and northings are stored in this file, so it skips over these two prompts. Specify that you wish to import fields 1 to 1 (you only want one field). Then for the 'Importation method' specify 'point in polygon', and for the 'Summary type' specify 'count'. Against the 'New data item name' enter 'Event count'. What you have now specified is that you wish to count the number of events (i.e. locations) in the 'Cardiff offenders' file that fall into each of the quadrats you have set up in the new file and you wish to assign the name 'Event count' to the data item for the new file that results. Now use <CR> to start the import. The process may take a little while, but eventually you should see a spreadsheet appear, with the new data item that you have created displayed. Use `Save` from the `File` menu to save this data file. Then use `Exit` from the `File` menu to return to the familiar 'top level' of INFO-MAP.

You should now have a new INFO-MAP data set consisting of a set of (8×8) quadrat counts for events in the 'Cardiff offenders' file. You can now analyse this as you wish. First try a `Choropleth map` of the counts to get a rough idea of what you have. Next calculate the mean quadrat count by using `Calculate/display worksheet` from the `Analysis` menu with the formula:

```
work[1]=mean({1})
```

Then calculate the variance of the quadrat counts into another worksheet item:

```
work[2]=stdev({1})**2
```

and the 'index of cluster size':

```
work[3]=work[2]/work[1]-1
```

What does the value imply about the distribution of 'Cardiff offenders'? Finally, calculate the test statistic for the 'index of dispersion' test:

```
work[4]=(nonmis({1})-1)*work[2]/work[1]
```

Compare this with tabulated percentage points of the χ^2_{63} distribution. Is it an 'extreme' value in the positive tail of the distribution? What conclusions can you draw from this?

We could have used a random sample of the quadrat counts in the above procedure, instead of all of them. For example using `Calculate` from the `Data` menu we can set up a new variable as:

```
samplects=if(rand()<=0.75,{1},miss())
```

This will give approximately a 75% random sample of the quadrat counts into a new variable. If this is variable 2 in the file, then repeating the above calculations replacing variable 1 with variable 2 will give the corresponding results for the random sample of quadrats. Note that the number of degrees of freedom for the χ^2 distribution would then be `nonmis({2})-1`.

You might then go on to calculate an estimate of the intensity of events over the estate based on the random sample of quadrats by calculating a worksheet item:

```
work[5]=mean({2})/area()[1]
```

Note here that all the quadrats have the same area, so we just use that of the first of them in the file. Can you work out a way to calculate a confidence interval for your intensity estimate?

Exercise 3.5

Load the 'Bodmin tors' data and again create a `Dot map`. Calculate a Clark–Evans test statistic and evaluate this for departure from complete spatial randomness. You may do this as follows. First, calculate a new variable of nearest neighbour distances:

```
nearneigh=dist([near(1)])
```

We assume that this new variable will be variable 2. Next, use `Calculate worksheet` in the `Analysis` menu to calculate items that are needed to form the significance test. First, calculate an estimate of intensity (events per unit area) as:

```
work[1]=count()/area()
```

where count() gives the number of events. Next, estimate the mean and variance for the mean nearest neighbour distance using the following:

```
work[2]=1/(2*(work[1]**0.5))
work[3]=(4-3.14159)/(4*work[1]*3.14159*count())
```

Note that the first expression involves taking the square root of the estimated intensity. Lastly, calculate the standardised normal deviate for the Clark–Evans test as:

```
work[4]=(mean({2})-work[2])/(work[3]**0.5)
```

where 'mean({2})' will give the mean nearest neighbour distance, $\bar{w}$. What are your conclusions from these results?

Can you modify this procedure, incorporating the `perim()` function described in the *User's Reference*, in order to perform the corrections suggested in the text that are appropriate when *all* nearest neighbour distances are being used?

You could also experiment with using the `rand()` function described in the *User's Reference*, to calculate Clark–Evans on a random subset of nearest neighbour distances, instead of the full set. For example

```
nearneigh=if(rand()<=0.5,dist([near(1)]),miss())
```

will give approximately a 50% random sample of nearest neighbour distances.

Exercise 3.6

It is possible to perform formal simulation tests on estimated K functions within INFO-MAP. It is also possible to estimate such functions within spatial subsets of a full data set. Load the data on 'Cardiff offenders' and create a `Dot map`. The overall $\hat{L}()$ for these data is shown in the text. Choose `Zoom/Restrict` within the `File` menu and select a small subset of the study region to 'zoom'. You do this by marking one corner of an appropriate rectangular region by a 'double click' and then 'dragging' a window away from the chosen corner, fixing the final position with another 'double click'. This defines the 'source' area for the zoom. You then similarly define a rectangular 'target' area (usually a moderate sized area roughly in the centre of the working screen). In this case use the 'raster' zoom option (there are no polygons).

Once the zoom is completed, you may redraw the `Dot map` and estimate the K function for just this sub-region, specifying the number of simulations you require for the envelopes. Do not specify a large number since the computations can take some time; twenty should be sufficient to illustrate the idea. By using this idea you can explore other sub-regions of this data set or indeed, sub-regions of other data sets. Note that the same principle, of selecting sub-regions for study, can be used with forms of analysis other than just the K function.

4

Further methods for point patterns

Here we consider various more specialised topics concerned with the analysis of 'point patterns'. We start by considering the case where more than one type of 'event' is observed and interest lies in relationships between the patterns of these different types. We then discuss the situation when a single type of 'event' occurs, but each occurrence is at a different time. Our interest is then in whether and how the spatial pattern varies over time. We conclude with a common problem: that of requiring to adjust our analysis of pattern for a covariate which is known to affect the rate at which 'events' occur; for example, geographical variations in the 'population' from which 'events' arise.

4.1 Introduction

Looking back on the case studies we introduced in the previous chapter, it is clear that several questions that arise from these examples are not directly addressed by the analysis methods we presented there.

Consider, for example, the data set on the locations of 'theft from property' offences in Oklahoma City, USA, in the late 1970s. These data comprise two distinct categories of events; one set refers to offences committed by whites, the other by blacks. The methods of the previous chapter would be appropriate to investigate either of the two patterns separately, or indeed both taken together, but what of possible interactions between the two types of events? We posed the question earlier of whether the two sub-groups of offenders have different 'activity spaces', based perhaps on different cognitive or mental maps of the city? Do the crimes committed by different groups display different spatial patterns and is there evidence that these patterns are related in any way? These are questions which the methods of the previous chapter fail to address directly.

Another example relates to the 188 cases of Burkitt's lymphoma, in the West Nile District of Uganda. These data, which relate to the time period 1961–75, include the age of the child involved and also the date of onset of the disease (measured in days elapsed since January 1st 1960). As discussed previously, interest in analysing these data does not relate to looking for spatial clustering (we would expect this *a priori*), but rather in assessing evidence for space–time clustering. Are cases that are near each other in geographic space also 'near' each other in time? If so, this might be evidence in support of the hypothesis that suggests an infective aetiology for the disease. Here the methods presented in the previous chapter would appear to be of even less use than in the crime example.

A further situation where the point pattern methods discussed so far have only limited application, is when a pattern of potential interest is possibly obscured by a more dominant clustering that results from an underlying and obvious spatial covariate which needs to be allowed for in any analysis. Nowhere is this problem more acute than in the study of patterns in epidemiological data, where variations in the population at risk are nearly always present. The data set relating to 57 cases of cancer of the larynx, for part of Lancashire in Britain, collected over a 10 year period (1974–83), provide a good example. The question was whether these cases 'cluster' near the site of an old industrial waste incinerator. But in addressing this we need to correct for the natural background variation in population, either directly, or through use of the incidence of lung cancer data in the same area as a surrogate to 'mimic' the background distribution of population. We have not so far discussed any methods of analysis that attempt to deal with this kind of problem, although some of the alternative models to CSR that we introduced, such as the heterogeneous Poisson process, clearly relate to these situations.

In this chapter we extend some of the methods that we discussed in the last chapter in order to try and 'plug' some of these 'gaps'. We are aware that these are not the only deficiencies that might be identified in our methods, but we believe them to be some of the most important. We are also conscious that we will only be able to deal here with some of the simpler approaches to such problems, but there are limits as to how far we are able to go in this book.

4.2 The analysis of multiple types of events

In the analyses discussed in the previous chapter all the events that occurred were of one type. Suppose now that we have two or more types of events. The previous methods we have discussed enable us to analyse whether the occurrences of any one type of event taken separately (or all of them taken together) exhibit clustering or regularity; however there is now a further set of questions that may be of interest. Is pattern in the occurrences of one type of event related to that in the occurrences of another? Does the distribution of one 'explain' the distribution of the other? In general, we shall be looking for evidence of *independence* between types of events as opposed to either *attraction* or *repulsion*.

In the same way as CSR provided a basis from which to compare the degree of clustering or regularity in a *univariate* point pattern, independence forms the basis from which to assess evidence of repulsion or attraction in observed *bivariate* or *multivariate* point patterns. Independence implies that the overall pattern in events is made up from independent component processes, one for each of the types of events. Note that independence does not necessarily imply that any of these component processes need be CSR processes—there is no reason why two clustered processes should not be independent. Indeed, one could have independence between a clustered process and a process exhibiting regularity. In general, any of the constituent processes may be quite dissimilar in pattern, both from each other and from the composite pattern. Note further that independence is not the same hypothesis as that of a *random labelling* of events. The latter stipulates that the 'marking' of events is independent of their locations and is a uniform distribution over the number of types of event. This would only be equivalent to independence if all the component processes were CSR.

We do not discuss any specific theoretical models for multivariate point processes, except to say that it is possible to generalise our earlier Poisson, Poisson cluster and Cox processes to incorporate the idea of linked processes which either display attraction or repulsion—for example *linked Cox processes*, or *linked pair processes*. We concern ourselves here solely with methods to establish departures from independence rather than the fitting of specific models.

An extremely simple approach to the testing of independence in the spatial distribution of two types of events is based on counts of the number of each event in quadrats over $\mathcal{R}$ (either randomly scattered or in a regular grid in the case of a completely enumerated point pattern). It is common practice to present results in a 'presence–absence' form, that is to say a 2×2 table consisting of the number of quadrats, c_{ij}, which contain in turn: both types of events, neither, type 1s but no type 2s, and type 2s but no type 1s. So we might obtain:

		Type 2	
		Absent	Present
Type 1	Absent	c_{11}	c_{12}
	Present	c_{21}	c_{22}

This is then tested by the simple standard χ_1^2 test of independence in a 2×2 table, details of which may be found in any standard elementary text on statistics. That is, we compare:

$$X^2 = \frac{(c_{11}c_{22} - c_{12}c_{21})^2 \sum_i \sum_j c_{ij}}{\sum_j c_{1j} \sum_j c_{2j} \sum_i c_{i1} \sum_i c_{i2}}$$

with the percentage points of the χ_1^2 distribution, significantly large values indicating lack of independence between the two patterns.

The problem with such a simple approach is that, as usual with quadrat methods, a large amount of locational information present in the observed patterns is being ignored in the test. In our discussion of univariate spatial point patterns in the previous chapter two of the most powerful tools we introduced were the distribution functions of nearest neighbour distances and the K function. Both of these can also be used in tests of independence for multivariate point patterns.

The distribution function of nearest neighbour event–event distances for the univariate pattern is replaced in the multivariate case by a set of nearest neighbour distribution functions $G_{ij}(h)$. $G_{ij}(h)$ is the probability that the distance from a randomly chosen type i event to the nearest event of type j is less than or equal to h. For nearest point–event distribution functions, $F_j(h)$ is the probability that the distance from a randomly chosen point to the nearest event of type j is less than or equal to h. Estimates of these functions $\hat{F}_j(h)$ and $\hat{G}_{ij}(h)$ can be obtained from an observed mapped point pattern using a natural extension of the method given in the previous chapter for the corresponding univariate distributions. If required, edge corrections can also be included in the estimates of the type described there.

If component point processes in a multivariate point pattern are independent, then the distribution of nearest neighbour distances to events of type j from an origin of measurement should be the same, whether the origin is a randomly chosen point or a randomly chosen type i event, $(i \neq j)$. That is:

$$G_{ij}(h) = F_j(h) \quad i \neq j$$

This suggests an obvious approach to exploring for independence, namely to plot the estimates $\hat{F}_j(h)$ and $\hat{G}_{ij}(h)$ on a single diagram with h varying between 0 and the maximum nearest neighbour event–event or point–event distance. The plot is then examined for evidence of marked variation between the two estimated distributions.

A crude test of independence in a bivariate pattern can be obtained without the full estimation of $\hat{G}_{ij}(h)$ or $\hat{F}_j(h)$, used above. A number of alternative approaches have been suggested, but a simple one to implement is to take a random sample of points in $\mathcal{R}$ and measure the nearest neighbour point–event i distances and nearest neighbour point–event j distances corresponding to this sample of random points. Distances in each of these samples are then replaced by their rank (position in order of size), within their respective sample. As a result, one is left with a set of pairs of point–event distance ranks. These may then be tested for independence using a standard procedure such as that based on Spearman's or Kendall's rank correlation coefficient. We do not give details of such procedures here, which may be found in any text on 'distribution free' or 'non-parametric' statistics. One advantage of this crude test for independence is that it can be applied in the sampling of point patterns as well as to fully mapped patterns. An edge correction is advisable, it being usual to discard from the two samples any distances where random points are nearer to the boundary of $\mathcal{R}$ than to a neighbouring event of either type.

For mapped data sets, analyses based upon the K function are generally more powerful than those which use nearest neighbour distances. In the case of a multivariate point pattern where the intensity of type j events is λ_j, $K_{ij}(h)$ is defined as:

$$\lambda_j K_{ij}(h) = E(\#(\text{type } j \text{ events } \leq h \text{ from an arbitrary type } i \text{ event}))$$

$K_{ii}(h)$ is clearly just the univariate K function for events of type i. $K_{ij}(h), i \neq j$, is known as the *cross K function*.

Under the assumption of independence between types of events the locations of type j events should be random with respect to those of type i events, regardless of whether the spatial distribution of either type i or type j events is clustered, regular, or random, when considered separately. Thus under independence the expected number of type j events within a distance h of a randomly chosen type i event is $\lambda_j \pi h^2$. Hence theoretically under independence:

$$K_{ij}(h) = \pi h^2$$

Under repulsion $K_{ij}(h)$ tends to be less than πh^2, whereas under attraction $K_{ij}(h)$ tends to be greater than πh^2. This suggests the obvious approach of comparing the empirical cross K function, $\hat{K}_{ij}(h)$, estimated from the observed data, with its theoretical value, as a test of independence.

In the case of a bivariate point pattern with n_1 type 1 events and n_2 type 2 events over a study region $\mathcal{R}$, with area R, $K_{12}(h)$ may be estimated by analogy with the edge corrected method used in the univariate case. Let d_{ij} be the distance between the ith type 1 event and the jth type 2 event and let $I_h(d_{ij})$ be 1 if $d_{ij} \leq h$, 0 otherwise. Further, consider a circle centred on the ith type 1 event, passing through the jth type 2 event and let w_{ij} be the proportion of the circumference of this circle which lies within $\mathcal{R}$. Then a suitable edge corrected estimator for $K_{12}(h)$ is

$$\tilde{K}_{12}(h) = \frac{R}{n_1 n_2} \sum_{i=1}^{n_1} \sum_{j=1}^{n_2} \frac{I_h(d_{ij})}{w_{ij}}$$

Theoretically $K_{12}(h) = K_{21}(h)$ but this will not necessarily be the case for their corresponding estimates. One suggestion is therefore that a better estimate for $K_{12}(h)$ is given by

$$\hat{K}_{12}(h) = \frac{(n_2 \tilde{K}_{12}(h) + n_1 \tilde{K}_{21}(h))}{n_1 + n_2}$$

The use of K functions for multivariate point patterns is analogous to that described for univariate point patterns in the previous chapter. $\hat{L}_{ij}(h)$ is plotted against h where:

$$\hat{L}_{ij}(h) = \sqrt{\frac{\hat{K}_{ij}(h)}{\pi}} - h$$

Plots of $\hat{L}_{11}(h)$, $\hat{L}_{22}(h)$, $\hat{L}_{12}(h)$ on the same axes simultaneously reveal the tendency for the individual component patterns to depart from CSR as well as any tendency for attraction (positive peaks in the plot of the cross L function) or repulsion (negative troughs in the plot of the cross L function) between the component patterns, regardless of whether their individual patterns are either random, clustered or regular.

In the univariate case, upper and lower simulation envelopes of the K function under CSR provided a means to assess significant departures of $\hat{K}(h)$ from its theoretical value. These still apply for $\hat{K}_{ii}(h)$ or $\hat{K}_{jj}(h)$. The theoretical value of $K_{ij}(h)$ $i \neq j$ under independence does not depend on CSR in the component patterns and therefore no assumptions can be made about models for either of the component patterns in order to provide a basis for development of simulation envelopes. The separate component event patterns need to be preserved in their observed form in any simulations. One way of achieving this is by simulations which involve random shifts of the whole of one component pattern relative to the other. In the case of a rectangular $\mathcal{R}$ these shifts are performed using a toroidal idea—events moved outside of $\mathcal{R}$ by a shift 'reappear' in $\mathcal{R}$ in their new position in the opposite corner. For irregular regions the boundary is also shifted by the same amount and then attention focused only on that sub-region of $\mathcal{R}$ which also lies within the shifted study region. Using these sort of techniques one may obtain simulation envelopes to assess the significance of peaks or troughs in the $\hat{L}_{ij}(h)$ plots.

To illustrate these ideas let us consider the data on the offences committed by white and black residents in Oklahoma City. Visually, (Figure 4.1) the two distributions do look rather different, with those offences committed by the whites showing limited evidence of clustering but those for blacks showing considerable spatial aggregation in the centre of the study region. Perhaps this reflects a more limited 'spatial search' for crime opportunity among black offenders.

However, in this section we are concerned with the hypothesis of spatial interaction (attraction or repulsion) between these two types of events, rather than the rather obvious difference in their individual patterns. To assess this 'interaction' hypothesis we estimate $\hat{L}_{12}(h)$, where the subscript 1 represents the black offenders and 2 the white. When we plot this function, together with the corresponding envelopes derived from random toroidal shifts (Figure 4.2), we find no significant evidence of spatial interaction between the spatial distributions of offences committed by blacks and those committed by whites.

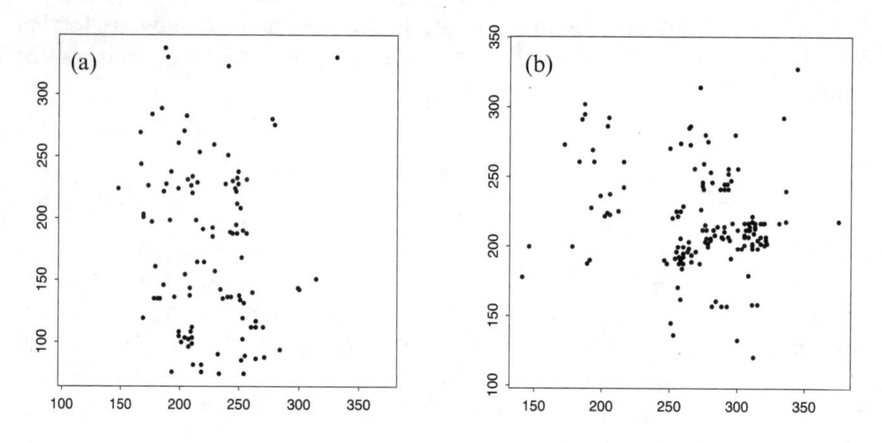

Fig. 4.1 Oklahoma City offences by (a) whites and (b) blacks

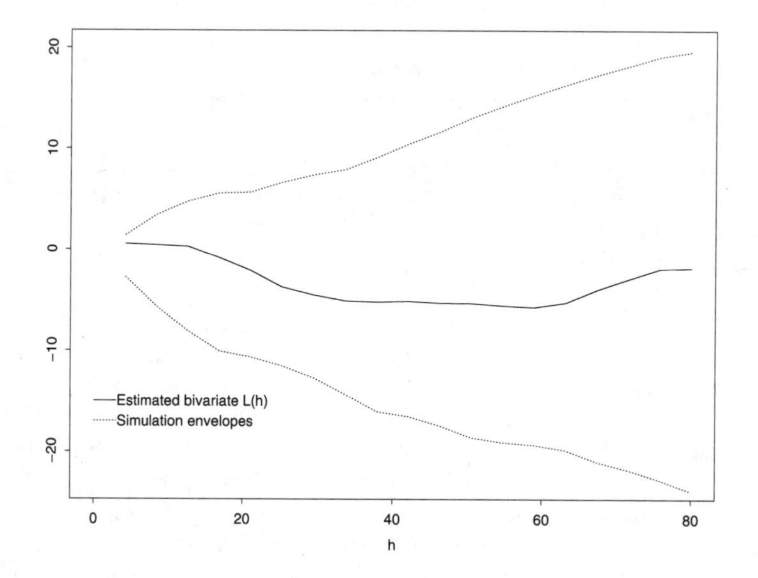

Fig. 4.2 $\hat{L}_{12}$ for black and white offences

4.3 Space–time clustering

In the discussion of the analysis of multivariate point patterns above, we were concerned with a 'labelling' of events in terms of different types. A different kind of 'labelling', which is of interest in some applications, particularly in an epidemiological context, is when the time of occurrence of events has been recorded. We may then be interested in whether space and time interact—that is, whether events cluster in space when time is allowed for. This would be particularly important in studying a dynamic process, since the pattern at any fixed period of time would not necessarily be very informative about the way that the pattern had 'grown' through time.

One simple space–time interaction test is due to the epidemiologist George Knox. Knox's test is based on a consideration of all $n(n-1)$ ordered pairs that can be formed from an observed set of n events. For each of these pairs we can measure both their spatial separation and the time interval between them. Suppose S pairs are 'close in space' and T pairs are 'close in time'. In other words, both space and time are divided into two classes ('close' and 'not close'), and pairs of events are assigned to one of the four cells of a contingency table, according to whether the pair is, or is not, close in space and/or time. Then, under the assumption of independence of space and time the number of pairs X that would simultaneously be both 'close' in time and space will have an approximate Poisson distribution with mean $ST/n(n-1)$ (assuming the $n(n-1)$ pairs are regarded as being independent of one another). We therefore reject independence in favour of clustering in time and space if our observed value of X is 'unusually large' when compared with this theoretical distribution. It is left to the analyst to decide in any particular application how to construct the zero–one proximity measures 'close in space' and 'close in time'. In practice, analyses are usually repeated for ranges of distance and time intervals that are thought to be of relevance.

An extension to Knox's test, devised by Nathan Mantel, is based on the following test statistic:

$$\sum_{i \neq j}\sum x_{ij} y_{ij}$$

where x_{ij} represents the 'distance' between events i and j, and y_{ij} is the corresponding 'time interval'. As an aside, note that if we let y_{ij} and x_{ij} each take the value 1 if the events are close in both space *and* time, and zero otherwise, then the above test statistic is simply the count of close space–time pairs used as the test statistic by Knox. In Mantel's case, numerical rather than categorical values of spatial and temporal separation are employed. Often in studies of space–time interaction, interest centres not on all pairs of events, but only on those which are 'reasonably' close. As a result, Mantel suggests using separation measures x_{ij} and y_{ij} given by:

$$x_{ij} = \frac{1}{(k_s + d_{ij})}$$

and

$$y_{ij} = \frac{1}{(k_t + t_{ij})}$$

where d_{ij} and t_{ij} are Euclidean distance and time interval respectively, between events i and j and k_s and k_t are arbitrary constants chosen to allow for events that have identical space or time coordinates. Mantel gives expressions for the mean and variance of the sampling distribution of his test statistic, $\sum\sum x_{ij}y_{ij}$, under the assumption of no space–time clustering. These can then be used to test observed values of the statistic. Details are provided in the references given at the end of this chapter.

The problem with the Knox test is that it requires the user to specify, *a priori*, a set of distance and time intervals. One difficulty with Mantel's approach is the need to specify values for the constants in the expressions for time and distance separation. In order to avoid these somewhat arbitrary decisions another approach to the investigation of space–time interaction could use the K function, discussed previously.

One simple approach, if there are sufficiently few time periods, is to regard events at different times as different types of events and base analysis upon $K_{ij}(h)$ to detect attraction (clustering in time) as opposed to independence. A more widely applicable method to detect space–time interaction, based on K functions, is that which was applied in the analysis of the data on Legionnaires' disease, discussed in Chapter 1. The approach is as follows. Recall first that the definition of the K function is the expected number of events within a distance h of an arbitrary event, divided by the intensity, λ. We may extend this idea to a bivariate 'space–time' K function, $K(h, t)$, defined as the expected number of events within distance h and time interval t of an arbitrary event, now scaled by the expected number of events per unit area *and* per unit time. We may obtain an estimate of this function, say $\hat{K}(h, t)$, with an edge correction, by extensions to the method used previously. If the processes operating in time and space are independent (no space–time interaction) then $K(h, t)$ should be just the product of two K functions, one in space, $K_S(h)$, and the other in time, $K_T(t)$, both of which may also be estimated separately from the data. Thus, a test for space–time interaction may be based on the observed differences:

$$\hat{D}(h, t) = \hat{K}(h, t) - \hat{K}_S(h)\hat{K}_T(t)$$

If there is space–time interaction in the data we will see evidence of raised values of $\hat{D}(h, t)$; this will be clarified by a three-dimensional plot of $\hat{D}(h, t)$ against distance and time. One way to devise a more formal assessment of the significance of the observed values of $\hat{D}(h, t)$, is to perform m simulations, in each of which the n events are randomly labelled with the observed n time 'markers'. We can thus obtain m estimates $\hat{D}_i(h, t)$, $i = 1, \ldots, m$. The actual observed sum of $\hat{D}(h, t)$ over all h and t, is then compared with the empirical frequency distribution of m such sums, each of which is obtained from one of the $\hat{D}_i(h, t)$. An 'extreme' value of the observed sum when compared with this distribution would indicate evidence of space–time interaction.

Applying these methods to the Burkitt's lymphoma data the $\hat{D}()$ plot, shown in Figure 4.3, provides visual evidence of space–time clustering, while the formal test of significance using simulation generates a set of simulated summed $\hat{D}_i()$ values, of which only one exceeds the observed value (Figure 4.4). These results imply that there is significant space–time clustering in the data, possibly giving support to the hypothesis that the lymphoma is the result of some infectious agent.

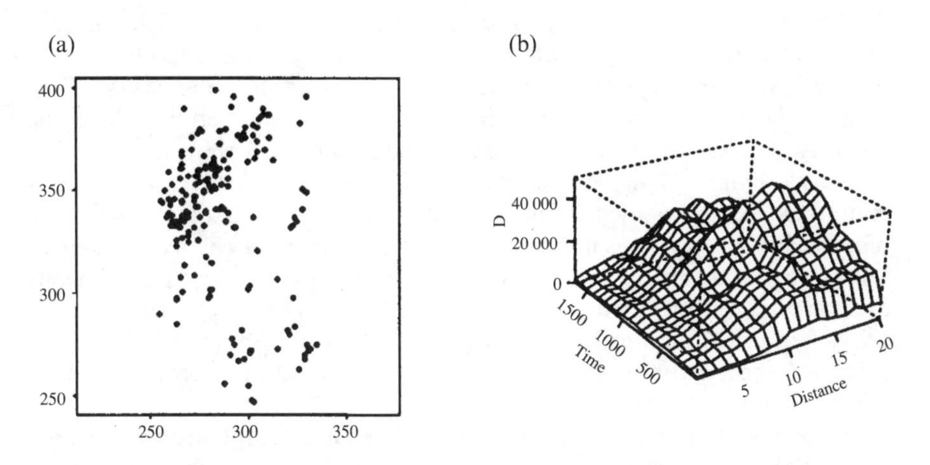

Fig. 4.3 (a) Distribution and (b) $\hat{D}(h, t)$ for Burkitt's lymphoma

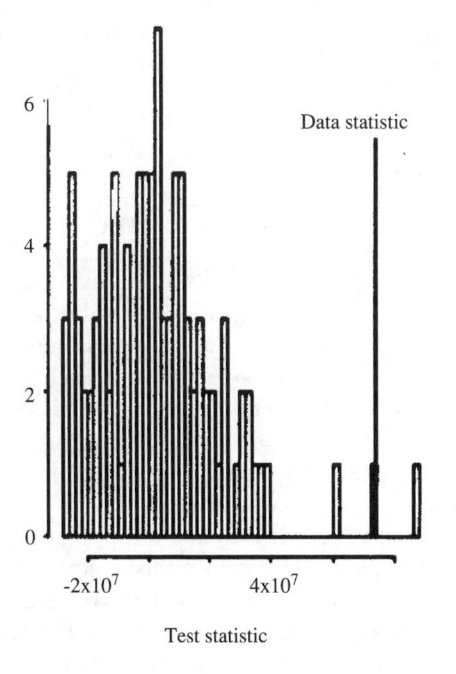

Fig. 4.4 Testing Burkitt's lymphoma data for space–time interaction

4.4 Correcting for spatial variations in the population at risk

As discussed earlier, in some applications where the hypothesis of clustering is of interest, comparison with CSR is not sensible because of an expectation that there will be some degree of natural spatial variation in the intensity of the process. This would be the case in an epidemiological setting where the intensity of cases of disease would be expected to vary with population density. Another example might be if we wished to investigate possible clustering of crime data, such as residential burglaries, where we know that the distribution of residential properties is not homogeneous. In such cases one really wants to test clustering against a base hypothesis which consists of a heterogeneous Poisson process with varying intensity $\lambda(\boldsymbol{s})$ rather than CSR. The heterogeneity arises from the natural variation in background population (in an epidemiological setting) or in the distribution of residential properties (in a crime setting).

In the previous chapter we discussed kernel methods to obtain an estimate of $\hat{\lambda}(\boldsymbol{s})$. An obvious question is whether we can use such methods in attempting to 'correct' for underlying variations in 'population at risk'.

Suppose in the first instance that we have direct information on variations in an appropriate 'at risk population' in our study area $\mathcal{R}$. The most usual form that such information would take would be the numbers of individuals in this population in census districts covering $\mathcal{R}$. In other words, we have, say, m population observations y_j, each associated with a fixed area $\mathcal{A}_j$. Our previous kernel estimate of event intensity over $\mathcal{R}$ (ignoring the edge correction for simplicity) was:

$$\hat{\lambda}_\tau(\boldsymbol{s}) = \sum_{i=1}^{n} \frac{1}{\tau^2} k\left(\frac{(\boldsymbol{s} - \boldsymbol{s}_i)}{\tau}\right)$$

This is an estimate of 'events per unit area'. An intuitive way to convert to an estimate of 'events per unit population', say $\hat{\rho}_\tau(\boldsymbol{s})$, is by dividing $\hat{\lambda}_\tau(\boldsymbol{s})$ by an estimate of the population density at $\boldsymbol{s}$. A crude estimate of this may be obtained by simply converting absolute population in each $\mathcal{A}_j$ into population density by dividing by the corresponding area A_j, and attaching to $\boldsymbol{s}$ the population density of that district in which $\boldsymbol{s}$ lies. Examination of the resulting $\hat{\rho}_\tau(\boldsymbol{s})$ over $\mathcal{R}$ then gives some indication of variations in intensity per unit population, which is the real question of interest.

Such an approach uses only a crude estimate of population density at $\boldsymbol{s}$ and an obvious development would be to improve this in some way. Suppose that each of the m population values, y_j, are assumed to be associated with m specific points $\boldsymbol{s}'_j$ in $\mathcal{R}$. These might for example be the 'centroids' of the corresponding areas $\mathcal{A}_j$ or the approximate 'centre of population' in that area. Then the possibility exists of a kernel estimate of population density, say $\hat{\lambda}'_\tau(\boldsymbol{s})$, using these points. We obtain this from a natural extension to our previous kernel estimate as:

$$\hat{\lambda}'_\tau(s) = \sum_{j=1}^{m} \frac{1}{\tau^2} k\left(\frac{(s - s'_j)}{\tau}\right) y_j$$

on the basis of the argument that we are now estimating 'population per unit area' as opposed to 'points per unit area'.

We can now replace our previous estimate of 'events per unit population', $\hat{\rho}_\tau(s)$, by one based on the ratio of the kernel estimates for intensity of events and population density respectively, obtaining:

$$\hat{\rho}_\tau(s) = \frac{\sum_{i=1}^{n} k\left(\frac{(s - s_i)}{\tau}\right)}{\sum_{j=1}^{m} k\left(\frac{(s - s'_j)}{\tau}\right) y_j}$$

where we have assumed that the same bandwidth, τ, and kernel, $k()$, are used in the estimate of both event intensity and population density; clearly, this is not necessary and it would be a simple modification to use different values in each. In fact, the use of different bandwidths may well be sensible. We are concerned here with the ratio of two kernel estimates of intensity. It does not necessarily follow that 'good' estimates of the numerator and the denominator will automatically lead to a 'good' estimate of their ratio. For instance, relatively small changes in the denominator (the estimate of population density) in regions where its value is small will produce dramatic and unacceptable variations in the ratio of the two kernel estimates. For such reasons it may be preferable to deliberately 'over-smooth' the kernel estimate of population intensity when estimating the ratio, by choice of a larger bandwidth than would be appropriate if we were just interested in an estimate of the population intensity alone.

The assumption that population values are 'centred' at specific points may seem debatable. The interpolation of variations in population density from a set of census districts in this way is undoubtedly rather questionable; we will have more to say about such techniques in Chapter 7. An alternative approach to the use of direct population information might be to use observed events from another spatial process which was felt 'representative' of population variations (a 'control' process). This is essentially the idea suggested in relation to the data set on cancer of the larynx, where it was considered that the incidence of lung cancer in the same area might be used as a surrogate to 'mimic' background population distribution.

In this situation, the application of the above kernel ratio ideas is even more straightforward. Suppose that the 'control' process consists of m events occurring at points s'_j in $\mathcal{R}$. Our previous kernel estimate of population density is replaced by a simple kernel estimate of the intensity of the control process, and 'events per unit population', $\hat{\rho}_\tau(s)$, is now estimated by:

$$\hat{\rho}_\tau(\mathbf{s}) = \frac{\sum_{i=1}^{n} k\left(\frac{(\mathbf{s} - \mathbf{s}_i)}{\tau}\right)}{\sum_{j=1}^{m} k\left(\frac{(\mathbf{s} - \mathbf{s}'_j)}{\tau}\right)}$$

Ex 4.4

where again we have assumed that the same bandwidth and kernel are used in the estimate of both event intensity and control intensity, and obvious modifications apply if this is not the case. The comments made earlier concerning the advisability of 'over-smoothing' the kernel estimate of the intensity of 'controls' also apply here.

There is also a somewhat different approach which can be used where a 'control' process is available to act as a surrogate for variations in population at risk. This employs some of the ideas we have previously discussed in relation to multivariate point patterns. Suppose that n_1 cases of a disease have been observed in a study area $\mathcal{R}$. Suppose further that we also have a random sample of the locations in $\mathcal{R}$ of n_2 'control' individuals from the population at risk. We now have $n = n_1 + n_2$ events in $\mathcal{R}$ of two types, 'cases' and 'controls'. If there is no clustering of 'cases' relative to 'controls' this amounts to the 'cases' being just a random sample from the pattern of both cases and controls. Thus the hypothesis we wish to test is that of random 'labelling' of cases and controls. Note that this hypothesis requires no assumptions about the specific form of the underlying component processes of either cases or controls.

As stated earlier the random 'labelling' hypothesis is not the same as that of the independence hypothesis between types of events, with which we were concerned earlier in our discussion of multivariate point patterns. However, we can use K functions to examine the random 'labelling' hypothesis as well as that of independence.

Under random 'labelling' the pattern of either the cases or the 'controls' taken separately represents random 'thinning' of the combined spatial point process. From their definition, K functions are invariant under random 'thinning' and so it follows that under random 'labelling' we would theoretically expect:

$$K_{11}(h) = K_{22}(h) = K_{12}(h)$$

This suggests that a useful way of investigating departures from random 'labelling' is to assess the significance of differences amongst estimates of these functions. If 'cases' are type 1 events and 'controls' type 2, then we can use a plot of $\hat{K}_{11}(h) - \hat{K}_{22}(h)$ against h to assess departure from random 'labelling'. Peaks then represent spatial clustering of cases over and above the natural environmental spatial clustering of controls.

Upper and lower simulation envelopes for assessing the significance of peaks in this plot are developed from estimates of $\hat{K}_{11}$ and $\hat{K}_{22}$ in repeated simulation using the fixed $n_1 + n_2$ locations but randomly assigning 'case' labels to n_1 of these locations.

We may illustrate the idea with reference to the data on larynx and lung cancers. The distributions are shown in Figure 4.5 and the results are given in Figure 4.6. There is slight evidence that the larynx cancers are more 'dispersed' than the lung cancers.

We can also apply the same idea to the data on the offences committed by white and black residents in Oklahoma City, the distributions of which were shown earlier in Figure 4.1. Before, we looked at the hypothesis of spatial interaction (attraction or repulsion) between these two types of events, whereas now we are concerned with differences in the individual patterns. In other

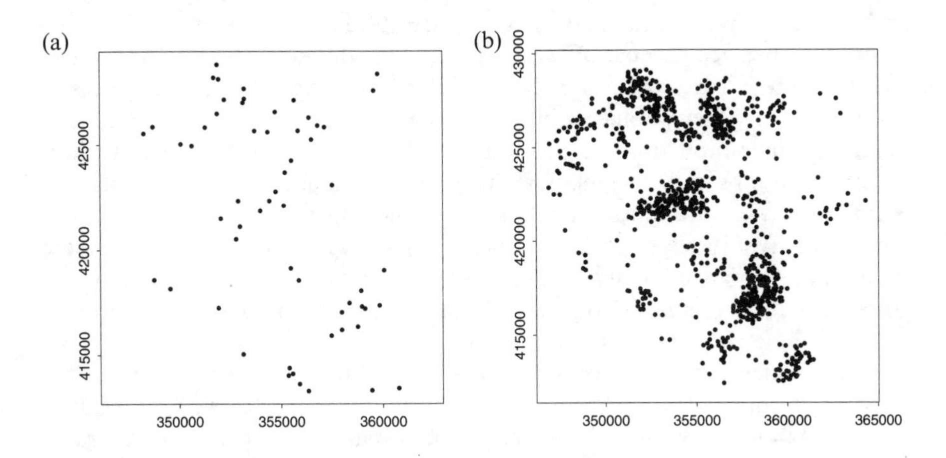

Fig. 4.5 Distribution of (a) larynx and (b) lung cancers

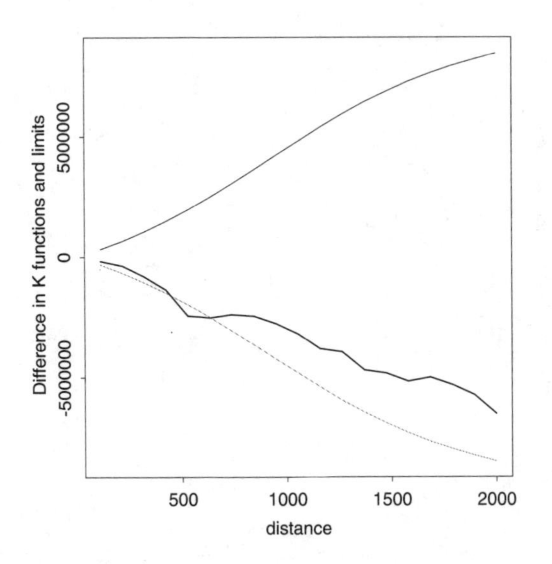

Fig. 4.6 Difference between *K* functions for larynx and lung cancers

words, are the black crimes just a random subset of the overall pattern of both black and white crimes? The results are shown in Figure 4.7 and confirm strongly the visual difference between the two patterns, with those offences committed by blacks showing considerably more spatial aggregation.

An alternative approach to the problem of detecting clustering when there is a heterogeneous background population from which events arise, is to use test statistics based on inter-event distances. These methods are similar in nature to the use of the distribution functions $G_{ij}()$ and $F_j()$, of nearest neighbour distances used earlier in our discussion of bivariate point patterns, except in this case the hypothesis considered is one of random labelling rather than independence. We assume that we have a set of 'cases' and 'controls'. We then count the observed number of events that are 'neighbours' of 'cases' and which are themselves 'cases'; in other words, the test statistic is the count of 'case'–'case' pairs. By 'neighbour' we might typically take the nearest event to that under consideration, though the test statistic can, in principle, use various other definitions of a 'neighbour'. In order to evaluate the test statistic for significance, we require to know how we would expect this statistic to behave under the hypothesis of random labelling. Expressions are available for the mean and variance of such counts under this hypothesis as well as an approximate sampling distribution. Details may be found in the references given at the end of this chapter.

One problem with this sort of approach is that it is based simply on the existence or otherwise of a 'case' 'neighbour' for any given 'case'. As a result, if the 'neighbour' does indeed happen to be a 'case', but it is a considerable distance away, we get a contribution to the test statistic that is given the same

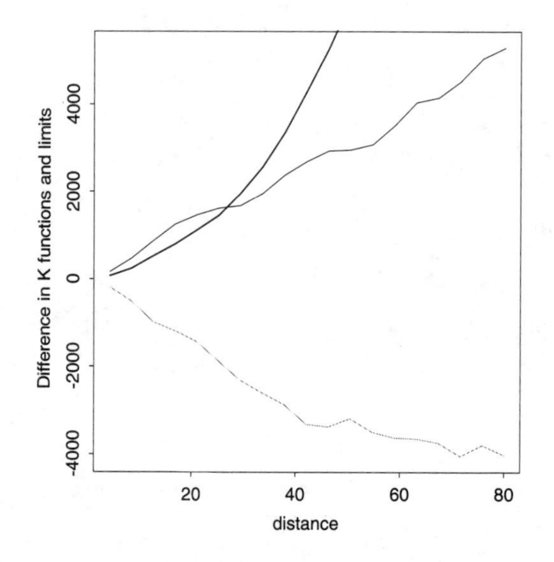

Fig. 4.7 Difference between *K* functions for black and white offences

weight as 'case'–'case' 'neighbours' separated by a very short distance. Modifications to the approach exist which try to circumvent this problem; again we do not give details here, but the reader is referred to the readings at the end of the chapter.

4.5 Clustering around a specific point source

The methods outlined in the previous section allow us to detect the presence or absence of clustering when there is a heterogeneous 'at risk' population. That is, we are able to ascertain whether a set of events (such as cases of a disease) have a tendency to be spatially aggregated when compared with the population distribution. But suppose we want to know if there is clustering around a *particular* point source (or maybe sources), or even clustering along a linear source, as opposed to just clustering in general. We are then interested in detecting *raised incidence* or elevated risk, around such a source, or sources. Of course, one simple way to explore this would be to examine the ratio of the two kernel estimates, $\hat{\rho}_\tau(\boldsymbol{s})$, as described in an earlier section of this chapter. If this ratio revealed an unusual 'peak' we might wish to conduct investigations in the vicinity of that peak in order to ascertain whether there was anything of note nearby. But we emphasise that this is merely an exploratory device, not an explicit test of a hypothesis concerning a particular suspected point source.

Before seeing how we might set up a more formal model, let us consider some examples of where this may be useful. First, we can refer again to the example on larynx and lung cancer. As described earlier in Chapter 3, one problem of interest here centres on whether there is an elevation in the risk of cancer of the larynx around a suspected point source of pollution, namely an old industrial waste incinerator. Another example might be to detect whether there is raised incidence of childhood cancer along high voltage power lines (as has been hypothesised). A further case might be the possibility of elevated risk of respiratory disease along busy main roads.

Several approaches to this modelling problem have been suggested, but we focus here on that developed in two ways by the statistician Peter Diggle. Some of our references at the end of the chapter discuss alternative approaches. The first model we consider is a heterogeneous Poisson model, where the intensity, $\lambda(\boldsymbol{s})$, of one type of event (hereafter 'cases') is expressed as a multiplicative function of the intensity of background population ('controls') together with distance from the suspected point source of pollution. Formally, if $\boldsymbol{s}_0$ represents the location of the suspected source, and h is the distance of an arbitrary point $\boldsymbol{s}$ from this source, then the model proposed is:

$$\lambda(\boldsymbol{s}) = \rho\lambda'(\boldsymbol{s})f(h;\boldsymbol{\theta})$$

where ρ is a scaling parameter (representing the ratio of numbers of 'cases' to 'controls'), $\lambda'(\boldsymbol{s})$ represents the background population intensity, as in our earlier section, and $f()$ is a distance decay function involving a vector of parameters $\boldsymbol{\theta}$, which describes how the incidence varies with distance h from the suspected source, $\boldsymbol{s}_0$. If there is no elevated risk around the suspected source, then we expect $f(h; \boldsymbol{\theta}) = 1$ for all h.

More specifically, we might postulate a particular functional form for $f(h; \boldsymbol{\theta})$ as, say:

$$f(h; \boldsymbol{\theta}) = 1 + \theta_1 \mathrm{e}^{-\theta_2 h^2}$$

where θ_1 and θ_2 are parameters to be estimated. The background population intensity $\lambda'(\boldsymbol{s})$ may be estimated by using the kernel estimation methods described earlier. Maximum likelihood methods are then used to estimate the parameters θ_1 and θ_2. The null hypothesis is that there is no distance decay effect, which effectively amounts to testing whether the parameter θ_2 is zero. If this is the case, we can see from our model that $\lambda(\boldsymbol{s})$ is independent of h and therefore not dependent on distance from the suspected source $\boldsymbol{s}_0$; it is simply equal to the intensity of controls, scaled by a constant.

The advantage of explicitly specifying a form for $f()$ in this way is that other covariates can also be introduced into the model, if required. That is, $f()$ can be replaced by a function that represents multiplicative risk factors other than simply distance from the point source. For example, assuming the data were available, we might include covariates relating to the population characteristics at different distances from the point source. This allows for the fact that the population around the source, because of its socio-economic composition or age–sex structure, might be more susceptible to the disease for reasons that have nothing to do with the source. In addition, extensions to consider multiple possible sources are also possible, making this model formulation an extremely general one.

A slightly different approach avoids the need for a kernel estimate of the background population intensity $\lambda'(\boldsymbol{s})$. Instead, an alternative model is proposed, with the following form:

$$p(\boldsymbol{s}) = \frac{\rho f(h; \boldsymbol{\theta})}{(1 + \rho f(h; \boldsymbol{\theta}))}$$

Here, $p(\boldsymbol{s})$ denotes the probability that an event at location $\boldsymbol{s}$ is a 'case', rather than a 'control'. Again, parameters are estimated by maximum likelihood. Additional covariates may also be included in $f()$, so that we might relate the probability of an event being a 'case' to factors such as smoking behaviour, occupation, and so on. An extension to multiple sources is again possible.

We may get some flavour of these kind of 'raised incidence' models by applying one of them to the data on larynx cancer. Here, we assume that lung cancer serves as an adequate surrogate for background population distribution (the larynx cancers as 'cases', the lung cancers as 'controls'). The distribution of both sets of events has already been illustrated in Figure 4.5. The old incinerator is located approximately at grid reference (355000,414000). If we fit

the first of the two raised incidence models described above using the form for $f()$ given, then we get estimates of $\hat{\theta}_1 = 33.69$ and $\hat{\theta}_2 = 1.11$. The standard error of $\hat{\theta}_2$ confirms θ_2 to be significantly different from zero, suggesting that there is indeed raised incidence around the specified pollution source (the incinerator). It is of interest to contrast this conclusion to that obtained earlier, when comparison of K functions for the same data gave no evidence of *overall* clustering of larynx cases relative to lung cancers. Here the hypothesis being tested is more specific to a particular point source and the clustering test is correspondingly more sensitive.

4.6 Summary

In this chapter we have built on the basic point pattern analysis techniques given in the previous chapter, extending them to cope with various situations where information other than just the location of events needs to be taken into account in analysis. We started by looking at the situation where the relationship between patterns in two or more different types is examined. We followed this by a discussion of methods to investigate space–time interaction. We then went on to consider techniques designed to analyse pattern in a set of observed events when the background population from which they have arisen cannot reasonably be assumed to be homogeneous throughout the study region. Finally, we looked at various models relating to detecting raised incidence of events around a specific point source of interest.

Some of the methods discussed in this chapter are rather specialised, having perhaps most applicability in an epidemiological context. For example, space–time tests were originally developed in such a context, and it is to the distribution of diseases such as childhood cancers that such methods have found the broadest range of application. However, methods for examining space–time clustering have also been put to genuine practical use in other areas, for example by British police forces in the investigation of the clustering of crime data. Other methods outlined here have also been put to effective use in investigations of whether there is clustering of one type of event (a particular disease) around one or more point sources of pollution. We would argue that, while some of the techniques are quite demanding, they provide a rich collection of techniques for certain kinds of spatial data analysis and are a useful addition to the tool kit of many specialists who deal with 'event' data.

In concluding this part of the book on point pattern analysis it is perhaps appropriate to make some general remarks in relation to the subject as a whole. The usefulness of analysis methods for point patterns has been questioned by some researchers in recent years. One reason has been that some of rather simple tests, described in Chapter 3, (e.g. simple 'nearest neighbour' tests), are admittedly of little value in many applications, especially in human geography. This is because the assumption of underlying homogeneity is usually unrealistic. For example, to test for departures from randomness in the distribution of urban centres hardly makes much intuitive sense; there will be obvious clustering in areas of higher population density. However, the sorts of methods outlined in this chapter, which are less well known and appreciated, can be brought to bear effectively on such issues, since they allow for such heterogeneity. Indeed, in some applications there may be much to commend using such methods, as opposed to summarising event data by estimating occurrence rates in what are usually arbitrary areal units, and then analysing these rates using analysis methods appropriate to area data, such as those we describe later in Part D. In epidemiology, for example, there is often a tendency automatically to use standardised morbidity or mortality rates within 'convenient' areas, even though more specific point locational data may be available. Of course, it is appreciated that there will be many circumstances when, for various reasons, analysis within an area framework will be unavoidable.

We now leave the subject of 'point patterns' and, in the next section of the book, turn our attention to the analysis of a different type of spatial data, that which arises when a 'spatially continuous' attribute has been sampled at point sites in a region.

4.7 Further reading

An overview of many of the methods we have described in this chapter is to be found in:

Diggle, P.J. (1993) Point process modelling in environmental epidemiology, in Barnett, V. and Turkman, K.F. (eds.) *Statistics for the Environment*, John Wiley, Chichester, Chapter 4.

On bivariate point patterns see:

Upton, G. and **Fingleton, B.** (1985) *Spatial Data Analysis by Example: Volume 1: Point Pattern and Quantitative Data*, Wiley, Chichester, Chapter 4.

Diggle, P.J. (1983) *Statistical Analysis of Spatial Point Patterns*, Academic Press, London, Chapters 6–7.

On space–time methods the reader is directed to:

Knox, E.G. (1964) Epidemiology of childhood leukaemia in Northumberland and Durham, *British Journal of Preventive and Social Medicine*, 18, 17–24.

Mantel, N. (1967) The detection of disease clustering and a generalised regression approach, *Cancer Research*, 27, 209–20.

Siemiatycki, J., Brubaker, G. and **Geser, A.** (1980) Space–time clustering of Burkitt's lymphoma in East Africa: analysis of recent data and a new look at old data, *International Journal of Cancer*, 25, 197-203.

Thomas, R. (1992) *Geomedical Systems: Intervention and Control*, Routledge, London, 200–215.

Diggle, P.J., Chetwynd, A.G., Haggkvist, R. and **Morris, S.** (1993) Second-order analysis of space–time clustering, Department of Mathematics, Lancaster University, Technical Report.

There is quite a large literature on correcting for spatial variations in population at risk. The 'distance-based' approach using nearest neighbours is described in:

Cuzick, J. and **Edwards, R.** (1990) Spatial clustering for inhomogeneous populations (with discussion), *Journal of the Royal Statistical Society, Series B*, 52, 73–104.

Research on elevated risk or raised incidence around a point source has also attracted a large literature. In addition to Diggle (1993), cited above, some useful references are:

Bithell, J.F. and **Stone, R.A.** (1989) On statistical methods for analysing the geographical distribution of cancer cases near nuclear installations, *Journal of Epidemiology and Community Health*, 43, 79–85.

which discusses a particular 'isotonic regression' approach. See also

Diggle, P.J. and **Rowlingson, B.S.** (1994) A conditional approach to point process modelling of elevated risk, *Journal of the Royal Statistical Society, Series A*, to appear.

which deals with the elevated risk models that we have discussed.

The larynx cancer data were originally analysed in:

Diggle, P.J., Gatrell, A.C. and **Lovett, A.A.** (1990) Modelling the prevalence of cancer of the larynx in part of Lancashire: a new methodology for spatial epidemiology, in Thomas, R.W. (ed.) *Spatial Epidemiology*, Pion, London.

but for a full review (and rejection!) of the hypothesis that larynx cancer is clustered around incinerators see:

Elliott, P.J., Hills, M., Beresford, J., Kleinschmidt, I., Jolley, D., Pattenden, S., Rodrigues, L., Westlake, A. and **Rose, G.** (1992) Incidence of cancer of the larynx and lung near incinerators of waste solvents and oils in Great Britain, *The Lancet*, 339, 854–58.

For details of how to perform various kinds of point pattern analysis within a GIS environment (specifically, using ARC/INFO) see:

Gatrell, A.C. and **Rowlingson, B.S.** (1993) Spatial statistical modelling within a Geographical Information Systems framework, in Fotheringham, A.S. and Rogerson, P. (eds.) *Spatial Analysis and GIS*, Taylor and Francis, London.

Most of the results reported in this chapter were obtained using the software SPLANCS, which we described in Chapter 2. See:

Diggle, P.J. and **Rowlingson, B.S.** (1993) SPLANCS: spatial point pattern analysis code in S-Plus, *Computers and Geosciences*, 19, 627–55.

4.8 Computer exercises

Here are suggested exercises that you can try on ideas discussed in this chapter using INFO-MAP and our example data sets. These exercises have been referenced at appropriate points in the chapter by numbered symbols in the margin.

Exercise 4.1

Open the 'Oklahoma City property crimes' file and load the data. You can draw dot maps of either 'white crime' or 'black crime'. You could also map both at once by creating a new data item such as

```
both=if(miss({1}),1,2)
```

and then dot mapping the resulting variable. You should change the legend by using `Scaling` in the `Map` menu and choosing the 'discrete' option, before producing the dot map, in order to show only two classes.

Now try mapping one type of crime and producing an $\hat{L}()$ for that distribution, using `K-function` from the `Analysis` menu with the 'univariate' option. Specify zero simulations. Whilst this is still displayed use `Cut to clipboard` in the `Edit` menu. 'Double click' on one corner of the plot and 'drag' a box to encompass it. Another double click then copies this plot to the 'clipboard'. Next, map the second type of crime and derive an $\hat{L}()$ for this distribution, using the same maximum distance and again the 'univariate' option. Whilst this is still displayed use `Paste from clipboard` in the `Edit` menu and 'drag' the retrieved plot to a suitable screen position, double clicking to finalise its display. You should then have $\hat{L}()$ functions for both types of crime displayed simultaneously. What conclusions can you draw from their comparison?

Try experimenting with these data using the 'bivariate' option provided when `K-function` is selected, to produce a $\hat{L}_{12}()$ function. Use only small maximum distances with this option as it will take some considerable time to compute. You could reproduce the simulation envelopes for assessing the significance of 'peaks' or 'troughs' and formally compare $\hat{L}_{12}()$ with the result given in the text.

Exercise 4.2

Open the 'Burkitt's lymphoma in Uganda' file and load the data. Map the time of onset of the disease. When looking for space–time clustering we wish to examine the correlation between events which are 'close' in time and 'close' in space. A crude approach would be to calculate a correlation between the time of onset of a case and the time of onset of 'neighbouring' cases. This may be done by calculating a new variable as

```
nearby=({1}[near(1)]+{1}[near(2)]+{1}[near(3)])/3
```

For each event the new variable contains not the time of onset for that event but the average time of onset for the three nearest neighbouring events. You could now plot time of onset against the new variable, or work out the correlation between them.

To try out the more formal Knox statistic we need to define first what we take as being 'close' in both space and time. In the first instance we will take 'close' in space as being a separation of less than 4 in the grid referencing system applying in this file. For time we will take 'close' as being a time separation of less than 180 days. We can now proceed to calculate the statistic. Start by assigning the value zero to the three worksheet items `work[1]`, `work[2]` and `work[3]`. Use `Calculate/display worksheet` from the `Analysis` menu with the formula:

```
i=1,3:work[i]=0
```

Then, still using the `Calculate/display worksheet` option, calculate the components of the Knox test with three successive formulae as follows, noting that calculations may take a little while):

```
i=1,count():work[1]=work[1]+sum(if(dist([i])<4,1,0))

i=1,count():work[2]=work[2]+sum(if(abs({1}[i]-{1})
                    <180,1,0))

i=1,count():work[3]=work[3]+sum(if(abs({1}[i]-{1})<180 and
                    dist([i])<4,1,0))
```

Convince yourself that these formulae are computing the required quantities as described in the text! One can then compare `work[3]` with the tail probability of a Poisson distribution whose mean is:

```
work[1]*work[2]/(count()*(count()-1))
```

What conclusions do you come to?

You may like to try exploring similar ideas but using subsets of the data, based either on age, or some 'window' within the 14 year overall time period. One of the exercises in Chapter 2 considers how to restrict a data set, using the `Zoom/Restrict` option in the `File` menu, so that only a subset of cases remains and this is probably a simpler way to proceed than including additional logical expressions into the above formulae. Note that if you do restrict the data set in this way, you need to load the original file again to get back to the full data set.

Exercise 4.3

In order to try out the idea of the ratio of kernel estimates for intensity of events and population density open the 'Lancashire lung and larynx cases' file and load the data. Calculate a new variable for the density of the population at risk by the formula

```
density={1}/area()
```

Produce a shaded map (using the `Choropleth` option) of the new 'population density'. Switch on the `Superimpose` option in the `Map` menu and dot map the larynx cases. Visually examine the correspondence between the population density and the distribution of cancers. What do you observe?

Switch off the `Superimpose` option and use `Kernel map` from the `Analysis` menu and choose the 'ratio of intensities' option. Specify the variable 'larynx' as the numerator and 'Total population' as the denominator. You then need to specify a bandwidth for the numerator kernel estimate (that is, intensity of larynx cases), followed by a bandwidth for the denominator (density of population). The subsequent kernel estimate will take some time to compute, but should finally result in a kernel estimate of 'relative risk' for the larynx cancer, relative to the crude 'population at risk' represented by assuming all population in an area to be located at the 'centre' of the area.

Exercise 4.4

This is very similar to the previous exercise with the lung and larynx data, except that now we use the distribution of lung cancer cases as possibly a better estimate of 'population at risk'. We also try applying the technique within a subset of the whole region.

First look at the distribution of both lung and larynx cases separately to get a visual impression of the two distributions. Now 'zoom' into a spatially defined subset of data. Draw a `Dot map` of the lung cancer cases. They are particularly dense around the urban area roughly centred on grid reference (354000,422000). You can use `Grid references` within the `Map` menu to display the grid references of locations in the region. Choose `Zoom/Restrict` from the `File` menu and define a rectangular 'source area' to be rescaled, which is roughly centred on (354000,422000) and about 6 km square (the grid references here are in metres); One of the exercises in Chapter 2 gives more detail on exactly how to do this. Rescale your defined area to a moderately large area in the centre of the screen by using a 'raster' zoom. You should then be left with a temporary 'zoomed' map and the data set will have been restricted to the 'source area' that you defined; any observations outside this area have been removed from the data set.

Now try a `Dot map` of both the lung and larynx cancer cases; the distributions within the area of interest will now be able to be seen in much greater detail. Finally, apply the kernel estimate of ratio of intensities, using the larynx cases as numerator and this time the lung cases as denominator. Use a small bandwidth for the lung cases (there are many of these) and a larger bandwidth for the larynx cases (since you will probably only have very few within your defined area). The resulting estimate will take some time to compute and this might be a good time to take a coffee break!

Perhaps try the same idea using other subsets of the data, or even using the whole region, in which case a lunch break (or, alternatively, multiple cups of coffee) would probably be sensible!

Exercise 4.5

Open the 'Oklahoma City property crimes' file used in an earlier exercise. Previously, we estimated $\hat{L}_{11}()$, $\hat{L}_{22}()$ and $\hat{L}_{12}()$ for these data, looking at the hypothesis of independence in the two point patterns. Now examine the hypothesis of 'random labelling'. Try to repeat the analysis described in the text by using `K-function` from the `Analysis` menu but this time using the 'Difference between univariate' option. This computes and plots $\hat{K}_{11}() - K_{22}()$ according to the two data items you specify. In this case take the first variable as 'black crime' and the second as 'white crime'. Specify a small maximum distance over which to estimate this difference as it will take some considerable time to compute the result.

You could reproduce the simulation envelopes for assessing the significance of 'peaks' or 'troughs' in this function of the differences and compare your findings informally with those given in the text. Use a small number of simulations to speed up the computations as much as possible.

Exercise 4.6

It is not possible to fit any 'raised incidence' models in INFO-MAP, but it is of some interest to try some simpler, informal ways of exploring for elevated risk. Open the 'Lancashire lung and larynx cases' file and load the data. By using `Open` in the `Overlay` menu superimpose the location of the incinerator referred to in the text. By using `Grid reference` in the `Map` menu find the approximate grid reference of the incinerator.

Suppose this is (354850,413550). We now wish to calculate the mean distance of all lung cases from the incinerator and to compare this with the mean distance of the larynx cases from the incinerator. The argument here is that if the distribution of both types of events in the study region in general is similar and also in the vicinity of the incinerator, then these mean distances should also be similar. To calculate these two mean distances use the `Calculate worksheet item` in the `Analysis` menu. Firstly enter the formula:

```
work[1]=mean(if(miss({2}),miss(),((east()-354850)**2
        +(north()-413550)**2)**0.5))
```

This rather complicated expression should give the mean distance of lung cases from the incinerator. Try to work out exactly how it works. Then use:

```
work[2]=mean(if(miss({3}),miss(),((east()-354850)**2
        +(north()-413550)**2)**0.5))
```

to get the corresponding mean distance of larynx cases from the incinerator. Compare the two means. Can you work out how to perform a 't-test' of the difference in these two means? What comments do you have concerning whether this is a sensible 'test' of clustering around the incinerator?

Part C

The Analysis of Spatially Continuous Data

5

Introductory methods for spatially continuous data

Here and in the following chapter we consider the analysis of data which are 'spatially continuous'. Our objectives are significantly different from those in Part B. There, our observations were the locations of 'events', and we were interested in any pattern in such locations; here, we concentrate on understanding the spatial distribution of values of an attribute over the whole study region, given values at fixed sampling points. Our objective is to model the pattern of variability and to establish any factors to which this might relate. Ultimately, we may wish to use such models to obtain good predictions of values at points where the attribute has not been sampled. Such methods are relevant to many studies in the geosciences, such as soil science, climate study, hydrology, mining geology, and so on.

5.1 Introduction

In Part B we were concerned with the analysis of spatial 'point patterns', where our data consisted of the locations of a series of 'events' occurring in some study region $\mathcal{R}$. We also considered generalisations of this type of situation where additional information was available relating to the 'events', such as a 'labelling' of different types, a time of their occurrence, or background information on variations in the population from which they arose. However, in all these cases our interest remained in possible patterns in the *locations* of the events. Where additional information became involved it was of interest only in so much as it might help us to 'explain' or otherwise improve our analysis of the pattern in the locations.

That objective should be distinguished from the one which we consider in this and the following chapter, and indeed for the remainder of the book. It is now attribute values over $\mathcal{R}$ which we study, based on observed values at a pre-defined and fixed set of locations. Previously, we were interested in patterns in the locations of observations; now we are interested in patterns in attribute values. The locations are now simply sample sites at which attribute values have been recorded.

Here in Part C the focus of our discussion will be the analysis of an attribute which is conceptually *spatially continuous* over $\mathcal{R}$, and whose value has been sampled at particular fixed point locations $\boldsymbol{s}_i$. (Again we shall maintain the notation introduced earlier, where $\boldsymbol{s}_i = (s_{i1}, s_{i2})^T$, is a (2×1) vector, representing respectively the 'x' and 'y' coordinates of the ith location.) Typical examples of such spatially continuous data might be geological measures on an ore deposit such as mineral grade, or the concentration of some pollutant, or rainfall measurements, or soil salinity and permeability, and so on. This type of spatially continuous data is often referred to as *geostatistical* data. Clearly, much data arising in the fields of geography, geology and the environmental sciences is of this nature.

We distinguish this from the situation considered subsequently, in Part D of the book, where the attribute values are not considered spatially continuous, but instead relate only to a finite set of areas or zones that partition the study region. Perhaps a useful analogy here is with processes in time. Some processes evolve continuously—for example, temperature during the day can be sampled at any time. Other processes only evolve at discrete fixed intervals—it makes little sense to think of many economic measures as evolving continuously, because by their definition they can only be sampled say monthly, or quarterly. Currently we intend to consider the spatial equivalent of continuous processes in time; later, in Part D, we will be concerned with the spatial equivalent of fixed time period development, the areas being analogous to the time periods. We should acknowledge in passing that it may be perfectly possible to regard certain types of data under either framework—we might treat pollution measurements as spatially continuous for the purpose of predicting pollution levels at sites where we do not have measurements; however, we might also form the average of pollution measurements for a set of administrative health districts, so as to analyse such data in conjunction with health indices available only at that spatial level.

Statistically then, the situation we are considering in the remainder of this chapter and the next, is one where a series of observations y_i, $i = 1, \ldots, n$, on a spatially continuous attribute, have been recorded at corresponding spatial locations, $\boldsymbol{s}_i$, in the study region $\mathcal{R}$. Occasionally we shall refer to our observations collectively as the $(n \times 1)$ vector $\boldsymbol{y} = (y_1, \ldots, y_n)^T$. The measurements, y_i, are assumed to be observations on a spatial stochastic process $\{Y(\boldsymbol{s}),\ \boldsymbol{s} \in \mathcal{R}\}$, which varies in a spatially continuous way over $\mathcal{R}$ and has been sampled at fixed points. Note that strictly we should probably refer to our observed data values as $y(\boldsymbol{s}_i)$, since they are observations on the random variables $Y(\boldsymbol{s}_i)$. However, we shall mostly use the simpler notation y_i. Where it is convenient we may also refer to the random variable $Y(\boldsymbol{s}_i)$ simply as Y_i, or collectively for all sample sites by the $(n \times 1)$ vector $\boldsymbol{Y} = (Y(\boldsymbol{s}_1), \ldots, Y(\boldsymbol{s}_n))^T$.

The main objective of our analyses will be to infer the nature of spatial variation in the attribute over the whole of $\mathcal{R}$, from the sampled point values.

We may seek description in terms of a smooth surface which captures large scale global trends; or alternatively, we might also wish to study aspects of local variability, particularly if the primary objective is one of accurate interpolation or prediction of the value of the attribute at points other than $\mathbf{s}_i$. Essentially these objectives correspond to the notions discussed in Chapter 1 of modelling first order variation in the mean value of the process, $E(Y(\mathbf{s}))$, which we shall refer to as $\mu(\mathbf{s})$; or of modelling second order variation or spatial dependence between $Y(\mathbf{s}_i)$ and $Y(\mathbf{s}_j)$ for any two locations, $\mathbf{s}_i$ and $\mathbf{s}_j$, in $\mathcal{R}$; that is, $COV(Y(\mathbf{s}_i), Y(\mathbf{s}_j))$.

In general, as with 'point patterns' previously, our approach will involve a mixture of methods, some designed to examine large scale heterogeneity in the mean value of the attribute, $\mu(\mathbf{s})$, over $\mathcal{R}$, others where we shall examine areas within which we assume the process to be stationary or isotropic (see Chapter 1) and look for spatial dependence, in this case through the use of the *variogram* or the *covariogram*, which, as we will see, attempt to capture the covariance structure of the process. When it comes to proposing possible statistical models in order to try to 'explain' any effects we may detect, we shall usually think of these as consisting of two 'components'; a first order component, representing large scale variations in $\mu(\mathbf{s})$; and a stationary second order component representing small scale spatial dependence in the process. The reader should bear in mind that, as in the case of 'point pattern' analysis, there will inevitably be cases where spatial distribution explained in terms of dependence with the assumption of homogeneity, could possibly arise also from heterogeneity in mean value. Ultimately, our division into the two components, although guided by our analyses, will be to some extent arbitrary.

In line with the structure established in Part B, we begin by outlining some case studies that we shall use throughout this part of the book. We then move on to consider methods, under the general headings of visualisation, exploration and modelling.

5.2 Case studies

Throughout this and the following chapter, we shall use a selection of data sets involving spatially continuous variables, to illustrate various forms of analyses. We have provided copies of these data sets on disk. They include:

- Rainfall measurements in California
- Rainfall measurements in central Sudan
- Temperatures for weather stations in England and Wales
- Groundwater levels in Venice
- Radon gas levels in Lancashire
- PCBs in an area of south Wales
- Geochemical data for north Vancouver Island, Canada
- South American climate measures.

We begin by describing all of these data sets in more detail; full references to the sources of the data are given at the end of this chapter.

We have already encountered the first of these data sets in Chapter 1, that concerned with the spatial variability of rainfall in California. We described some analyses there and refer the reader back to that discussion. Recall that the data consist of recordings of average annual precipitation at a set of 30 monitoring stations, distributed across the state of California and shown in Figure 1.1. For these same sites we have measures of altitude, latitude, and distance from the coast, each of which is a possible covariate that might explain the variation in precipitation. With these data our interest is perhaps less in making spatial predictions of rainfall than in trying to explain spatial variation using the available covariates.

The next example seems, on the surface, very similar, in that it too deals with spatial variability of precipitation, but this time in central Sudan. Here, however, we have no covariates. The problem is to try to describe the spatial variation and perhaps to make estimates of precipitation in areas where there are no monitoring stations. Resources for collecting such basic information are obviously limited, particularly in the developing world. The data we include consist of measurements of total annual precipitation in 1942, 1962 and 1982, recorded at 31 sites. The sites are unevenly distributed across the region, with a preponderance of monitoring stations around the capital city of Khartoum, at the confluence of the White Nile and the Blue Nile (Figure 5.1).

Understanding the distribution of rainfall in central Sudan is important, particularly because the area faces some of the most severe population

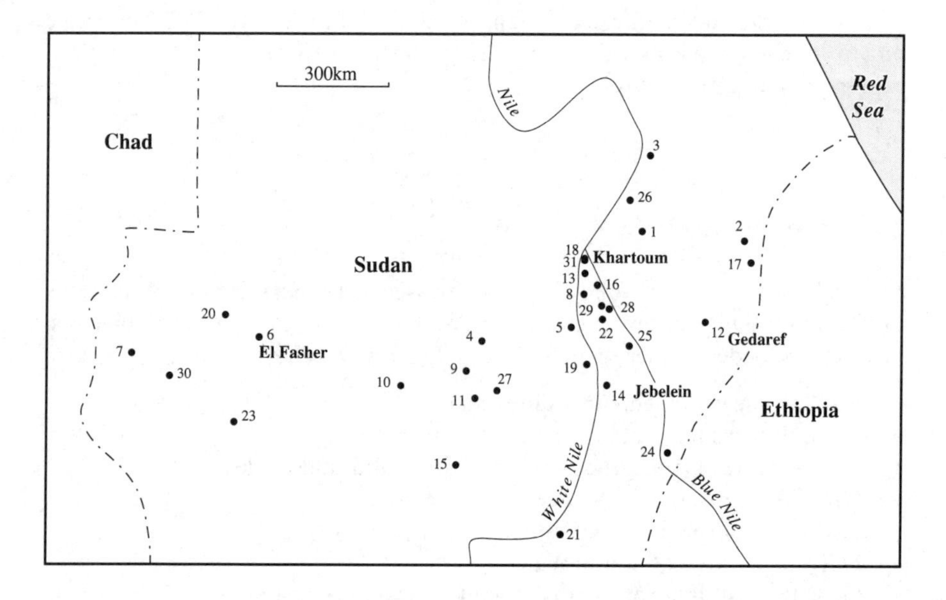

Fig. 5.1 Rainfall measuring sites in Sudan

pressures of any part of the Sahel region. The intensity of crop cultivation is considerable, and this, coupled with severely limited rainfall resources, has led to the desertification so characteristic of much of the wider region.

Although we include here rainfall totals for only three years it is well known that precipitation is highly variable from year to year. Further, there is, of course, very great seasonal variability in rainfall, with the wettest months generally in July and August, though this seasonality itself varies geographically. We restrict attention to seeing how annual rainfall varies across space and how the nature of this spatial variability has changed over two twenty-year periods. Some climatologists have detected no real evidence for long-term reduction in rainfall, the droughts in recent years being thought of as natural fluctuations. Further, we shall be interested to see to what extent rainfall is highly localised in space. We should appreciate that when considering issues of desertification what matters more is not so much annual precipitation as seasonal and diurnal variation. However, the spatial localisation is certainly also of interest.

A further set of climate data which we include relate to England and Wales. Data on the mean daily temperature in August 1981 and August 1991 were extracted from the Monthly Weather Report, for a set of 48 stations distributed across the country. The main criterion for choosing sites was to ensure a reasonable geographical spread, but we also chose only those stations for which values were available in both years. In addition, we have also included the elevation of each site. We shall be interested in seeing whether we can obtain a good description of the geographical variation in temperature. In the simplest case we shall want to see the extent to which temperature variation can be explained solely in terms of geographical location. But we will then wish to see if further explanatory power is achieved by adding in elevation as a further covariate. Given that we have data for two years we might also want to investigate whether the character of spatial variation in one year is different from that in another.

Some parts of the world rely heavily on groundwater for supplies of both drinking water and water for industrial and commercial use and our next example relates to data on levels of such groundwater. In Venice withdrawals from several aquifers (the rock formations that contain water) at different depths have often been heavy, and have led to major problems of land subsidence. The result has been that the local population has been exposed to the risk of flooding from the Adriatic Sea. In order to be able to control the pumping from wells, hydrologists require accurate maps of the subsurface levels of groundwater (the so-called 'piezometric' surface).

The data we include for the Venice region come from a series of sparsely distributed boreholes (Figure 5.2). The groundwater levels have been measured in 1973 and 1977. In 1973 data were measured for 40 sites; however, data for only 35 of these sites are available for 1977. Since 1973 withdrawals of groundwater from the major industrial area (Porto Marghera) have declined markedly, leading to a lower risk of subsidence and of saltwater intrusion into the freshwater aquifer. As with other data sets considered in this chapter we

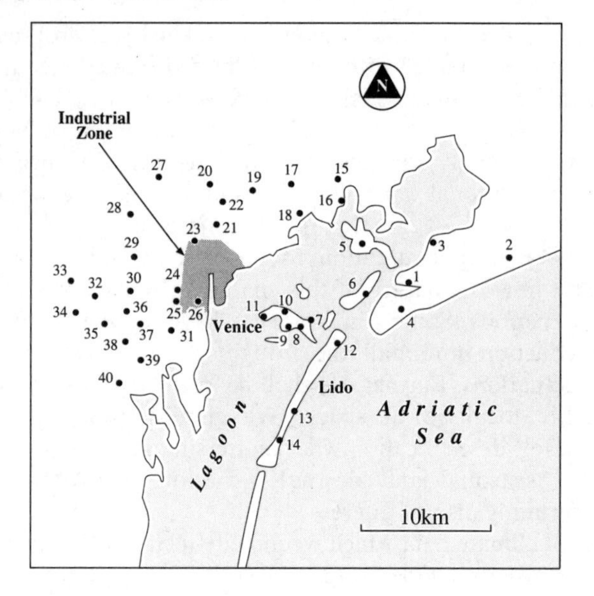

Fig. 5.2 Boreholes at which groundwater levels were measured in Venice 1973

would like to be able to describe the nature of spatial variation as accurately as possible, in order to provide estimates of groundwater levels at locations that have not been sampled. We also wish to know how reliable these estimates are.

Radon-222, commonly called just radon, is a naturally occurring radioactive gas produced by the decay of trace quantities of uranium. Released to the atmosphere it is harmless but when trapped within buildings it can accumulate and is considered to be a serious risk factor for lung and possibly other cancers. As a result, local authorities in many parts of the world are monitoring the gas in homes where the risk is thought to be raised. For example, some areas of South West England comprise uranium-bearing granitic rocks, and radon levels recorded in many properties there are extremely high. However, it is known that levels of radon are very highly variable in space, with levels in one property bearing little relationship to values in adjacent properties; this is thought to be a function of building materials, degree of ventilation and insulation, and so on. It is clearly important to try to characterise this degree of spatial variability. The data we include here relate to Lancashire, an area not thought to be at particular risk. Data are available for 344 homes and were measured in 1989 (Figure 5.3). For reasons of confidentiality the locations are defined with a resolution of 100 metres and have been randomly shifted. As a result, while any results obtained are a faithful description of spatial variation it is not possible to identify individual properties!

In general, the picture is of relatively low radon values to the west and south, higher values to the north and east. This reflects the fact that the area to the

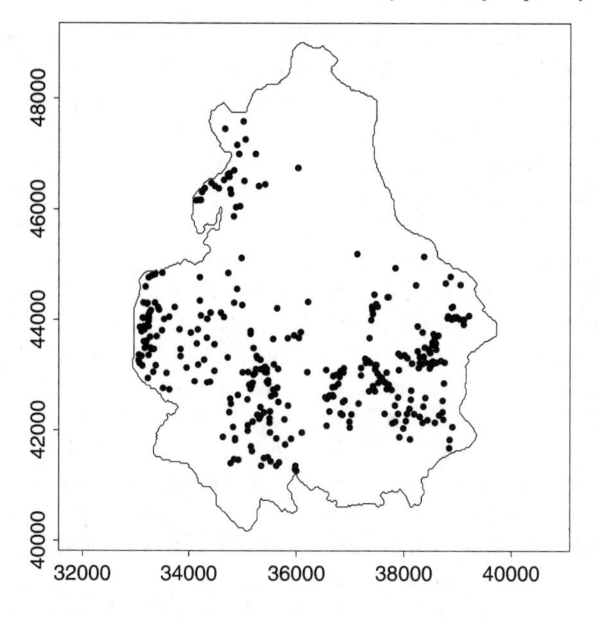

Fig. 5.3 Radon measurement sites in Lancashire

west (known as the 'Fylde') is covered by glacial deposits which trap the radon gas and prevent it rising to the surface. Further north and east there are no such surficial deposits and the gas escapes from the sandstones and limestones to the surface. The distribution of sample sites reflects population distribution. Population is very sparsely distributed to the north-east of the study region (the 'Trough of Bowland') and most of the sites are clustered in the major centres of population: Blackpool, Lancaster, Preston, Blackburn, and Burnley. With these data interest lies in understanding such broad regional trends in more detail, and examining if there is smaller scale variation superimposed on such trends. It might be possible to interpolate the data to provide a regional map of radon levels, or identify areas where sampling is 'thin' and needs to be improved.

The next case study relates to data on environmental pollution in soil samples. The data collection arose as a result of concerns over possible risks associated with the contamination of the environment with polychlorinated biphenyls (PCBs) in a small area of South Wales. Near the town of Pontypool is located a large plant for the incineration of chemical wastes (including PCBs) at very high temperatures. There had been worries that some of these substances had been escaping into the surrounding environment, possibly contaminating soil and vegetation. Data on 70 sites within an area of about 6 square kilometres are included here (Figure 5.15a). The soil samples were taken in late 1991. Measuring minute concentrations of PCBs is notoriously difficult and there were painstaking efforts to do this as accurately as possible, with different laboratories cross-checking the measurements. The research report from which the data are taken lists different sets of data for the sites; we have used the standardised data for the sum of seven different types of PCB (known as 'congeners'). We shall, however, simply refer to the data as PCB measurements. The analytical problem is to try to characterise the pattern of spatial variability; are there locally elevated concentrations around the incineration plant?

A further set of data that we shall find of interest in a spatially continuous setting are taken from the National Geochemical Reconnaissance performed by the Geological Survey of Canada. The particular study area is that part of Vancouver Island north of latitude 50 and west of longitude 126. In the small data set included here we have 916 sites (stream locations) at which five elements (part of a much larger set), have been measured; these are: zinc (Zn), copper (Cu), nickel (Ni), cobalt (Co), and manganese (Mn). The coverage is quite dense over the study area; in general, the reconnaissance programme aims to achieve a sampling density of approximately 1 sample per 13 km^2. As with the radon and PCB data, we shall be interested in characterising the nature and scale of spatial variation in geochemistry.

Broad-scale, multi-element reconnaissance programmes are carried out in many countries, since an understanding of spatial variation is clearly of some economic importance. There is quite a long tradition in geochemistry of mapping and analysing this kind of data, particularly in order to try to identify unusual features, or anomalies, that may suggest potential mineralisation. We comment later on the visualisation of spatially continuous data, but it is useful to point out here that geochemists are not only concerned with mapping single elements. They frequently wish to derive multi-component maps, which show the simultaneous variation of a number of elements. Three elements are usually selected; for instance, we might use the combination: zinc, copper, lead. The fact that interest may be as much in combinations of elements as opposed to separate ones, implies that an understanding of geochemical data is often a multivariate problem. We may want to obtain estimates of the distribution of a single element; but we might wish to look at an ensemble, using the sorts of methods we consider in Chapter 6.

Geochemical data are not of use solely in mineral exploration and resource evaluation. Depending on the particular trace elements being sampled, such data are also of value in an environmental context. High precision geochemical maps can be used to identify areas with raised levels of toxic elements such as lead, cadmium and arsenic. They might also be used to identify areas that are deficient in some elements. Some diseases in animal populations have been correlated with mineral deficiencies. For example, sheep can suffer from a disease called 'swayback', which has been linked to copper deficiency. We might then use a regional geochemical map to identify areas of crop cultivation where the soil could be treated to alleviate the problem. In some cases we might even wish to look for associations with human health problems.

One feature of geochemical data is that they often tend to be log-normally distributed; at least, the distribution will often be highly skewed, with a preponderance of generally low values, together with a much smaller set of extreme values. This is a common feature of many data sets which involve measurement of concentrations of some description (such as that on radon and PCBs). It will therefore often be advisable to transform the data before analysis, a logarithmic transformation being the obvious choice.

During this century several different schemes for climate classification have been devised, many because of the view that such schemes are relevant to human activity. A major constraint, however, in devising a reasonably objective classification concerns the type of data recorded at climate stations. Schemes that rely solely on temperature and precipitation, both of which can be measured with simple equipment, are going to be most useful in characterising climates across the global land surface. Because we are wanting to use a variety of both temperature and precipitation measures we need multivariate methods to help devise our classification. So with our next data set we are interested less in the interpolation of a single variable, as with the Sudan rainfall data, but more in using the available sample information to devise a multivariate classification.

The data relate to 76 climate stations in South America (Figure 6.1). Stations were selected that were all under 200 metres above sea level, in order to remove the effects of elevation. There are 16 climate variables, all of which are concerned with particular aspects of temperature and precipitation. The full list of variables is given in Table 5.1 and includes information on seasonality of climate and not simply annual totals or means. Clearly, we can expect broad regional differences, separating the tropical low latitude regimes in much of Amazonia from the middle latitude steppe conditions in parts of Argentina, for example. But to what extent can we paint a richer picture of climatic variability? We shall be using these data not so much in the present chapter but in Chapter 6, where we explore the role of multivariate methods.

Table 5.1 Description of climate variables for South America

1. Average annual temperature
2. Average daily January temperature
3. Maximum January temperature
4. Minimum January temperature
5. Average daily July temperature
6. Maximum July temperature
7. Minimum July temperature
8. Average annual precipitation
9. Average January precipitation
10. Average July precipitation
11. Average annual number of days precipitation > 1 mm
12. Average number of days in January precipitation > 1 mm
13. Average number of days in July precipitation > 1 mm
14. Temperature range (January–July)
15. Precipitation ratio (July/January)
16. Rain days ratio (July/January)

5.3 Visualising spatially continuous data

Having outlined the sort of data that we might be interested in analysing, let us now consider what kinds of visual representations we can obtain of such data before we go on to consider more sophisticated ways of exploring spatial variation. The simplest type of map that we can produce from data of this type is one in which the data value is written alongside the sampled location. However, this will not look very elegant or informative! A better solution is to use a symbol at each site, the nature of which carries useful information about the data value. We might use a variety of symbol types to represent different classes of variation; however, a preferred solution is to use *proportional symbol* maps.

Different geometric symbols can be used in this way, but commonly, and most simply, circles are used, where the area of the circle is proportional to the data value at that location. Alternatively, we might use rectangles, the height of which represents value. Although more difficult to produce, spheres are sometimes used to represent a variable such that the volume of the sphere is proportional to the data value (see Figure 2.1 for an example). Regardless of what symbol we use there are some important cartographic issues that arise in such mapping. The main problems concern how the map user perceives data values, since it has long been known that people tend to underestimate areas on maps. For example, it seems sensible to draw circles whose radii are proportional to the square root of the data values. In practice, however, cartographers introduce a correction factor to compensate for this perceptual underestimation, so that instead of taking square roots (or an exponent of 0.5) they use an (empirically derived) exponent of 0.57. Other problems concern the overlap of symbols, since with many locations the map will appear cluttered. Clever cartographic software allows for 'halos' around symbols to alleviate this problem.

As an aside, proportional symbol maps are widely used in contexts other than that of spatially continuous data. For example, we might map retail sales for stores in this way, even though this variable is in no way spatially continuous. Or, we could take data such as total population, representing an aggregate value within an areal unit, and map it using proportional symbols; we return to such uses later, when dealing with area data in Chapter 7.

If we use proportional circles to represent one data item we may then fill the circles with colours (or shading) to indicate the classified value of another variable. However, this too makes demands on the map reader. Another strategy is to shade the symbols with classified values of the same variable used to derive the symbol. In cases where many data values are present and proportional symbols overlap to a large extent, it might even be useful to do away with proportional symbols altogether and use small symbols of fixed size, shaded or coloured according to such a classification of data values. We need to say something about such classification since both the number of classes we use and the type of class intervals we select will determine the 'message' that our map communicates. Precisely the same issues concerned with class interval selection arise in analysing data collected for areal units, and we will refer back to this discussion in Chapter 7, where we consider that type of data.

There are no hard and fast rules about numbers of classes; clearly, this is largely a function of how many data values we have. For example, if we have only a small sample of 20 or 30 sites (as for the Sudan rainfall data) it hardly makes sense to use seven or eight classes; however, with perhaps two or three hundred measurements (as in the example of the radon data) a set of seven or eight classes is likely to prove informative. As a general rule of thumb some statisticians recommend a number of classes of $(1 + 3.3 \log n)$, where n is the number of observations.

As for class interval selection, there is a wide variety of schemes that may be considered. We do not review these in detail here, but simply comment on some possibilities. 'Equal intervals' are self-explanatory; they are valuable where data are reasonably uniformly distributed over their range, but if the data are markedly skewed they will give large numbers of values in just a few classes. This is not necessarily a problem, since unusually high (or low) values are easily picked out on the map. An extension of this scheme is to use 'trimmed equal' intervals where the top and bottom of the frequency distribution (for example, the top and bottom 10 per cent) are each treated as separate classes and the remainder of the values are divided into equal intervals. A way of ensuring that an equal number of observations fall into each class is to base class intervals on the percentiles of the distribution, that is, to use 'quantiles'; for example, selecting five classes will produce a quintile map. Another idea might be to use 'standard deviates' where the classes are based on intervals distributed around the mean in units of standard deviation. Finally, of course, we could let cartographers define their own class intervals. For instance, if we are mapping income data we may want to fix certain intervals (such as a 'poverty line'). In general, there is no substitute for a careful examination of the distribution of data values before selecting class intervals. As should be clear, we can obtain a huge variety of maps simply by varying the class intervals. This turns out to be of particular significance if mapping data for areal units and we will return to this point again in Chapter 7.

Ex 5.1

We can try out some of these ideas and get some initial sense of spatial variation in the data that comprise some of our case studies by visualising them using proportional symbols. Regardless of which year is taken the Sudan rainfall data give the same general impression when mapped, of higher levels to the south-east of the study area (Figure 5.4). This reflects the fact that the southern part of the study area is rather more humid. Further west and north we are moving into arid and semi-arid areas.

Groundwater levels in the Venice region may be mapped for both 1973 and 1977. In 1973 (Figure 5.5), note the cluster of negative values at sites 7–10 (in the city of Venice: see Figure 5.2 for the specific locations) and in the industrial region to the north west of Venice (sites 23–26). As noted earlier, considerable volumes of groundwater were being extracted in these areas. By 1977, while there are still some negative values these are lower in magnitude. If we examine

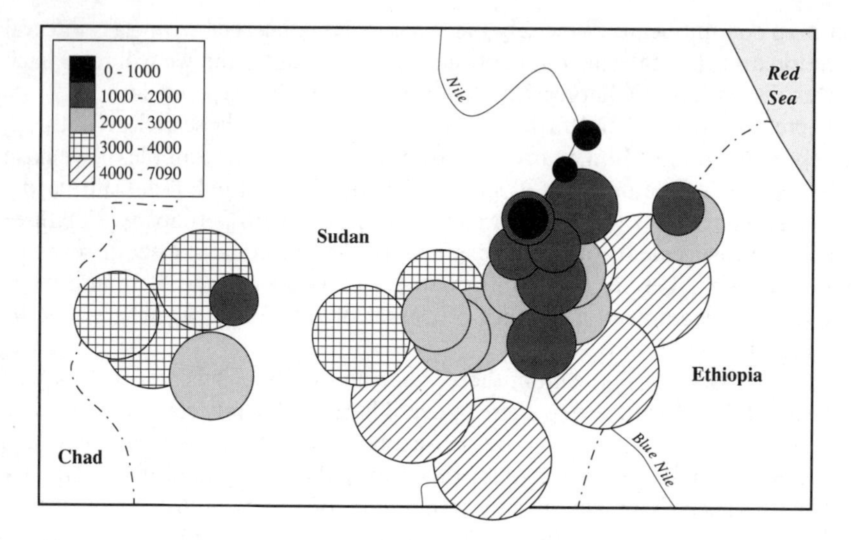

Fig. 5.4 Proportional symbol map of Sudan rainfall in 1982

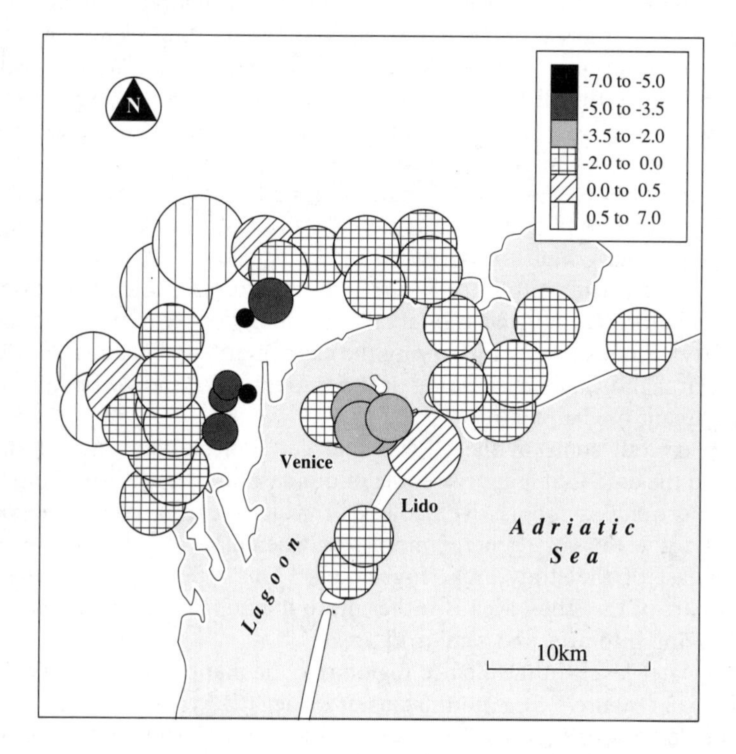

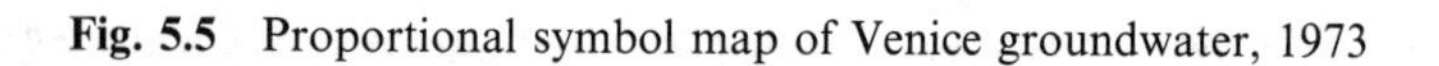

Fig. 5.5 Proportional symbol map of Venice groundwater, 1973

the relationship between 1977 and 1973 values there is a close correlation, but some sites with negative values in 1973 had become positive in 1977.

As mentioned earlier, some environmental data have highly skewed distributions and transformation is necessary before such data can sensibly be visualised in the sort of ways we have discussed. This would certainly be true, for example, if we wished to produce proportional symbol maps for either the PCB or radon data, where a logarithmic transformation is advisable.

These sorts of mapping tools are useful preliminary visualisations of data, but, typically, when we wish to look at variation in spatially continuous data we really need to use maps that *show* such continuity. For instance, the temperature maps that we see on television or in newspapers show not the values at weather stations but an interpolated or 'contoured' surface of temperature variation. How we might obtain such maps is the subject of the next section.

5.4 Exploring spatially continuous data

In this section we consider methods involving various summary statistics or plots which may be derived from the observed data and used informally to investigate hypotheses of interest or suggest possible models. Some of these are more concerned with investigating first order effects in the process, while others address the possibility of spatial dependence or second order effects.

We start by discussing various simple approaches to estimating how $\mu(\boldsymbol{s})$, the mean value of the attribute of interest, varies across the study region. The 'surfaces' that result from such techniques may be viewed as contour maps for purposes of interpretation. In a sense, the methods discussed here are equivalent to those we considered in relation to the estimation of the intensity of a point process, in Chapter 3. We prefer to think of these methods as exploratory techniques useful mainly in examining global trends in variation. It is tempting to use some of the methods for prediction and indeed this is often done. However, it should be appreciated that the justification for doing so is rather weak. In particular, they do not involve an explicit statistical model for the data under consideration and make no attempt to incorporate explicitly the possibility of spatial dependence. As a result, none of the methods provides any estimates of the errors that can be expected in the results. Such provisos make such methods very dubious for the purposes of prediction.

We then move on to discuss techniques for exploring spatial dependence in the data, through the *covariogram* and *variogram*. These are equivalent to the techniques used in Chapter 3 for examining second order properties of a spatial point process, such as the distribution of inter-event distances and the K function.

5.4.1 Spatial moving averages

A very simple way to estimate $\mu(\boldsymbol{s})$ is by the average of the values at neighbouring sampled data points. For example, $\mu(\boldsymbol{s})$ may be estimated as the unweighted average of the sample values at the three sampling points nearest to $\boldsymbol{s}$—a *three-point spatial moving average*. If this averaging is also applied at the sample points $\boldsymbol{s}_i$, then the resulting map will be smoother than the original observations and will indicate global trends in the data. The more points included in the moving average, the greater the smoothing will be. Patterns that initially show a lot of 'noise' can be 'cleaned up' in this way. The effect, of course, is to remove local place-to-place variation in the data.

The obvious problem with using this approach is that it does not allow for spatial variations in the distribution of sample sites. For instance, there is no discrimination between a site that is a considerable distance from its 'neighbours' (as defined by the averaging scheme used) and one which is very close to them. To counter this problem, we might want to use a weighted average of neighbouring points:

$$\hat{\mu}(\boldsymbol{s}) = \sum_{i=1}^{n} w_i(\boldsymbol{s}) y_i$$

where $\sum w_i(\boldsymbol{s}) = 1$ and $w_i(\boldsymbol{s}) \propto h_i^{-\alpha}$ or $w_i(\boldsymbol{s}) \propto \mathrm{e}^{-\alpha h_i}$ where h_i is the distance from $\boldsymbol{s}$ to $\boldsymbol{s}_i$ and α is a parameter with a value chosen to provide a suitable degree of smoothing. Usually $w_i(\boldsymbol{s})$ is chosen to be zero beyond some appropriate maximum distance, since, again, we might only want to examine a few neighbours. The advantage over the unweighted scheme is that we 'downgrade' the influence of neighbours that are some distance from the site under consideration. As we shall see, there are more sophisticated ways of performing this sort of weighting.

5.4.2 Methods based on tessellation

There are a number of estimation techniques for $\mu(\boldsymbol{s})$, which have been developed on the basis of *tessellation* (or tiling) of the observed sample locations $\boldsymbol{s}_i$. The most commonly used employ the *Delaunay triangulation*, also known as a *triangulated irregular network* (TIN).

Given n distinct locations in a planar region $\mathcal{R}$ we can assign to each location, $\boldsymbol{s}_i$, a 'territory' consisting of that part of $\mathcal{R}$ which is closer to $\boldsymbol{s}_i$ than to any other of the locations. This construction is referred to as the *Dirichlet tessellation* of the locations in $\mathcal{R}$ (Figure 5.6). The 'tiles' so constructed are sometimes referred to as *Voronoi* or *Thiessen polygons*.

Except possibly along the boundary of $\mathcal{R}$ each Thiessen polygon is a convex region. Locations $\boldsymbol{s}_i$ and $\boldsymbol{s}_j$ whose Thiessen polygons share a common boundary can be thought of as being 'contiguous'. The lines joining all such pairs of 'contiguous' locations define a triangulation of the locations called the *Delaunay triangulation*. The Delaunay triangulation and the Dirichlet tessellation can be thought of as two sides of the same coin (Figure 5.6). One way to form the Dirichlet tessellation is from polygon sides which are the

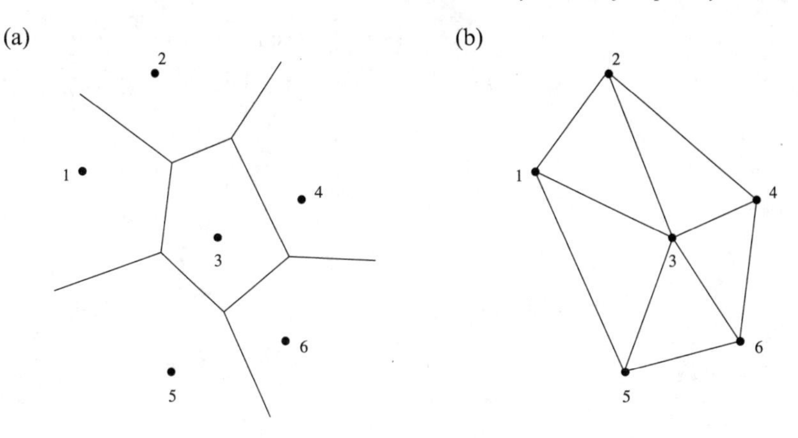

Fig. 5.6 (a) Dirichlet tessellation and (b) Delaunay triangulation

perpendicular bisectors of the edges of the Delaunay triangulation and polygon vertices which are the corresponding circumcentres.

The argument for using such tessellations of the sample sites $\boldsymbol{s}_i$, when estimating $\mu(\boldsymbol{s})$ at a unsampled site, $\boldsymbol{s}$, arises by analogy with curve fitting in one dimension. It is common in fitting a curve to a set of data points in one dimension to use a series of simple functions, such as polynomials, each of which is fitted to successive groups of data points and constrained to give some degree of continuity at their joins. This is the basis of what is known as *spline smoothing*, a common example being the use of *cubic splines*, successive cubic functions. The problem with applying this idea to the spatial or two-dimensional case is that the sample points $\boldsymbol{s}_i$ do not have a natural ordering—they do not divide $\mathcal{R}$ into obvious sub-regions over which to fit two-dimensional splines. One obvious way to produce a natural partitioning of $\mathcal{R}$ is to base it on a 'good' tessellation of the observed points. The Delaunay triangulation provides such a tessellation in that the triangles produced are as close to equilateral as possible. Other triangulations can be defined with 'good' properties of various other kinds, but we do not go into details here.

Once the set of non-overlapping triangles is defined, each vertex of which represents a site or sampled location, we may imagine a vertical line being constructed at each site, the height of which is proportional to the value at that site. The triangles on the base map may then be projected to give a set of tilted planes (Figure 5.7a). A simple way to proceed is then to assume that the surface we wish to estimate can be approximated by these tilted planes. Given that the heights are known at the three vertices of each of the planes, we can use simple geometry to estimate the value at any point on the map.

Once we have, in principle, estimated a value at any location on the map, we can then go on to draw *isolines* (also known as *isarithms*), lines joining points of equal value. We need to select the number of isolines we want and to define the

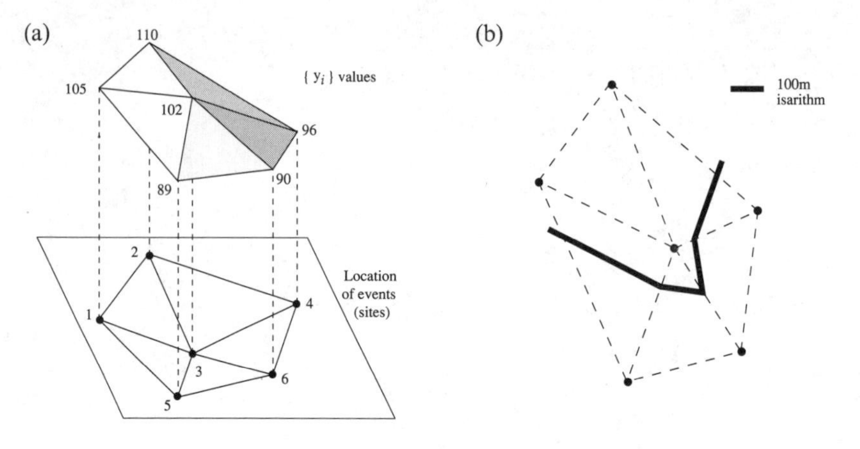

Fig. 5.7 Interpolation using a TIN

contour intervals (where we use the word 'contour' by analogy with the mapping of height above sea level). What the number of contours and contour intervals should be, raises the same sort of issues as in class interval selection for proportional symbol maps discussed earlier. Once these decisions are taken we can 'thread' an isarithm through the data points, initially as a set of straight line segments (Figure 5.7b).

We emphasise that our aim here is not cartographic excellence—we are not interested in 'elegant' contouring! We are using the contouring as an exploratory device to examine variations in $\mu(\mathbf{s})$. In particular, note that the contours obtained in Figure 5.7 are angular rather than smooth; this reflects the nature of the TIN 'model' that is used in the interpolation. Too much generalisation and 'cleaning up' of the contours to give smoothly varying lines can prove somewhat misleading, since the contours may be extrapolated well beyond the spatial range of sample locations. Note in particular that the Delaunay triangulation is only defined over the 'convex hull' formed by the data points and will not necessarily cover the edges of $\mathcal{R}$.

We can illustrate these ideas by 'contouring', in this way, one of the data sets we have already examined, that of groundwater in Venice. A TIN contour map of the 1973 levels is given in Figure 5.8 and clearly highlights the low values in the industrial area on the mainland.

There are many variations on the basic idea of TIN interpolation that we have described. The simple approach of using a series of planes each of which corresponds to a Delaunay triangle being fitted to the observed y_i at the corners, can be extended by fitting quadratic surfaces to the tiles and constraining them to have certain amounts of continuity at the joining edges. Another related approach with several variations falls under the heading of *natural neighbourhood interpolation*. This is based on repeated Dirichlet tessellations. Essentially, a weighted average of neighbouring sample points is used to interpolate the value at $\mathbf{s}$:

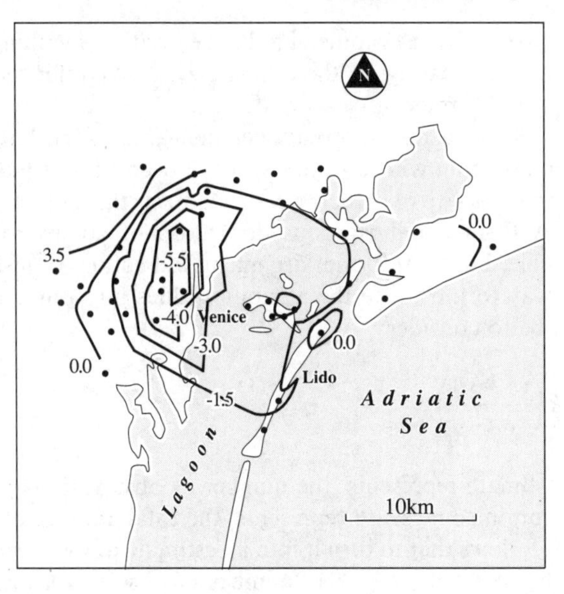

Fig. 5.8 TIN contours of Venice groundwater in 1973

$$\hat{\mu}(\mathbf{s}) = \sum_{i=1}^{n} w_i(\mathbf{s}) y_i$$

However, unlike the case in our earlier discussion of spatial moving averages, the weights $w_i(\mathbf{s})$ are now the proportion of the area of the Dirichlet tile around $\mathbf{s}_i$ which is 'stolen' by the tile around $\mathbf{s}$, when a Dirichlet tessellation is performed on $(\mathbf{s}, \mathbf{s}_1, \ldots, \mathbf{s}_n)$ as opposed to just $(\mathbf{s}_1, \ldots, \mathbf{s}_n)$. A new tessellation needs to be performed for each interpolated point; however since the new tessellation involves just one extra point, efficient methods may be developed to modify the base tessellation on $(\mathbf{s}_1, \ldots, \mathbf{s}_n)$ in each case, rather than compute an entirely fresh tessellation.

5.4.3 Kernel estimation

In the case of the analysis of 'point patterns' in Part B a flexible exploratory approach to the estimation of intensity, $\lambda(\mathbf{s})$, or 'events per unit area', over $\mathcal{R}$ was provided by kernel estimation. Ignoring edge corrections our basic kernel estimate was:

$$\hat{\lambda}_\tau(\mathbf{s}) = \sum_{i=1}^{n} \frac{1}{\tau^2} k\left(\frac{(\mathbf{s} - \mathbf{s}_i)}{\tau}\right)$$

where $k()$ was a bivariate probability density function (*the kernel*) which was symmetric about the origin, and $\tau > 0$ (*the bandwidth*) determined the amount of smoothing—essentially the radius of a disc centred on $\boldsymbol{s}$ within which points $\boldsymbol{s}_i$ will contribute 'significantly' to the estimate $\hat{\lambda}_\tau(\boldsymbol{s})$. We refer the reader back to the discussion in Chapter 3 for more details.

We are now interested not in 'events per unit area', $\lambda(\boldsymbol{s})$, but in the mean value $\mu(\boldsymbol{s})$ of an attribute whose values y_i, have been sampled at locations $\boldsymbol{s}_i$. The question arises as to whether we can adjust the kernel technique to estimate $\mu(\boldsymbol{s})$. A formal and rigorous derivation of an appropriate kernel estimate is possible, but an informal argument will suffice to justify it here.

An intuitive way to introduce the attribute values into our previous kernel estimate would be to consider:

$$\sum_{i=1}^{n} \frac{1}{\tau^2} k\left(\frac{(\boldsymbol{s}-\boldsymbol{s}_i)}{\tau}\right) y_i$$

If the original estimate represents 'the number of observations per unit area', then this extension in some sense represents 'the total amount of the attribute per unit area'. It follows that to turn it into an estimate of the mean value of the attribute we need to divide it by the 'number of observations per unit area'. This suggests that an appropriate kernel estimate for $\mu(\boldsymbol{s})$ would be:

$$\hat{\mu}_\tau(\boldsymbol{s}) = \frac{\sum_{i=1}^{n} k\left(\frac{(\boldsymbol{s}-\boldsymbol{s}_i)}{\tau}\right) y_i}{\sum_{i=1}^{n} k\left(\frac{(\boldsymbol{s}-\boldsymbol{s}_i)}{\tau}\right)}$$

where τ^{-2} in the denominator and the numerator have been cancelled. At points $\boldsymbol{s}$ where the denominator is 0 the numerator must also be 0 and by convention $\hat{\mu}_\tau(\boldsymbol{s})$ is set to 0 at these points.

Corrections for edge effects have been ignored so far, but since the form of $\hat{\mu}_\tau(\boldsymbol{s})$ is a ratio of two kernel estimates, involving the same set of $\boldsymbol{s}_i$, we do not need to incorporate any edge correction since it would appear identically in both the numerator and the denominator and cancel.

As discussed in Chapter 3, the choice of the specific form of the kernel is of secondary importance relative to the choice of an appropriate bandwidth in terms of significantly affecting the resulting estimate. A typical choice for $k()$ might be the quartic kernel which we have already discussed in some detail in Chapter 3 (see Figure 3.3), that is:

$$k(\boldsymbol{u}) = \begin{cases} \frac{3}{\pi}\left(1-\boldsymbol{u}^T\boldsymbol{u}\right)^2 & \textit{for } \boldsymbol{u}^T\boldsymbol{u} \le 1 \\ 0 & \textit{otherwise} \end{cases}$$

The effect of increasing the bandwidth τ is to expand the region around $\boldsymbol{s}$ within which observed values y_i influence the estimate at $\boldsymbol{s}$. For very large τ, $\hat{\mu}_\tau(\boldsymbol{s})$ will appear flat and local features will be obscured; if τ is small then $\hat{\mu}_\tau(\boldsymbol{s})$ tends to a collection of spikes centred on the $\boldsymbol{s}_i$.

Note that the kernel estimate $\hat{\mu}_\tau(\boldsymbol{s})$ is really just a more sophisticated version of the weighted moving average scheme discussed earlier. That is, for a given τ, it is essentially just a weighted average of the sample data points $\sum w_i(\boldsymbol{s})y_i$, where the weights

$$w_i(\boldsymbol{s}) = \frac{k\left(\dfrac{(\boldsymbol{s}-\boldsymbol{s}_i)}{\tau}\right)}{\sum_{j=1}^{n} k\left(\dfrac{(\boldsymbol{s}-\boldsymbol{s}_j)}{\tau}\right)}$$

depend upon τ, upon $\boldsymbol{s}$ through the distance between $\boldsymbol{s}$ and $\boldsymbol{s}_j$ *and* upon the local intensity of sample points in the neighbourhood of $\boldsymbol{s}$ through the denominator. Sample observations at a given distance from $\boldsymbol{s}$ obtain more weight in regions of $\mathcal{R}$ where sample points are sparse than where they are dense.

For any chosen kernel and bandwidth, values of $\hat{\mu}_\tau(\boldsymbol{s})$ can be examined at locations on a suitably chosen fine grid over $\mathcal{R}$ to provide a useful visual indication of the variation in the mean value of the attribute over $\mathcal{R}$. The bandwidth τ can be used to vary the level of 'smoothness' of this estimate, as described previously. As also discussed there, methods exist which attempt to determine the optimal bandwidth for any particular data set, and for 'adapting' the value of τ over $\mathcal{R}$ to reflect the local density of sample points. We refer the reader back to our section on kernel estimation in Chapter 3 for a fuller discussion.

Ex 5.3

5.4.4 Covariogram and variogram

The exploratory methods discussed so far in this chapter have been concerned with first order variations in attribute values, in other words with estimating the way in which the mean or expected value of the process varies in the study region. We now turn our attention to methods designed more explicitly to explore the spatial dependence of deviations in attribute values from their mean, the second order properties.

In Chapter 3 we used the K function or *reduced second moment measure* as a tool to summarise and analyse second order properties of a point pattern. In the current situation the equivalent tool is the *covariance function* or *covariogram*. Before looking at this formally, what do such functions represent in an intuitive sense?

Recall that, in elementary statistics, we speak of covariance as measuring the extent to which two variables vary together. For instance, if, as one increases, so too does another, we say there is positive covariance and this is estimated as the sum of cross-products of deviations of observations from the respective means of the two variables. If we divide this covariance by the product of the two standard deviations we obtain an estimated correlation coefficient. In a spatial context, as we have discussed in Chapter 1, the same ideas apply, except that we are interested not in the covariance or correlation between two variables but rather the way in which the deviations of observations from their mean value at different locations on the map co-vary or are correlated. Because we know that many variables show 'spatial persistence' (for example, the deviation from mean rainfall measured at one location is likely to be similar to that recorded a kilometre away, but less likely to co-vary with that twenty kilometres away) we can anticipate typically observing positive covariance or correlation at short distances and lower covariance or correlation at greater distances for many spatially continuous phenomena.

More formally, if we have a spatial stochastic process $\{Y(\mathbf{s}),\ \mathbf{s} \in \mathcal{R}\}$ where we denote $E(Y(\mathbf{s}))$ as $\mu(\mathbf{s})$ and $VAR(Y(\mathbf{s}))$ as $\sigma^2(\mathbf{s})$ then the covariance of this process at any two particular points $\mathbf{s}_i$ and $\mathbf{s}_j$ is defined as:

$$C(\mathbf{s}_i, \mathbf{s}_j) = E((Y(\mathbf{s}_i) - \mu(\mathbf{s}_i))(Y(\mathbf{s}_j) - \mu(\mathbf{s}_j)))$$

with the corresponding correlation defined as:

$$\rho(\mathbf{s}_i, \mathbf{s}_j) = \frac{C(\mathbf{s}_i, \mathbf{s}_j)}{\sigma(\mathbf{s}_i)\sigma(\mathbf{s}_j)}$$

Notice that $C(\mathbf{s}, \mathbf{s}) = \sigma^2(\mathbf{s})$.

Such a process is said to be *stationary* if $\mu(\mathbf{s}) = \mu$ and $\sigma^2(\mathbf{s}) = \sigma^2$ (that is, the mean and variance are independent of location and constant throughout $\mathcal{R}$) and, in addition,

$$C(\mathbf{s}_i, \mathbf{s}_j) = C(\mathbf{s}_i - \mathbf{s}_j) = C(\mathbf{h})$$

This means that $C(\mathbf{s}_i, \mathbf{s}_j)$ depends only on the vector difference, $\mathbf{h}$, between $\mathbf{s}_i$ and $\mathbf{s}_j$ (that is, direction and distance of separation) and not on their absolute locations. $C(\mathbf{h})$ is often referred to as the *covariance function* or the *covariogram* of the process and the corresponding correlation $\rho(\mathbf{h})$ as the *correlogram*. Note that $C(0) = \sigma^2$.

The process is said to be *isotropic* if the dependence is purely a function of the distance between $\mathbf{s}_i$ and $\mathbf{s}_j$ and not the direction, that is, dependent only on the length of the vector $\mathbf{h}$, which we shall denote h. Then, $C(\mathbf{s}_i, \mathbf{s}_j) = C(h)$. If this holds, then clearly $\rho(\mathbf{s}_i, \mathbf{s}_j) = \rho(h)$ as well.

A weaker assumption than that of stationarity is *intrinsic stationarity* which is defined through a constant mean and a constant variance in the differences between values at locations separated by a given distance and direction. That is:

$$\begin{aligned} E(Y(\mathbf{s}+\mathbf{h}) - Y(\mathbf{s})) &= 0 \\ VAR(Y(\mathbf{s}+\mathbf{h}) - Y(\mathbf{s})) &= 2\gamma(\mathbf{h}) \end{aligned}$$

The function $\gamma(\mathbf{h})$ is known as the *variogram*. Strictly it is the semi-variogram but the prefix 'semi' is often omitted and we shall follow this convention in our subsequent references to it (the factor 2 is simply added for convenience so that for large separations $\gamma(\mathbf{h})$ is equal to σ^2, rather than twice this).

For a stationary process the covariogram, correlogram and variogram are directly related by:

$$\rho(\boldsymbol{h}) = \frac{C(\boldsymbol{h})}{\sigma^2}$$
$$\gamma(\boldsymbol{h}) = \sigma^2 - C(\boldsymbol{h})$$

So, for a stationary spatial process the covariogram, correlogram and variogram provide similar information in a slightly different form (Figure 5.9). The covariogram and correlogram have the same shape, with the correlogram being scaled so that its maximum value is 1. The variogram also has the same shape as the covariogram, except that it is 'inverted'. While the covariogram starts from a maximum of σ^2 at $\boldsymbol{h} = 0$ and decreases to 0, the variogram starts at 0 and increases, at a distance referred to as the *range*, to a maximum of σ^2, often referred to as the *sill*.

In this section we are interested in using the covariogram, or variogram, simply as exploratory devices to examine spatial dependence in the observed data. Later we will see how these functions play a major role in the modelling of such data. In order to estimate $C(\boldsymbol{h})$ or $\gamma(\boldsymbol{h})$ for an observed spatial process it is in practice necessary to make some sort of stationarity assumption—we need to assume that the behaviour of the process in some parts of the space is the same as that in others, otherwise we do not have the repetition required to form an estimate of the second order properties.

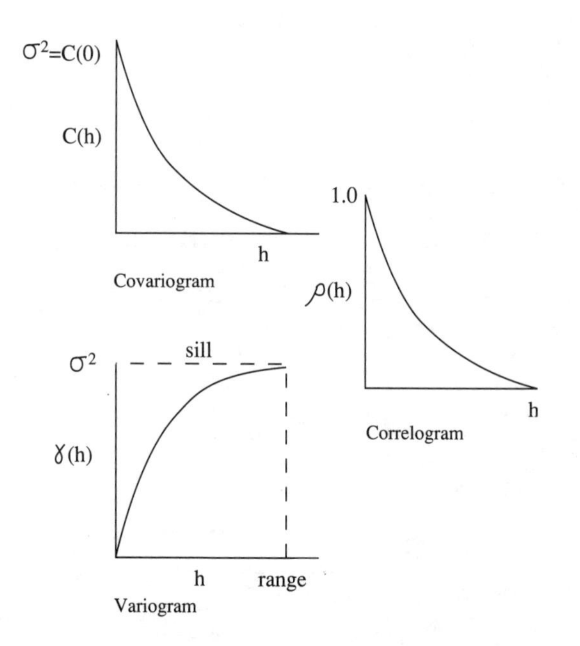

Fig. 5.9 Covariogram, variogram and correlogram

It is a matter of judgement whether to assume stationarity or intrinsic stationarity. Often, we will need to assume isotropy as well, in order to obtain a description of the covariance structure which is sufficiently simple to work with. In general, statisticians and geographers have tended to prefer to work with the covariogram or correlogram, whereas geologists and environmental scientists have favoured the variogram. It should be noted that estimates of the variogram are in general more robust to minor departures from stationarity in the form of a first order trend in the process.

The natural sample estimator of the variogram is:

$$2\hat{\gamma}(\boldsymbol{h}) = \frac{1}{n(\boldsymbol{h})} \sum_{\boldsymbol{s}_i - \boldsymbol{s}_j = \boldsymbol{h}} (y_i - y_j)^2$$

where the summation is over all pairs of observed data points with a vector separation of $\boldsymbol{h}$ and $n(\boldsymbol{h})$ is the number of these pairs. As $\boldsymbol{h}$ is varied, a set of values is obtained, so giving a sample variogram (Figure 5.10). Of course in practice, for irregularly spaced sample points, there will rarely be enough observations with an exact vector separation of $\boldsymbol{h}$ and so a series of intervals are used and the estimator calculated on the basis of separations which lie in these intervals, that is, which are 'close' to $\boldsymbol{h}$. In general this will involve specifying a tolerance on both distance and direction. Usually, disjoint intervals are chosen.

Under the assumption of isotropy the variogram is estimated over all directions for a given distance separation h:

$$2\hat{\gamma}(h) = \frac{1}{n(h)} \sum_{|\boldsymbol{s}_i - \boldsymbol{s}_j| = h} (y_i - y_j)^2$$

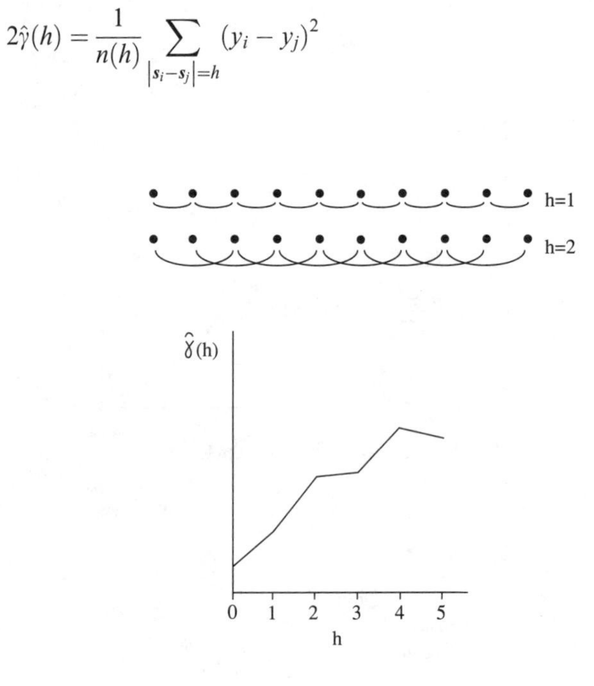

Fig. 5.10 Typical sample variogram

The equivalent estimate for the sample covariogram of a stationary process is:

$$\hat{C}(\boldsymbol{h}) = \frac{1}{n(\boldsymbol{h})} \sum_{\boldsymbol{s}_i - \boldsymbol{s}_j = \boldsymbol{h}} (y_i - \bar{y})(y_j - \bar{y})$$

where $\bar{y}$ is the mean of all the observed sample values and similar remarks apply concerning tolerances around $\boldsymbol{h}$ and simplification under the assumption of isotropy, as in the variogram case.

Note that in general the theoretical relationship between variogram and covariogram for a stationary process will not necessarily hold for the estimates. That is, in general, $\hat{\gamma}(\boldsymbol{h}) \neq \hat{C}(0) - \hat{C}(\boldsymbol{h})$.

Although theoretically $\gamma(0) = 0$, sampling errors and small scale variability may often cause sample values with small separations to be quite dissimilar. This causes a discontinuity at the origin of the sample variogram. This is often referred to as the *nugget effect*. Clearly a variogram consisting of pure nugget effect, that is, horizontal except at the origin, corresponds to a process with no spatial dependence.

Notice that in general $n(\boldsymbol{h})$ will increase as h, the length of the vector separation $\boldsymbol{h}$, increases, so that the reliability of the sample estimate of variogram or covariogram decreases as h decreases. Unfortunately, it is the more local behaviour which is likely to be of practical interest. In other words, in the area in which we are most interested, we have the least reliability in our estimate. One pragmatic suggestion is to attempt to smooth the variability caused by small numbers of data pairs at short distances by using a modified estimate of $\gamma(\boldsymbol{h})$ as:

$$\tilde{\gamma}(\boldsymbol{h}) = \frac{\hat{\gamma}(\boldsymbol{h})}{\bar{y}^2(\boldsymbol{h})}$$

where $\bar{y}^2(\boldsymbol{h})$ is the mean of all the data values used to calculate $\hat{\gamma}(\boldsymbol{h})$ at different values of $\boldsymbol{h}$. $\tilde{\gamma}(\boldsymbol{h})$ is referred to as the *relative variogram*.

How do all these ideas concerning covariograms, variograms and their sample estimates relate to practical spatial data analysis? One would typically begin an exploratory analysis of second order properties or covariance structure by first estimating an isotropic variogram or covariogram. This estimate is based on more sample pairs than any directional variograms and so should be less erratic and have a more interpretable structure. It should also serve to establish the most appropriate h lags at which to compute estimated values. Subsequent analysis can then proceed to calculate directional variograms in two or three broad directions in order to explore for possible directional effects or *anisotropy*. Examination of the resulting plots allows a purely informal assessment of the degree of spatial dependence in the data and any particularly strong directional effects that may be relevant in relation to this.

An additional general exploratory technique which may be useful for gaining insight into covariance structure and to back up any informal inferences drawn from the variogram, is to plot $(y_i - y_j)^2$ (or $\sqrt{(y_i - y_j)^2}$) for all possible (i,j) pairs against the distance between them h. This *variogram cloud* may reveal extreme outlying points that are dominating the estimate of the sample variogram. It may also reveal skewness in the distribution of the differences at any lag which implies that the variogram will be a poor estimate of the true covariance structure at that lag. It may also be useful to examine simple scatter plots of y_i values against 'neighbouring' y_j values at various different spatial lags.

For a stationary process the sample variogram should rise to an upper bound, the *sill* referred to earlier, corresponding to σ^2. The distance at which this occurs is referred to as the *range* (see Figure 5.9). Failure to exhibit an upper bound will indicate some degree of non-stationarity in the process. For example, if the variogram rises as an unbounded, concave-upwards curve away from the origin, this may indicate a first order 'drift' or trend in the process. In this case the trend may be removed from the data using a trend surface model, and further analysis of second order effects carried out using the residuals from this model. In order to explain what this means we need to consider some models for global trends in spatially continuous data, which we will come on to in the next section. But before doing this let us apply some of these second order exploratory tools to real data.

The sample variogram obtained for the logarithms of the radon data is shown in Figure 5.11. Notice that even at very short distances there is still a high degree of variability. In particular, there is clear evidence of a substantial 'nugget effect', the sample variogram being virtually horizontal. We might expect this, given what we know about radon gas levels as discussed earlier. We noted then that they display considerable local spatial variability, with variations even between neighbouring properties depending upon their building materials, age, insulation, and so on. The variogram simply confirms this absence of spatial continuity. In particular, this implies that there is little point in attempting to exploit the sampled values in the locality of a site at which the value is unknown, in order to form estimates of levels at that site. In fact, noting the general global trends, pointed out when we introduced this data set, is probably about as far as we can go with such data in understanding the distribution of the highly variable radon gas levels.

In contrast, the sample variogram for one of the Canadian geochemical elements (nickel) shows much more spatial continuity (Figure 5.12). Again, we have used the logarithms because of the marked skewness in the raw data. Although there is some visual evidence of a nugget effect this is much less pronounced than for the radon data. The variogram increases gradually with distance, beginning to flatten out at a distance of about 1.5–2.0 kilometres.

5.5 Modelling spatially continuous data

So far in this chapter we have been concerned with exploring spatially continuous data in a fairly informal way. In this section we consider the

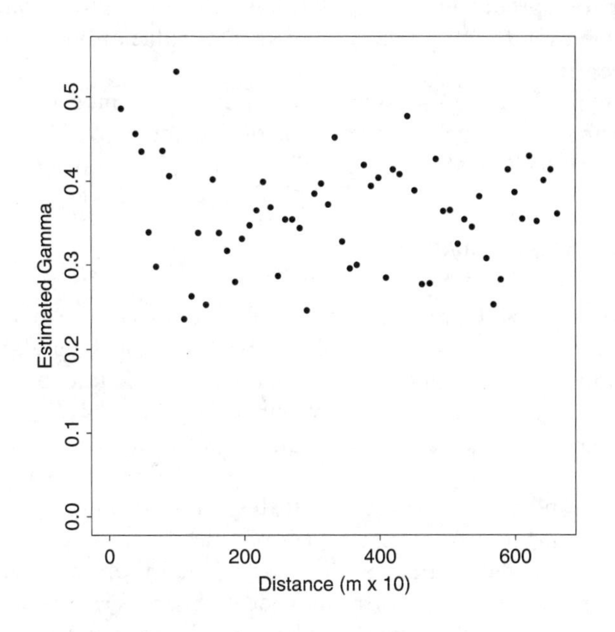

Fig. 5.11 Sample variogram for logarithms of radon levels

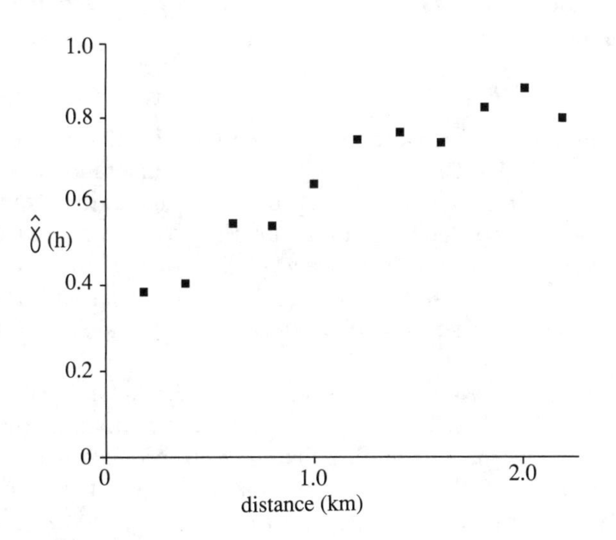

Fig. 5.12 Sample variogram for logarithms of nickel concentrations on north Vancouver Island

construction of specific models to 'explain' the observed variation in the attribute value over $\mathcal{R}$ and possibly to predict its value at points where values have not been recorded.

In the first instance we shall mostly be interested in models which involve first order variation in the mean value of the attribute. We then move on to consider how to include second order effects into such models.

5.5.1 Trend surface analysis

The methods discussed previously for exploring variations in $\mu(\boldsymbol{s})$ did not involve an explicit model. One simple approach to modelling global or large scale variations in the mean value of a spatially continuous process is *trend surface analysis*. This model has been widely, though not always appropriately, used by geographers, geologists, and others with an interest in explaining variations in spatially continuous data or interpolating values to sites others than those sampled. It involves the fitting of polynomial (or sometimes trigonometric) functions of the spatial coordinates (s_{i1}, s_{i2}), of the sample sites $\boldsymbol{s}_i$, to the observed data values at these sites, y_i, by ordinary least squares regression. Often covariates other than location are also included into the regression model in order to further understand or 'explain' spatial variations. In the first instance, we will present the method on the assumption that no such covariates are involved. Their inclusion is a straightforward extension, which we will comment on later.

The conventional multiple regression model that is used may be written using vector notation as:

$$Y(\boldsymbol{s}) = \boldsymbol{x}^T(\boldsymbol{s})\boldsymbol{\beta} + \epsilon(\boldsymbol{s})$$

where $Y(\boldsymbol{s})$ is the random variable representing our process at the point $\boldsymbol{s}$; $\boldsymbol{x}^T(\boldsymbol{s})\boldsymbol{\beta}$ is the trend (that is, the mean value, $\mu(\boldsymbol{s})$, of this random variable); and $\epsilon(\boldsymbol{s})$ is a zero mean random variable representing fluctuations from this trend. The $(p \times 1)$ vector $\boldsymbol{x}(\boldsymbol{s})$ consists of p functions of the spatial coordinates, (s_1, s_2), of the point $\boldsymbol{s}$. For a linear trend surface this is just $(1, s_1, s_2)^T$, whilst for a quadratic surface it is $(1, s_1, s_2, s_1{}^2, s_2{}^2, s_1 s_2)^T$; and so on for higher order surfaces. $\boldsymbol{\beta}$ is a $(p \times 1)$ vector of parameters to be estimated when the model is 'fitted' to the observed data.

The standard regression assumption is that the errors, $\epsilon(\boldsymbol{s})$, at different points $\boldsymbol{s}$, have constant variance, say σ^2, and that they are independent, so that their covariance is zero. This implies there are no second order effects present in the process $Y(\boldsymbol{s})$. Under this assumption we may 'fit' the model to our observed data, y_i, by *ordinary least squares*, deriving estimates $\hat{\boldsymbol{\beta}}$ and their associated standard errors, by using standard regression results:

$$\hat{\boldsymbol{\beta}} = (\boldsymbol{X}^T\boldsymbol{X})^{-1}\boldsymbol{X}^T\boldsymbol{y}$$

$$VAR\left(\hat{\boldsymbol{\beta}}\right) = \sigma^2(\boldsymbol{X}^T\boldsymbol{X})^{-1}$$

where $\boldsymbol{y} = (y_1, \ldots, y_n)^T$, is the vector of observed values at the sample sites $\boldsymbol{s}_i$; and $\boldsymbol{X}$ is an $(n \times p)$ matrix with row vectors $\boldsymbol{x}^T(\boldsymbol{s}_i)$, $i = 1, \ldots, n$. For example, for a quadratic trend surface:

$$\boldsymbol{X} = \begin{pmatrix} 1 & s_{11} & s_{12} & s_{11}^2 & s_{12}^2 & s_{11}s_{12} \\ 1 & s_{21} & s_{22} & s_{21}^2 & s_{22}^2 & s_{21}s_{22} \\ \vdots & \vdots & \vdots & \vdots & \vdots & \vdots \\ 1 & s_{n1} & s_{n2} & s_{n1}^2 & s_{n2}^2 & s_{n1}s_{n2} \end{pmatrix}$$

Details of this standard multiple regression model may be found in most elementary statistical texts, including considerations such as how to evaluate the 'fit' of the model to the observed data and test the significance of either individual β_i values or combinations of them taken together. For example, the variance, σ^2, of $\epsilon(\boldsymbol{s})$ in the above expression for $VAR(\hat{\boldsymbol{\beta}})$ would normally be unknown and would be estimated from the *residuals* of the model; that is, the differences between the observed values y_i at each sample site and those predicted by the model $\hat{y}_i = \boldsymbol{x}^T(\boldsymbol{s}_i)\hat{\boldsymbol{\beta}}$. The appropriate estimate of σ^2 is given by:

$$\hat{\sigma}^2 = \frac{\sum_{i=1}^{n}(y_i - \hat{y}_i)^2}{n - p}$$

Examination of the residuals is also used to assess the 'fit' of the model in various ways. In particular, a measure of overall 'goodness of fit' is provided by the *coefficient of determination*, sometimes referred to as the 'proportion of variation explained', or just 'R^2' for short. This is calculated as:

$$R^2 = 1 - \frac{\sum_{i=1}^{n}(y_i - \hat{y}_i)^2}{\sum_{i=1}^{n}(y_i - \bar{y}_i)^2}$$

Sometimes $\bar{R}^2 = 1 - (1 - R^2)(n - 1)/(n - p)$ is used instead, since this incorporates an adjustment to account for the number of explanatory variables in the model. 'F tests' may be used to compare the fits of alternative models, for example to ask whether a quadratic trend surface offers a significant improvement over a linear one. We do not reproduce further details here, but simply refer the reader to one of the many widely available texts on applied regression. Such texts do not usually comment specifically on spatial applications, but with the above assumptions on $\epsilon(\boldsymbol{s})$, there is essentially no difference between applying regression to spatial data and applying it to that which is not spatially referenced. Whether such assumptions are reasonable with spatial data is of course another question, a point which we will take up shortly.

Visually, what do such trend surfaces look like? A simple linear trend surface is represented, in three dimensions, as a tilted plane (Figure 5.13), projecting onto a base map as a series of linear (straight line) contours. In this case the vector of parameters $\boldsymbol{\beta}$ would have three elements whose estimates, $\hat{\beta}_1$, $\hat{\beta}_2$ and $\hat{\beta}_3$, relate respectively to the 'intercept' of the plane at the origin of the coordinate system, its slope in the coordinate direction s_1 and its slope in the coordinate direction s_2. A quadratic surface would have a 'dome' or 'trough' structure in three dimensions, the precise form depending, of course, on the values of the parameters $\boldsymbol{\beta}$ and whether these are positive or negative. Higher order surfaces may have several 'domes' and 'troughs' depending on the order of the surface.

As an exploratory tool the simple ordinary least squares regression model for $Y(\boldsymbol{s})$ may provide useful insight into overall trends in $\mu(\boldsymbol{s})$; but the underlying statistical assumptions are unlikely to be an appropriate representation of the process $Y(\boldsymbol{s})$ and results need to be treated with caution. The reason is that standard regression assumes that $\epsilon(\boldsymbol{s})$ are independent random errors. In other words, it assumes that there is only first order variation and no second order variation in the spatial stochastic process $Y(\boldsymbol{s})$. With most real spatial data this assumption is often violated and residuals from such a regression will be spatially correlated. In addition, it is possible that $VAR(\epsilon(\boldsymbol{s}))$ will not be constant over $\mathcal{R}$, a situation which is referred to as *heteroscedasticity*. As a result of one or both of these problems the conventional confidence intervals for the regression coefficients β_i and corresponding tests for goodness of fit and confidence intervals for predictions from the regression will be invalid if a trend surface is fitted to spatial data by ordinary least squares in the way described above.

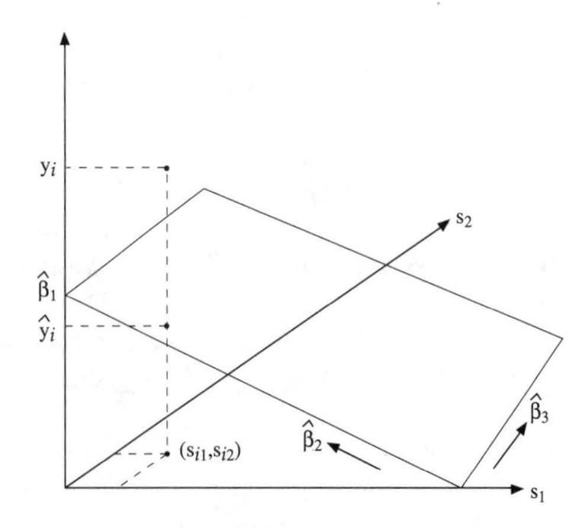

Fig. 5.13 A linear trend surface

Even if such trend surface analysis is simply regarded as an *ad hoc* approach to interpolation, rather than a formal statistical model, it has serious drawbacks. In most cases the model will be virtually useless for local prediction. The simple polynomial functions typically used in trend surfaces produce global smoothing which loses most of the local variability. When enough higher order terms are included in the model to retain some of this detail, the regression coefficients are highly correlated and their estimates correspondingly unstable and the fitted surface is very vulnerable to outliers or observational errors in y_i. In addition, local effects in one part of the region will influence the fit of a regression surface everywhere. Results are also affected by a clustered distribution of sample sites and also by the measurement units of locational coordinates. It is sometimes recommended that any given set of locational coordinates be scaled to a unit square, or to low values; otherwise, the raising of numerically large coordinates to higher powers causes problems in the subsequent matrix calculations.

This is not to say that simple least squares regression is not a useful tool in the analysis of spatially continuous data; rather that it ought to be treated with appropriate care. It is useful for indicating broad trends and relationships, without too much emphasis on formal hypothesis tests for parameter values and model fit, since these are likely to be rendered invalid by second order effects. The degree to which this may be the case can be assessed by examination of spatial dependence in the residuals from the model. Recall that these are the differences between the observed values y_i at each sample site and those predicted by the model $\hat{y}_i = \boldsymbol{x}^T(\boldsymbol{s}_i)\hat{\boldsymbol{\beta}}$.

We can use any of the exploratory techniques we have previously discussed for second order effects, such as the covariogram or variogram, to examine such residuals. Strictly speaking there are problems associated with doing this. If p parameters have been estimated in the regression, then the observed residuals are by mathematical necessity subject to p linear constraints. That is, the observed residuals are automatically correlated to some extent and so a bias is necessarily introduced into the estimate of a variogram or covariogram based on residuals. This problem is more theoretical than practical, and unlikely to be significant at short distances when n is reasonably large and the trend surface used is appropriate for the data. The degree of spatial dependence detected from such residual covariograms or variograms indicates just how suspect parameter estimates and standard errors from the ordinary least squares model are likely to be.

One area in which simple least squares regression is often useful is in assessment of the relationship between any spatial pattern in attribute values and other covariates whose values have been recorded at the sample locations $\boldsymbol{s}_i$. This simply involves a natural extension to the model above where the vector $\boldsymbol{x}(\boldsymbol{s})$ of spatial coordinates and functions of these, is augmented with other covariates. This requires no adjustment to any of our previous results. $\boldsymbol{X}$ still has rows $\boldsymbol{x}^T(\boldsymbol{s}_i)$, but we now simply regard these as consisting of the values of p explanatory variables, $(x_{i1}, \ldots, x_{ip})$, where we assume that in general some of these explanatory variables may be the coordinates (s_{i1}, s_{i2}) of $\boldsymbol{s}_i$ (and possibly their powers and cross products, etc.) and others will be the values of additional covariates recorded at these locations. We then have in our model a spatial trend as well as a dependence on the values of the other covariates. The appropriate individual elements of the parameter estimates $\hat{\boldsymbol{\beta}}$ can be used to assess the significance of the relationship with the different covariates. Residual covariograms will again give an indication of how 'valid' any significance tests are likely to be. Notice, in passing, that a model involving such covariates would only be of use in prediction if we know $\boldsymbol{x}(\boldsymbol{s})$, the vector of values of the explanatory variables at all points $\boldsymbol{s}$ where prediction is required. Clearly, this may not be the case when some of the elements of $\boldsymbol{x}(\boldsymbol{s})$ relate to covariates, rather than being solely functions of the spatial coordinates (s_1, s_2).

Ex 5.7

Previously, we mentioned that as well as residual spatial dependence, non-constant variance of $\epsilon(\boldsymbol{s})$ over $\mathcal{R}$ (heteroscedasticity) might also be a possible objection to use of the standard least squares regression model. This latter problem is one which also arises in many non-spatial applications of regression analysis and may be able to be addressed by either the use of *weighted regression* or a transformation of the response variable y_i. Weighted regression arises as a special case of *generalised least squares* which we will discuss in detail shortly in connection with formal modifications to ordinary regression in order to account for possible spatial dependence in residuals. The alternative, transformation of y_i to a scale, say y_i', more suitable for modelling, should always be considered in regression analysis, both to stabilise variance (i.e. the variance of the transformed variable is constant) and to linearise the relationship between $\mu(\boldsymbol{s})$ and the covariates when, for example, a multiplicative rather than an additive relationship is suspected. A useful set of transformations to consider is the *Box–Cox* family:

$$y_i' = \frac{(y_i^\lambda - 1)}{\lambda}, \quad \lambda \neq 0$$

$$y_i' = \log(y_i), \quad \lambda = 0$$

These bring together most of the transformations commonly used in practice, in terms of choice of the value of the single parameter, λ. In the case of spatially continuous data the logarithmic transformation from this family is often particularly useful. Firstly, it tends to stabilise unequal variances, when in the original data variance increases markedly with the mean value. In fact, the variance of the logarithm will be approximately constant when the variance of the original data is proportional to the square of its mean value. A second effect of the logarithmic transformation is to linearise the relationship between a variable and a covariate, when this relationship is exponential on the original scale. Such relationships are common when the original variable is a concentration of some substance. A third beneficial effect is to reduce the degree of 'skewness' in an original variable with a 'long tail' of extreme positive values, by compressing the upper part of the measurement scale.

However, it should be appreciated that, in general, transformations will not help to alleviate the other problem mentioned earlier that arises when regression is applied in a spatial context: that of possible second order effects or residuals being spatially correlated. We take this point up again in the next section, but before doing so it may be useful to illustrate some of these ideas by fitting some simple least squares trend surface models to two of the data sets we introduced earlier. When we examine the Sudan rainfall data shown earlier in Figure 5.1, both linear (Figure 5.14) and quadratic trend surfaces fitted to the 1982 figures show the predicted values rising to the south. Looking at residuals from the linear trend surface we note positive values to the east (around Gedaref, for instance), where rainfall is under predicted. To the south and central parts of the study area (around El Fasher and Jebelein, for example) there are some negative residuals, suggesting that rainfall totals are over predicted. A plot of residuals against predicted values (for either model) shows some evidence of heteroscedasticity and a transformation of the data is worth exploring. Repeating the analysis using the logarithm of rainfall reduces the heteroscedasticity. The residual map looks rather different, though the value at Gedaref remains high.

How can we assess the fit of this model? First, recall that we have assumed that the errors are spatially uncorrelated. A simple, informal, test of this is to examine the relationship between the residual at a station and that at its nearest spatial neighbour. A plot of the residuals against their nearest neighbours shows little evidence for spatial correlation. This done, how well do the linear and quadratic models fit the data? For the linear model on the Sudan data we obtain an R^2 value of approximately 52% and for the quadratic we get $R^2 \approx 62\%$. In this case although the quadratic surface is strictly a significantly better fit when the standard 'F' test is applied, the significance is marginal and the quadratic offers little improvement over the linear. The predominant trend in this data is the north–south effect which is adequately captured by the linear surface. We do not give details of the results of the same analysis using logarithms of the response variable, but the findings are very similar.

Turning to the PCB data discussed earlier in our case study section, recall that we have a possible source of environmental contamination in the centre of the map. The distribution of data values is very highly positively skewed and it is much more sensible to look at the logarithms of PCB concentrations. Mapping the logarithms of the data using contours based on a TIN reveals a very 'jagged' surface with a huge peak around the suspected site. *A priori* it seems not unreasonable to fit a quadratic surface to the data. Fitting a quadratic model to the logarithms and deriving residuals reflects the dome-like structure clearly (Figure 5.15). Further analysis reveals a group of negative residuals to the west and 'hot-spots' elsewhere.

Whilst the quadratic trend surface clearly highlights the huge peak in these data, it is not necessarily a tenable model. It is questionable whether a first order trend is really a sensible way to capture the obvious local 'spike' in concentrations we observe here. We might do better to attempt to model this via second order local effects, particularly if prediction of PCB values across this study area was our objective. We will come on to such methods in later sections of the chapter.

(a)

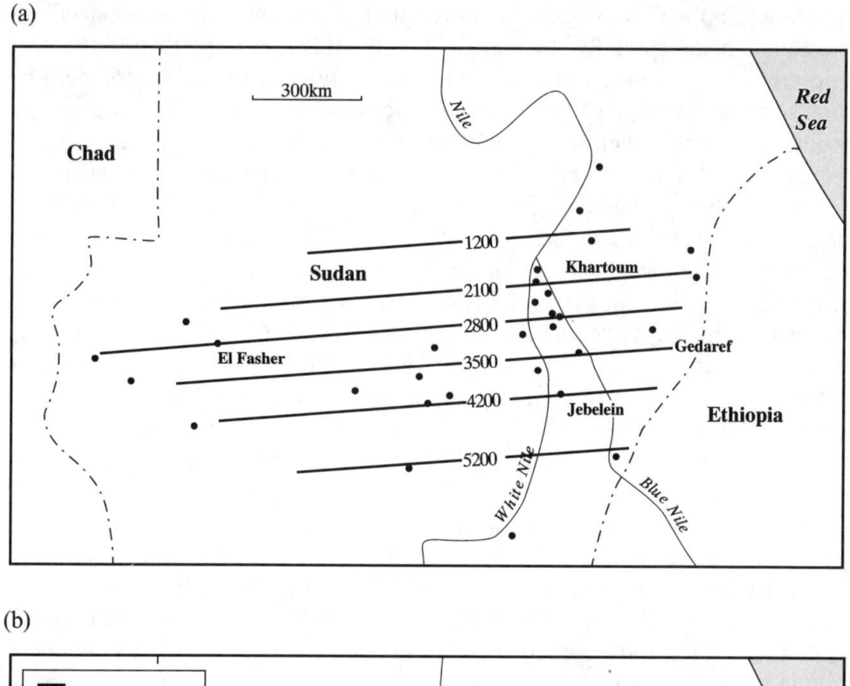

(b)

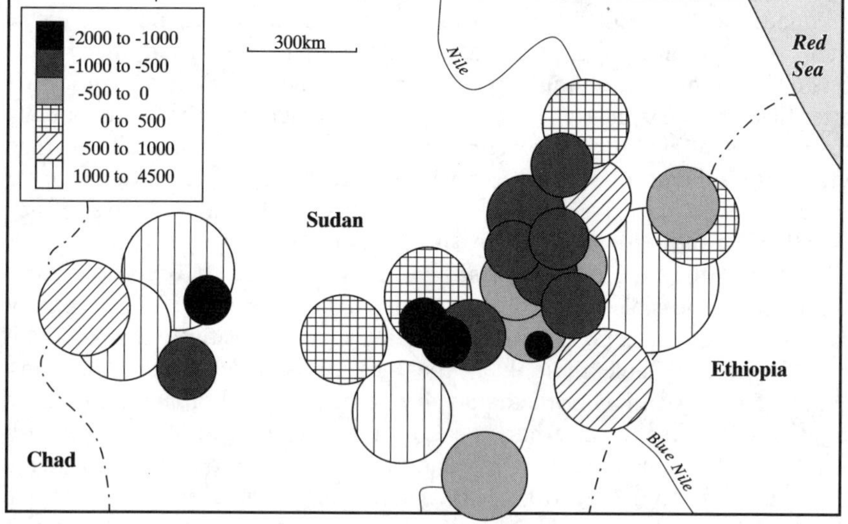

Fig. 5.14 (a) Linear trend surface and (b) residuals for Sudan rainfall, 1982

For an outline example of using covariates other than just location in ordinary regression models applied to spatially continuous data, we refer the reader back to our discussion of the analysis of the Californian rainfall measurements in Chapter 1, where covariates such as altitude and a rain shadow effect were included in the model.

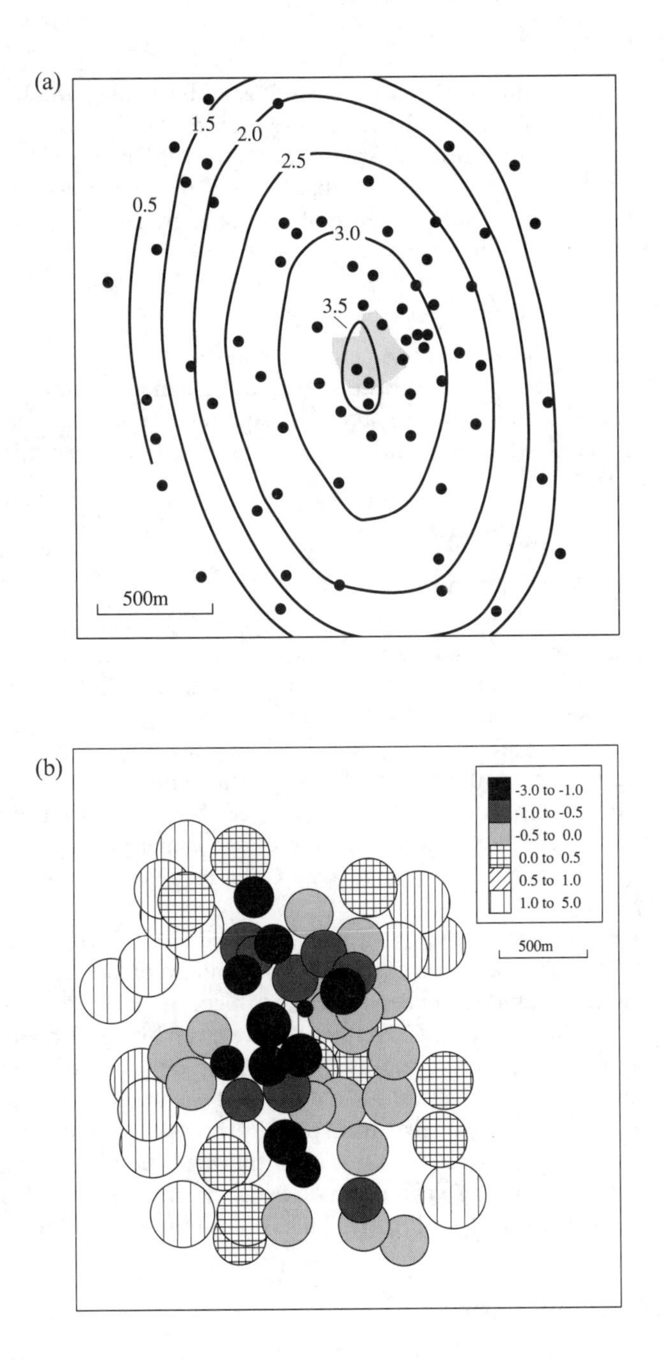

Fig. 5.15 (a) Quadratic trend surface and (b) residuals for log of PCBs

5.5.2 Generalised least squares

So far, the regression models that we have discussed have made the simple and possibly unrealistic assumption that there are only first order effects involved in the process $Y(\boldsymbol{s})$ and no residual spatial dependence. One way to avoid the implausibility of the assumptions of this basic trend surface regression model is to relax the assumption of independence of residuals, by the use of *generalised least squares*. The model then becomes:

$$Y(\boldsymbol{s}) = \boldsymbol{x}^T(\boldsymbol{s})\boldsymbol{\beta} + U(\boldsymbol{s})$$

Ex 5.8

where $U(\boldsymbol{s})$ are zero mean errors as before. However, they are not necessarily independent at different points $\boldsymbol{s}$; instead they have some covariance function $C()$. In this case the previous ordinary least squares estimates for $\boldsymbol{\beta}$ and the corresponding standard errors must be replaced by their generalised least squares equivalents:

$$\hat{\boldsymbol{\beta}} = (\boldsymbol{X}^T\boldsymbol{C}^{-1}\boldsymbol{X})^{-1}\boldsymbol{X}^T\boldsymbol{C}^{-1}\boldsymbol{y}$$
$$VAR(\hat{\boldsymbol{\beta}}) = (\boldsymbol{X}^T\boldsymbol{C}^{-1}\boldsymbol{X})^{-1}$$

with notation as previously, and where $\boldsymbol{C}$ is the $(n \times n)$ matrix of covariances, $C(\boldsymbol{s}_i, \boldsymbol{s}_j)$, between $U(\boldsymbol{s}_i)$ and $U(\boldsymbol{s}_j)$, for each possible pair (i,j) of the n sample sites. The diagonal elements of this matrix are $VAR(U(\boldsymbol{s}_i))$.

It is interesting to note that weighted least squares, to which we referred earlier, is a special case of this, where $\boldsymbol{C}$ is diagonal. This would be applicable to the situation where residuals were uncorrelated, but their variance was not constant over $\mathcal{R}$. Each observation in the regression is then weighted in inverse proportion to its variance, so giving less weight to observations with large variance.

Generalised least squares gives us a way of including both first and second order variation in our model and ensuring that standard errors and parameter estimates allow for spatial dependence. The problem is that in general we do not know the covariance matrix $\boldsymbol{C}$. Neither can we estimate the elements of this matrix directly from the data, since usually we have only one observation at each pair of sample locations $\boldsymbol{s}_i$ and $\boldsymbol{s}_j$. In general, if we are going to attempt to develop realistic statistical models for a spatial process that incorporate both first and second order variation then it is clear that we need to develop models of covariance structure that we can estimate from the data and which allow us to arrive at the elements of $\boldsymbol{C}$ indirectly. This is the subject of the next section.

5.5.3 Models for variograms

In our exploratory section we considered how to estimate a reasonably stable sample covariogram or variogram from the observed data values. Here we discuss some possible parametric models we might wish to fit to these in order to provide a smooth, continuous description of the covariance structure. The reason for doing this is that via such models we can derive an indirect estimate of the covariance matrix $\boldsymbol{C}$ of our previous section and will therefore be able to use generalised least squares in order to obtain a model for our spatially continuous data that incorporates both first order and second order effects.

Recall from our earlier discussion that sample variograms or covariograms only make sense as estimates of covariance structure under the assumption of some form of stationarity in the data to which they are applied. If such data are not stationary, the sample covariogram or variogram may be dominated by first order effects and provide virtually no information about second order variation in isolation. Even if there are no first order effects involved in a non-stationary process and so its mean value is constant, sample estimates of second order effects would still be useless, since in that case non-stationarity means that the covariance structure does not repeat itself in different parts of $\mathcal{R}$ and this invalidates the basis of our estimates, which involve 'averaging' observed relationships over different parts of $\mathcal{R}$. Consequently, we focus here only on models for the covariance structure of stationary processes. We have no reason to consider models for the covariance function of non-stationary processes, since we know that pragmatically we have no way of estimating the covariance structure of a non-stationary process from sample data in order to ascertain which of such models might be appropriate for such data.

What all this means in practice is that if the sample variogram or covariogram of the original data (or some appropriate transformation) reveals evidence of a 'drift' or trend in the process (for example, the sample variogram fails to rise to an upper bound), then we should first remove this trend by the regression methods discussed earlier and work instead with residuals when applying the modelling techniques for covariance structure which we are about to discuss. Throughout this section we will assume that, if necessary, this has been done, so that the sample variogram with which we are working relates to sample data for which the assumption of stationarity is viable.

The astute reader will notice a dilemma here. The statistically 'correct' method of estimating a trend in the presence of spatial dependence is generalised least squares; however this requires a knowledge of covariance structure, which, as we have said, can only be sensibly estimated if trend has been removed. One possible way of overcoming this problem is by iterating our modelling of covariance structure. We first remove the trend by ordinary least squares regression, obtaining residuals whose covariance structure we model. We then re-estimate the trend by generalised least squares using the estimated covariance structure, so obtaining better residuals upon which to re-estimate covariance structure; and so on, until stability is achieved. In practice this will be computationally very demanding and may well not be worth the effort. We need to bear in mind that the isolation of first and second order effects is to some extent arbitrary. Trend removal by ordinary regression will often suffice to give residuals which are adequate to provide us with a reasonable estimate of covariance structure, particularly if we use the more 'robust' variogram rather than the covariogram to arrive at this estimate, as we discuss below. From a practical point of view we probably do not need to iterate the modelling of covariance structure. If trend is present and has been removed by ordinary regression, then once the covariance structure has been estimated, the trend should be refitted by generalised least squares to 'clean up' the associated parameter estimates and standard errors, as we shall see in our next section. However, subsequent re-modelling of covariance structure is of questionable value.

Having established our interest in the covariance structure of stationary processes and ensured that this is what our sample estimate of covariance structure relates to, how do we go about developing appropriate models? We will work with variogram models, since in general the sample variogram provides more robust estimates of spatial dependence than the sample covariogram in the presence of any remaining minor departures from stationarity (recall that it only involves the assumption of intrinsic stationarity). If we require an equivalent covariogram model (which we will do in conjunction with generalised least squares) then, since stationarity is assumed, we may derive it directly from the variogram model by using the general theoretical relationship between the two which we gave earlier, that is: $\gamma(\boldsymbol{h}) = \sigma^2 - C(\boldsymbol{h})$, where $\sigma^2 = C(0) = \gamma(\infty)$ is the 'sill' of the variogram.

We need to begin by examining any constraints that any model for covariance structure must respect and which in turn place restrictions on the form of valid variogram models. Joining up the successive points of a sample variogram produces a variogram model, but not a valid one. Only certain mathematical functions are permissible models for covariance functions and so only certain functions are valid for variograms. Necessary and sufficient conditions for the covariance function of a general spatial process are those of symmetry, that is:

$$C(\boldsymbol{s}_i, \boldsymbol{s}_j) = C(\boldsymbol{s}_j, \boldsymbol{s}_i)$$

and non-negative definiteness, that is:

$$\sum_{i=1}^{n}\sum_{j=1}^{n} \alpha_i \alpha_j C(\boldsymbol{s}_i, \boldsymbol{s}_j) \geq 0$$

for all n, $\alpha_1, \ldots, \alpha_n$ and $\boldsymbol{s}_1, \ldots, \boldsymbol{s}_n$.

We are only interested in stationary processes, since as we have already noted, we need to assume some form of stationarity in order to be able to estimate a covariance function at all. The covariance functions for stationary or isotropic processes are a further restricted subset of the above functions. In practice the appropriate conditions are difficult to check and one way to guarantee that the relevant conditions are satisfied is to build our models from only a few families of functions that are known to be valid. This in turn results in corresponding families of valid variogram models. Amongst these, three of the variogram models that are most commonly used for stationary processes are:

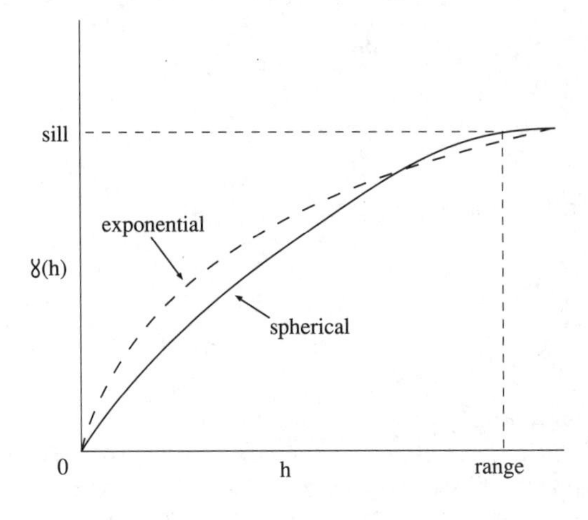

Fig. 5.16 Spherical and exponential variogram models

The *spherical model*:

$$\gamma(h) = \begin{cases} \sigma^2\left(\dfrac{3h}{2r} - \dfrac{h^3}{2r^3}\right) & \textit{for} \quad h \leq r \\ \sigma^2 & \textit{otherwise} \end{cases}$$

where r is the range and σ^2 is the sill.

The *exponential model*:

$$\gamma(h) = \sigma^2\left(1 - e^{-3h/r}\right)$$

where again r is the range and σ^2 the sill.

The *Gaussian model*:

$$\gamma(h) = \sigma^2\left(1 - e^{-3h^2/r^2}\right)$$

with r and σ^2 as above.

The shapes of the first two of these models are illustrated in Figure 5.16. Covariogram models, $C(h)$, corresponding to three common variogram models given are simply derived from the general relationship $\gamma(h) = \sigma^2 - C(h)$. In addition, a *nugget effect* can be introduced into any of them by the inclusion of a constant term, say a, on the understanding that this represents a discontinuity at the origin and that $\gamma(0)$ still remains zero. So the spherical model becomes:

$$\gamma(h) = \begin{cases} a + (\sigma^2 - a)\left(\dfrac{3h}{2r} - \dfrac{h^3}{2r^3}\right) & 0 < h \leq r \\ 0 & h = 0 \\ \sigma^2 & otherwise \end{cases}$$

and the exponential model becomes:

$$\gamma(h) = \begin{cases} a + (\sigma^2 - a)\left(1 - e^{-3h/r}\right) & h > 0 \\ 0 & h = 0 \end{cases}$$

Similar adjustments apply for the Gaussian model.

The ratio of the nugget effect to the sill is often referred to as the *relative nugget effect* and is usually expressed as a percentage. As discussed previously a variogram model consisting of pure nugget effect (so that $a = \sigma^2$, with a relative nugget effect of 100%) corresponds to a complete lack of spatial dependence.

Many other valid functional forms are possible for variogram models but are less often used in practice and we do not give details here. More importantly, we can construct more complex variogram models from the basic forms by 'mixing' them, using a model which is a linear combination of other models where all the weights involved are positive. In fact, these weights would need to sum to the 'sill' of the variogram. More than one type of model may be included, or the same type with differing parameter values. Any such combination is itself guaranteed to be a valid variogram model.

The models presented above are for isotropic processes. Models for anisotropic covariance structures are similarly defined, replacing their parameters by vector equivalents and the distance h by vector separation $\boldsymbol{h}$. In general the range may change with direction whilst the sill remains constant (*geometric anisotropy*), or the sill may change with direction whilst the range remains the same (*zonal anisotropy*). Mixtures of both of these effects are of course possible. The various directional models would first be established by fitting the above type of models along the identified axes of anisotropy. These models then need to be scaled and combined into a single vector model that is consistent in all directions. We do not discuss this process here, referring the reader to more specialist geostatistical texts for details.

Often, variogram models are fitted to the observed sample variogram 'by eye', although more formal fitting procedures can be developed. The distinguishing feature of the Gaussian model is its parabolic behaviour near the origin; it is the only one of these models with an inflexion (a flat 'S' shape) and provides a model for very spatially continuous phenomena. The exponential model rises more quickly to the sill than the spherical. For the exponential the tangent at the origin reaches the sill at about one fifth of the range, whilst for the spherical the corresponding proportion is about two thirds; the behaviour of both is fairly linear near the origin. These general characteristics may provide the basis of a choice between them. The range and sill can usually be guessed fairly easily from the sample variogram. Any nugget effect can be picked out by fitting a straight line to the first few points in the sample variogram and extrapolating this back to the origin. From there on, refinement becomes an exercise in curve fitting by adjusting these parameters and possibly trying mixtures of models. A good interactive graphics computer program is very useful. In particular, one should avoid a tendency to over fit sample variograms by trying to capture each and every kink. The objective of variogram modelling is to capture the basic structure of spatial dependence; complex models will almost certainly be an artefact of the inaccuracy of the sample variogram in the first place. In the next section we shall discuss one way of evaluating a chosen variogram model by using *cross validation.*

Let us return to the sample variogram we estimated earlier for the data on the distribution of nickel, illustrated in Figure 5.12, and try fitting a theoretical model. As before, we are using logarithms here. The results of fitting a spherical model are shown in Figure 5.17, with an estimate of the sill as 0.827 and of the range 2.13 kilometres. The nugget effect is 0.298, and if we divide this by the sill we get a relative nugget effect of 36%. This suggests that there is some degree of independence or 'random noise' in the data, but that there is clear spatial dependence at distances up to about 2 kilometres; at distances under 1 kilometre this is very marked.

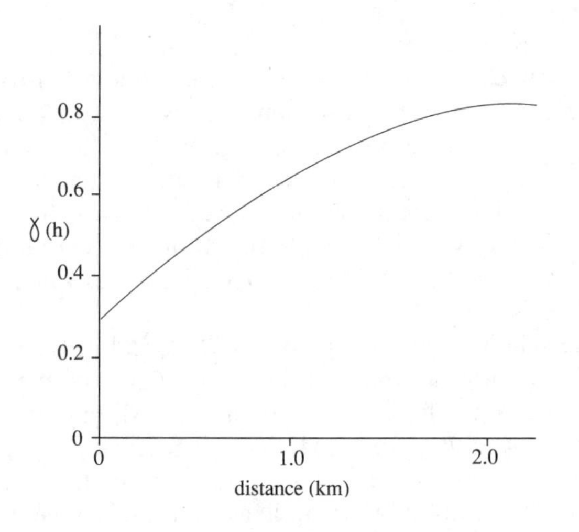

Fig. 5.17 Variogram model for logarithms of nickel concentrations

5.5.4 Generalised least squares revisited

We may now return to our discussion of generalised least squares and combine that with the above techniques for modelling covariance structure using the variogram. By doing this we can overcome the difficulties we encountered earlier and hence complete our development of a more realistic model for the spatial process $\{Y(\mathbf{s}),\ \mathbf{s} \in \mathcal{R}\}$ than that provided by ordinary least squares regression.

Earlier, we proposed a model for $Y(\mathbf{s})$ consisting of two components, namely:

$$Y(\mathbf{s}) = \mathbf{x}^T(\mathbf{s})\boldsymbol{\beta} + U(\mathbf{s})$$

where $\mathbf{x}^T(\mathbf{s})\boldsymbol{\beta}$ is a trend surface representing the mean, $\mu(\mathbf{s})$, a large scale or first order component, and $U(\mathbf{s})$ represents a local or second order component. We assume now that the second order component is a stationary zero mean, stochastic process with variance σ^2 and covariance function $C()$. The $(p \times 1)$ vector $\mathbf{x}(\mathbf{s})$ of values of explanatory variables at $\mathbf{s}$, which appears in the first order component, is taken to involve the spatial coordinates of $\mathbf{s}$ and possibly functions of these such as their powers and cross products. We could of course generalise this to the case where other covariates were included, although this would be of dubious value if the model is to be used for prediction, rather than just description, unless the values of these covariates are known at all points at which predictions are to be made. In the simplest possible case $\mathbf{x}(\mathbf{s})$ will simply be the scalar 1, corresponding to the situation where $\mu(\mathbf{s}) = \mu = \beta_1$, a constant over $\mathcal{R}$.

In practice, we can fit this model to the observed data, by first fitting the model using ordinary least squares regression, taking the residuals from this and using these to estimate a variogram model, $\hat{\gamma}()$, giving rise to an equivalent covariogram model $\hat{C}()$. This covariogram model allows us to construct an estimated covariance matrix between sample sites, $\hat{\mathbf{C}}$, with elements $\hat{C}(\mathbf{s}_i, \mathbf{s}_j)$. We can then refit the original model using generalised least squares with the estimated covariance matrix $\hat{\mathbf{C}}$. This 'corrects' the parameter estimates and standard errors of the ordinary least squares regression for second order effects. If felt necessary we could iterate the whole process until we arrive at stable estimates of $\hat{\boldsymbol{\beta}}$ and $\hat{C}()$; whereupon we have our final model for the process.

Of course, the validity of this final model will depend upon firstly the choice of an appropriate form of trend surface and secondly the choice of an appropriate variogram model. It has to be emphasised that the decomposition of a process into first and second order components in this way is, to some extent, arbitrary. In reality the two effects are confounded—certain types of second order variation can lead to trend like effects and so the two effects may not be able to be objectively distinguished for a 'real life' process. But if a decomposition model of the type described allows us to understand certain characteristics of the behaviour of the process and possibly to predict its behaviour then the modelling will have achieved its objectives.

Note here the mention of two possible objectives. If our primary interest is understanding and describing the nature of the variation in the observed attribute values and isolating any systematic large scale trend, then knowledge of the trend component and the form of the covariance structure of variations from this trend is as far as we need to pursue the modelling process. If, however, the primary interest is in prediction or interpolation of the attribute value at points where it has not been sampled, then we need to consider how to use our derived model for prediction purposes. This is the objective of the technique of *kriging* to which we devote considerable attention in the subsequent sections of this chapter.

5.5.5 Simple kriging

Suppose our primary interest does in fact lie in prediction of the values of a spatially continuous variable. Then the important point to appreciate is that if we have a model of the process as:

$$Y(\boldsymbol{s}) = \boldsymbol{x}^T(\boldsymbol{s})\boldsymbol{\beta} + U(\boldsymbol{s})$$

where $U(\boldsymbol{s})$ is a zero mean process with covariance function $C()$, then values of $U(\boldsymbol{s})$ are not entirely unpredictable. It follows that we can do better than simply predict a value of the process by its estimated mean value $\mu(\boldsymbol{s}) = \boldsymbol{x}^T(\boldsymbol{s})\hat{\boldsymbol{\beta}}$. We ought to be able to use our knowledge of the covariance function of the residual process $U(\boldsymbol{s})$ combined with the knowledge we have of its observed value at the sample points $\boldsymbol{s}_i$, to be able to add a local component to our prediction at $\boldsymbol{s}$ in addition to the mean. The technique by which this is achieved is commonly referred to as *kriging* (the name derives from the South African mining geologist D.G. Krige, who developed a preliminary version of the method, later to be refined by French geostatisticians).

We will consider first in this section the technique of *simple kriging*, where we assume that the first order component $\mu(\boldsymbol{s})$ in our model is known *a priori* and does not have to be estimated from the observed data. It can therefore be subtracted from our original sample observations y_i to provide a set of observed residuals u_i. Furthermore, we assume that the zero mean process, $U(\boldsymbol{s})$, upon which these are observations, has a known variance σ^2 and covariance function $C()$. Since $C()$ is known (rather than estimated) we do not have to assume $U(\boldsymbol{s})$ is a stationary process. Neither simple kriging nor any of the other kriging methods we will subsequently present depend inherently on the assumption of stationarity, although it will of course arise indirectly once $C()$ has to be estimated; as we have stressed before, such estimation is only possible under the assumption of some form of stationarity.

With the above assumptions, our problem in simple kriging reduces to that of wishing to find an estimate, $\hat{u}(\boldsymbol{s})$, for a value $u(\boldsymbol{s})$ of the random variable $U(\boldsymbol{s})$ at the location $\boldsymbol{s}$, given observed values, u_i, of the random variables, $U(\boldsymbol{s}_i)$, at the n sample locations $\boldsymbol{s}_i$. Once we have established a suitable estimate, we will form our prediction $\hat{y}(\boldsymbol{s})$ of the value of the random variable $Y(\boldsymbol{s})$ by simply adding $\hat{u}(\boldsymbol{s})$ to the known trend, $\mu(\boldsymbol{s})$, at the point $\boldsymbol{s}$.

It is intuitively sensible to consider estimates which are weighted linear combinations of the observed residuals u_i. In order to determine exactly what linear combination would be a 'best' estimate let us consider the following weighted linear combination of random variables:

$$\hat{U}(\mathbf{s}) = \sum_{i=1}^{n} \lambda_i(\mathbf{s}) U(\mathbf{s}_i)$$

This is simply a weighted sum of the n random variables $U(\mathbf{s}_i)$ at the sample sites $\mathbf{s}_i$. We denote the weights $\lambda_i(\mathbf{s})$, since different weights are to be applied to each of the $U(\mathbf{s}_i)$, and the weights used will be dependent on the particular location $\mathbf{s}$ at which we wish to predict the value (Figure 5.18).

Now since $\hat{U}(\mathbf{s})$ is a sum of random variables, it is itself another random variable. By choice of particular values for the weights, $\lambda_i(\mathbf{s})$, we can 'design' it to be a random value that is in some sense 'close' to the random value $U(\mathbf{s})$ whose value we wish to predict. The first thing to notice about $\hat{U}(\mathbf{s})$ is that its mean value is zero for any choice of weights, since, by assumption, the mean of each $U(\mathbf{s}_i)$ is zero and the weights are just constants. What this means is that 'on average' the value of $\hat{U}(\mathbf{s})$ will be zero, which is a desirable property given that we wish this random variable to be 'close' to $U(\mathbf{s})$ whose mean value is by definition also zero. Next we turn our attention to how much 'on average' we expect values of the random variable $\hat{U}(\mathbf{s})$ to differ from those of $U(\mathbf{s})$. One way of measuring this is by the *expected mean square error* between values of $U(\mathbf{s})$ and values of $\hat{U}(\mathbf{s})$. Bearing in mind that the random variables $U(\mathbf{s})$ and $U(\mathbf{s}_i)$ all have zero mean, this expected mean square error is:

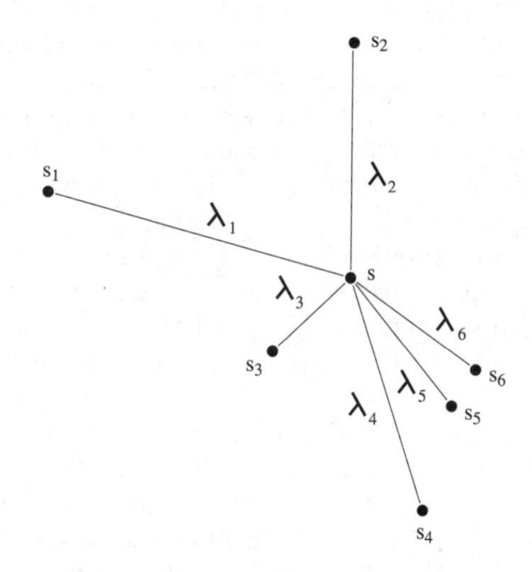

Fig. 5.18 Kriging weights at a location $\mathbf{s}$

$$E\left(\left(\hat{U}(\boldsymbol{s}) - U(\boldsymbol{s})\right)^2\right) = E\left(\hat{U}^2(\boldsymbol{s})\right) + E\left(U^2(\boldsymbol{s})\right) - 2E\left(U(\boldsymbol{s})\hat{U}(\boldsymbol{s})\right)$$
$$= \sum_{i=1}^{n}\sum_{j=1}^{n} \lambda_i(\boldsymbol{s})\lambda_j(\boldsymbol{s})C(\boldsymbol{s}_i, \boldsymbol{s}_j) + \sigma^2 - 2\sum_{i=1}^{n} \lambda_i(\boldsymbol{s})C(\boldsymbol{s}, \boldsymbol{s}_i)$$
$$= \boldsymbol{\lambda}^T(\boldsymbol{s})\boldsymbol{C}\boldsymbol{\lambda}(\boldsymbol{s}) + \sigma^2 - 2\boldsymbol{\lambda}^T(\boldsymbol{s})\boldsymbol{c}(\boldsymbol{s})$$

where $\boldsymbol{C}$ is the $(n \times n)$ matrix of covariances, $C(\boldsymbol{s}_i, \boldsymbol{s}_j)$, between all possible pairs of the n sample sites and $\boldsymbol{c}(\boldsymbol{s})$ is an $(n \times 1)$ column vector of covariances, $C(\boldsymbol{s}, \boldsymbol{s}_i)$, between the prediction point $\boldsymbol{s}$ and each of the n sample sites.

Since we wish $\hat{U}(\boldsymbol{s})$ to be 'close' to $U(\boldsymbol{s})$ we should choose the weights $\lambda_i(\boldsymbol{s})$ in order to minimise this mean square error. Differentiating with respect to the vector $\boldsymbol{\lambda}(\boldsymbol{s})$ in order to minimise gives the solution:

$$\boldsymbol{\lambda}(\boldsymbol{s}) = \boldsymbol{C}^{-1}\boldsymbol{c}(\boldsymbol{s})$$

from which it follows that

$$\hat{U}(\boldsymbol{s}) = \boldsymbol{\lambda}^T(\boldsymbol{s})\boldsymbol{U} = \boldsymbol{c}^T(\boldsymbol{s})\boldsymbol{C}^{-1}\boldsymbol{U}$$

where $\boldsymbol{U}$ denotes the $(n \times 1)$ column vector of random variables, $U(\boldsymbol{s}_i)$, at each of the n sample sites $\boldsymbol{s}_i$.

Furthermore, the minimised expected mean square error corresponding to this choice of weights is then:

$$E\left(\left(\hat{U}(\boldsymbol{s}) - U(\boldsymbol{s})\right)^2\right) = \sigma^2 - \boldsymbol{c}^T(\boldsymbol{s})\boldsymbol{C}^{-1}\boldsymbol{c}(\boldsymbol{s})$$

where it should be recalled that $\sigma^2 = VAR(U(\boldsymbol{s})) = C(\boldsymbol{s}, \boldsymbol{s})$. This minimised mean square error is often referred to as the *mean square prediction error*, or sometimes the *kriging variance*, which we shall denote σ_e^2. It provides the basis for deriving a confidence interval for our final prediction as described shortly.

To summarise, we have succeeded in deriving a linear combination $\hat{U}(\boldsymbol{s})$ of the random variables $U(\boldsymbol{s}_i)$, which has the same mean value as $U(\boldsymbol{s})$ and where the expected mean square error between the values of $U(\boldsymbol{s})$ and $\hat{U}(\boldsymbol{s})$ is minimised amongst all such linear combinations. What this means is that if we want to estimate a value of $U(\boldsymbol{s})$, then a 'best' guess would be to estimate a value of $\hat{U}(\boldsymbol{s})$ instead. We can do this in an obvious way. To obtain $\hat{u}(\boldsymbol{s})$, an estimated value of the random variable $\hat{U}(\boldsymbol{s})$, we simply replace $\boldsymbol{U}$ in our above expression for $\hat{U}(\boldsymbol{s})$ with our vector of observations $\boldsymbol{u}$, the vector of observed residuals u_i.

We complete the whole prediction process by adding our estimate $\hat{u}(\boldsymbol{s})$ to the known trend $\mu(\boldsymbol{s})$, to obtain our final prediction, $\hat{y}(\boldsymbol{s})$. Once we have obtained this prediction σ_e provides us with a standard error for prediction and hence a confidence interval. For example, if we make the typical assumption that prediction error is approximately normally distributed, then a 95% confidence interval for our prediction would be $\hat{y}(\boldsymbol{s}) \pm 1.96\sigma_e$. Note that since $\mu(\boldsymbol{s})$ is assumed known and subject to no prediction error, the standard error of prediction associated with $\hat{y}(\boldsymbol{s})$ is simply that associated with $\hat{u}(\boldsymbol{s})$.

This, then, is the technique of *simple kriging*. We can repeat the process for any point $\boldsymbol{s}$ where a prediction is required. Notice that for each new point $\boldsymbol{s}$ we require to calculate a new set of weights; but the only element in the calculation that changes is the vector $\boldsymbol{c}(\boldsymbol{s})$. The intensive computation involved is inversion of the $(n \times n)$ matrix $\boldsymbol{C}$ and this only has to be done once, since it is not dependent on the prediction point $\boldsymbol{s}$.

Before going any further, let us illustrate what all this means, with reference to an artificial, worked example. Consider the particular configuration of sites shown in Figure 5.19, with the site to be predicted marked as $\boldsymbol{s}$ and the others identified by the integers 1 to 5. We assume that the process, $Y(\boldsymbol{s})$, to be predicted has no trend but just a known constant mean value of 160.37. We can remove this known mean from the original sample observations y_i to give residuals u_i. The sample observations and residuals are as follows:

$$
\begin{array}{cc}
\textit{Sites} & \boldsymbol{y} \\
s_1 & \\
s_2 & \\
s_3 & \\
s_4 & \\
s_5 &
\end{array}
\begin{pmatrix} 122.0 \\ 183.0 \\ 148.0 \\ 160.0 \\ 176.0 \end{pmatrix}
\qquad
\begin{array}{c}
\textit{Sites} \\
s_1 \\
s_2 \\
s_3 \\
s_4 \\
s_5
\end{array}
\quad \boldsymbol{u} =
\begin{pmatrix} -38.37 \\ 22.63 \\ -12.37 \\ -0.37 \\ 15.63 \end{pmatrix}
$$

We also assume that we have been provided with a known covariance structure which happens to be isotropic and corresponds to an exponential variogram model having a 'sill' of 20.0 and a 'range' of 100 units and no nugget effect. From this it follows that the corresponding covariogram is:

$$C(h) = 20\mathrm{e}^{-3h/100}$$

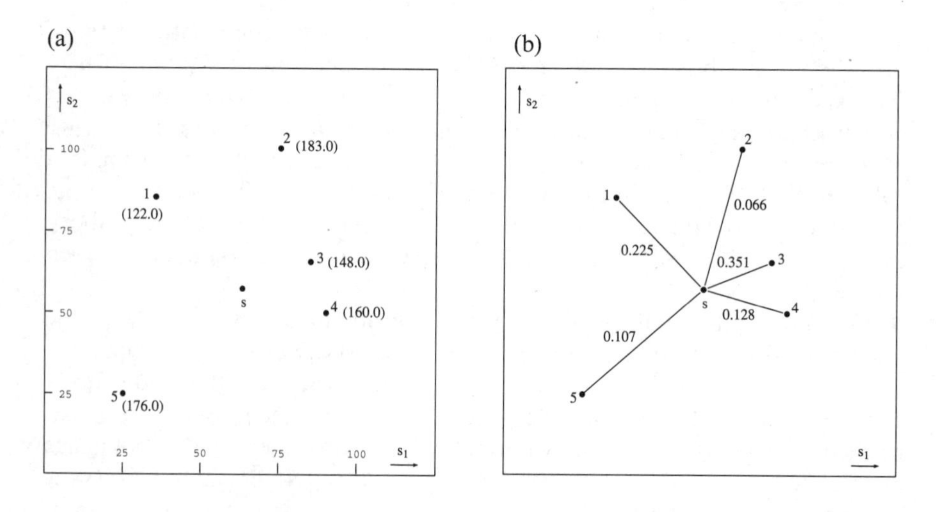

Fig. 5.19 An illustration of simple kriging

This covariance function contains all the information we need to obtain the estimated simple kriging weights $\lambda_i(\boldsymbol{s})$, $(i = 1, \ldots, 5)$. We can measure the distances between all sample sites, and between the sample sites and the location $\boldsymbol{s}$ and 'plug' these, as values of h, into the covariogram to derive the elements of the matrix $\boldsymbol{C}$ and the vector $\boldsymbol{c}(\boldsymbol{s})$ required in simple kriging. The results are shown below.

$$
\begin{array}{c}
\boldsymbol{C} \\
\begin{array}{cccccc}
\textit{Sites} & s_1 & s_2 & s_3 & s_4 & s_5 \\
s_1 & 20.00 & 4.571 & 3.970 & 2.828 & 3.228 \\
s_2 & 4.571 & 20.00 & 5.970 & 3.739 & 1.086 \\
s_3 & 3.970 & 5.970 & 20.00 & 12.45 & 2.300 \\
s_4 & 2.828 & 3.739 & 12.45 & 20.00 & 2.479 \\
s_5 & 3.228 & 1.086 & 2.300 & 2.479 & 20.00
\end{array}
\end{array}
\qquad
\begin{array}{c}
\boldsymbol{c}(\boldsymbol{s}) \\
\begin{array}{cc}
\textit{Sites} & s \\
s_1 & 6.895 \\
s_2 & 5.032 \\
s_3 & 10.15 \\
s_4 & 8.083 \\
s_5 & 4.079
\end{array}
\end{array}
$$

Notice that since the assumed form of the covariance function declines monotonically with distance, the elements in $\boldsymbol{C}$ are higher for sample sites that are close together, and lower for sites further away from each other. Also in $\boldsymbol{c}(\boldsymbol{s})$ those sites (such as 3 and 4) that are closest to $\boldsymbol{s}$ have the largest covariances with the site to be predicted. This does not necessarily mean that the simple kriging weights will also decline monotonically with distance, as we shall see.

Inverting $\boldsymbol{C}$ and multiplying by $\boldsymbol{c}(\boldsymbol{s})$ gives the required weights, shown below and in Figure 5.19b.

$$
\begin{array}{cc}
\textit{Sites} & \lambda(\boldsymbol{s}) \\
s_1 & 0.225 \\
s_2 & 0.066 \\
s_3 & 0.351 \\
s_4 & 0.128 \\
s_5 & 0.107
\end{array}
$$

Note that these simple kriging weights do not necessarily sum to unity, a point which we will return to later when we discuss kriging methods other than simple kriging. Note also, in passing, that the weights 'account' for possible 'redundancy' in the distribution of sample sites. Intuitively, we would not want those sample sites that are highly clustered to contribute, individually, the same amount of 'information' to the prediction as a site in a sparsely sampled area. The simple kriging weights at any point $\boldsymbol{s}$ automatically provide for this 'downweighting' of information carried by clustered sites. As a result, while sites 3 and 4 are both close to s the weights are by no means similar. Indeed, the weight for site 5, which is furthest from $\boldsymbol{s}$, is only slightly smaller than that contributed by site 4.

Our estimate $\hat{u}(\boldsymbol{s})$ is now:

$$\begin{aligned}\hat{u}(\boldsymbol{s}) &= 0.225 \times -38.37 + 0.066 \times 22.63 + 0.351 \times -12.37 \\ &\quad + 0.128 \times -0.37 + 0.107 \times 15.63 \\ &\simeq -9.86\end{aligned}$$

Hence our final predicted value is $\hat{y}(\boldsymbol{s}) = 160.37 - 9.86 \simeq 150.5$, when we add back the known mean.

The corresponding kriging variance is derived from

$$\begin{aligned}\sigma_e^2 &= \sigma^2 - \boldsymbol{c}^T(\boldsymbol{s})\boldsymbol{C}^{-1}\boldsymbol{c}(\boldsymbol{s}) \\ &= 20.0 - 6.92 \\ &= 13.08\end{aligned}$$

leading to an approximate 95% confidence interval of 150.5 ± 7.09.

We emphasise that this is a very simple worked example to illustrate how simple kriging works. Usually of course, we would have more than five sites and wish to predict more than one point! Notice that if $\boldsymbol{s}$ were further away from all the sample sites, $\boldsymbol{s}_i$, than the 'range' of the known variogram model, then $\boldsymbol{c}(\boldsymbol{s})$ would be identically zero and our estimate $\hat{y}(\boldsymbol{s})$ would simply be the known mean value $\mu(\boldsymbol{s})$. However, even if only one of the sample sites is within 'range' of $\boldsymbol{s}$ then more than one of the simple kriging weights may be non-zero—other sample sites get pulled into the prediction through a 'chaining' effect caused by their proximity to the site that is within the 'range'. Our final predicted value is somewhat less than the mean of the five observed values. This is intuitively sensible since the relatively high values in our sample data are further away from $\boldsymbol{s}$ than the lower values.

So much for simple kriging. Unfortunately it really isn't a very useful technique in practice. The reason is that it depends upon prior knowledge of both the trend in the process $\mu(\boldsymbol{s})$ and its covariance function $C()$. In reality both of these are likely to be unknown and have to be estimated from the data. In the next section we take up the modifications to the method which are necessary to enable this to be incorporated into the prediction.

5.5.6 General spatial prediction

So far, our discussion of spatial prediction has been restricted to simple kriging, which limits us to the case where the mean or trend of the process $Y(\boldsymbol{s})$, is known. We have not said anything yet about prediction when the first order or trend component, $\mu(\boldsymbol{s}) = \boldsymbol{x}^T(\boldsymbol{s})\boldsymbol{\beta}$, in our general model of $Y(\boldsymbol{s})$, has to be estimated from the data. We know how to derive this estimate, through generalised least squares, but how does this affect our technique for prediction from the model? We have also assumed that the covariance function of the process $C()$ is known. In practice this would have to be estimated from the data as well. Fortunately, we also know how to do this, using our techniques for fitting variogram models. We realise that this will involve us in assuming some form of stationarity and that if the first order component is anything other than constant over $\mathcal{R}$, we will need to estimate it and remove it from the original observations, working with the residuals to estimate the variogram. We have suggested a possible iterative scheme which may be of use in that case. How can we include this process into the prediction method?

From a practical point of view, all we have observed is $\boldsymbol{y}$, an $(n \times 1)$ column vector of observations, y_i, on the corresponding vector of random variables, $Y(\boldsymbol{s}_i)$, at each of the n sample sites, $\boldsymbol{s}_i$. We will now try to put together all we know about variogram modelling, trend surface estimation and simple kriging to give us an overall strategy for spatial prediction.

1. We first fit a suitable trend surface model:

$$Y(\boldsymbol{s}) = \boldsymbol{x}^T(\boldsymbol{s})\boldsymbol{\beta} + \epsilon(\boldsymbol{s})$$

   by ordinary least squares so obtaining $\hat{\boldsymbol{\beta}}$ and residuals, $\boldsymbol{y} - \boldsymbol{x}^T(\boldsymbol{s}_i)\hat{\boldsymbol{\beta}}$.
2. We then estimate the sample variogram from the current residuals and fit a suitable model to this, so obtaining a corresponding covariogram model $\hat{C}()$, from which we may derive the matrix $\hat{\boldsymbol{C}}$ having elements $\hat{C}(\boldsymbol{s}_i, \boldsymbol{s}_j)$.
3. We then estimate the trend surface model:

$$Y(\boldsymbol{s}) = \boldsymbol{x}^T(\boldsymbol{s})\boldsymbol{\beta} + U(\boldsymbol{s})$$

   by generalised least squares using the estimated covariance matrix $\hat{\boldsymbol{C}}$, obtaining a revised $\hat{\boldsymbol{\beta}}$ and revised residuals $\hat{\boldsymbol{u}} = \boldsymbol{y} - \boldsymbol{x}^T(\boldsymbol{s}_i)\hat{\boldsymbol{\beta}}$.
4. We iterate steps 2 and 3 if felt necessary until reasonable stability is obtained in $\hat{\boldsymbol{\beta}}$ and $\hat{C}()$; whereupon we derive final residuals, $\hat{\boldsymbol{u}}$, calculate a final estimate of $\hat{\boldsymbol{C}}$ and then go on to invert this matrix, obtaining $\hat{\boldsymbol{C}}^{-1}$.
5. For any location $\boldsymbol{s}$, at which a prediction is required, we first form $\hat{\boldsymbol{c}}(\boldsymbol{s})$, the estimated column vector of covariances, $\hat{C}(\boldsymbol{s}, \boldsymbol{s}_i)$, between location $\boldsymbol{s}$ and the n fixed sample locations $\boldsymbol{s}_i$. We then predict $Y(\boldsymbol{s})$ by analogy with simple kriging as:

$$\begin{aligned}\hat{y}(\boldsymbol{s}) &= \boldsymbol{x}^T(\boldsymbol{s})\hat{\boldsymbol{\beta}} + \hat{u}(\boldsymbol{s}) \\ &= \boldsymbol{x}^T(\boldsymbol{s})\hat{\boldsymbol{\beta}} + \hat{\boldsymbol{c}}^T(\boldsymbol{s})\hat{\boldsymbol{C}}^{-1}\hat{\boldsymbol{u}}\end{aligned}$$

6. We calculate the mean square prediction error as

$$\begin{aligned}\sigma_e^2 = &\left\{\left(\boldsymbol{x}(\boldsymbol{s}) - \boldsymbol{X}^T\hat{\boldsymbol{C}}^{-1}\hat{\boldsymbol{c}}(\boldsymbol{s})\right)^T\left(\boldsymbol{X}^T\hat{\boldsymbol{C}}^{-1}\boldsymbol{X}\right)^{-1}\left(\boldsymbol{x}(\boldsymbol{s}) - \boldsymbol{X}^T\hat{\boldsymbol{C}}^{-1}\hat{\boldsymbol{c}}(\boldsymbol{s})\right)\right\} \\ &+ \left\{\hat{\sigma^2} - \hat{\boldsymbol{c}}^T(\boldsymbol{s})\hat{\boldsymbol{C}}^{-1}\hat{\boldsymbol{c}}(\boldsymbol{s})\right\}\end{aligned}$$

   where $\hat{\sigma}^2 = \hat{C}(\boldsymbol{s}, \boldsymbol{s})$. The second term in the above expression is just the mean square prediction error given earlier for simple kriging. The first term is an extra component which arises because now the trend surface has also been estimated from the data. We do not give details of the derivation of this result here.

7. Steps 5 and 6 may be repeated for as many locations s as are required.

The above strategy provides a model-based approach to prediction which allows for the explicit identification and estimation of first order trend $\mu(s)$ in $Y(s)$ as well as the incorporation of second order effects. Notice that the whole process is a generalisation of our original ordinary least squares regression model. If the variogram model consists of pure 'nugget' effect, so that $\hat{C}()$ is zero everywhere except at the origin where it has the value $\hat{\sigma}^2$, then the prediction becomes simply $\hat{y}(s) = x^T(s)\hat{\beta}$, and

$$\sigma_e^2 = \hat{\sigma}^2\left(x^T(s)\left(X^TX\right)^{-1}x(s) + 1\right)$$

which is the usual mean square prediction error for ordinary least squares multiple regression.

Before going further it is perhaps instructive to return to the simple numerical example presented earlier in relation to simple kriging and apply the more general method. We use again the five sites shown in Figure 5.19 and consider prediction at the indicated site s. Previously we assumed that the process to be predicted, $Y(s)$, had just a known constant mean value. In other words $\mu(s) = \mu$. We now assume that there is still a constant mean value μ but that this needs to be estimated. This corresponds to the case in our above trend model where $\mu(s) = \mu = \beta_1$ and $x(s)$ is simply the scalar 1.

We continue to assume that we have been provided with same known isotropic covariogram as before, so that matrix C^{-1} is as previously calculated. We have then effectively skipped the first three steps in our general prediction approach. The final generalised least squares estimate of μ in step 4 is then calculated from $\hat{\mu} = \left(X^TC^{-1}X\right)^{-1}X^TC^{-1}y$, where in this case X is just the (5×1) column vector $(1, 1, 1, 1, 1)^T$ and $y = (122.0, 183.0, 148.0, 160.0, 176.0)^T$. This gives the result $\hat{\mu} = 160.369$, the same as the originally quoted known mean (we admitted this was an artificial example!). Hence the final residuals would be those that we quoted before. This implies that the prediction step 5 would be identical to that presented before for simple kriging and the final prediction would again be would be 150.5. However, in step 6 the earlier mean square prediction error is now adjusted slightly to reflect the estimation of μ as:

$$\sigma_e^2 = \left\{\left(1 - X^T\hat{C}^{-1}\hat{c}(s)\right)^T\left(X^T\hat{C}^{-1}X\right)^{-1}\left(1 - X^T\hat{C}^{-1}\hat{c}(s)\right)\right\} + \{13.08\}$$

which gives $\sigma_e^2 = 0.107 + 13.08 = 13.19$ and leads to an approximate 95% confidence interval of 150.5 ± 7.12—in this case a very minor change to the earlier result.

Returning to the general discussion, we need to look at some problems associated with our overall suggested scheme. Firstly, although the covariance function $C()$ is estimated from the data, no allowance for this is made in σ_e^2. The above strategy effectively assumes that $C()$ is estimated without error. Studies have shown that in practice different choices of variogram model, or use of different 'range' and 'sill' parameters within a family of models, can have quite significant effects on predictions from the model. Choice of different trend surface models will also have a considerable influence on predictions. For example, whether one takes $\mu(\mathbf{s})$ to be constant over $\mathcal{R}$ (corresponding to the choice $\mathbf{x}(\mathbf{s}) = 1$), or linear $(\mathbf{x}(\mathbf{s}) = (1, s_1, s_2)^T)$, or quadratic $(\mathbf{x}(\mathbf{s}) = (1, s_1, s_2, s_1^2, s_2^2, s_1 s_2)^T)$, can dramatically alter predictions. Choice of a suitable trend component is not straightforward, since it is a question of how much of the observed variation in the process can be judged 'trend' and how much arises from second order effects. In practice this is always an *ad hoc* division.

One way to evaluate such issues is by *cross validation*. Each sample data point y_i is deleted in turn and its value predicted from the rest of the data, using the chosen variogram model and the chosen form of the trend component. This results in an observed set of n prediction errors between the predicted and true values at each sample site. These prediction errors can then be studied in order to evaluate the overall prediction process and possibly to identify necessary adjustments to either the variogram model or the form of the trend.

A further problem concerns the computational difficulty involved in inversion of the $(n \times n)$ matrix $\mathbf{C}$. If n is large then this can cause problems in both computation time and required computer memory. In fact, since this is a symmetric positive definite matrix, very efficient inversion algorithms may be applied and continuing computer developments mean that memory is increasingly less and less of a problem. Nevertheless, the relative difficulty of inverting a large matrix has led to the practice of applying the above scheme to sample sites within moving 'local search neighbourhoods' about the prediction point $\mathbf{s}$, rather than to the full data set of the n sample sites. In practice, such search neighbourhoods would normally be defined on the basis of a fixed number of sample sites to include, or possibly a fixed radius about the point $\mathbf{s}$. For each point, $\mathbf{s}$, to be predicted a local matrix $\mathbf{C}$ and vector $\mathbf{c}(\mathbf{s})$ are formed from just the sample sites in the local search neighbourhood. Prediction then proceeds as if these were the only samples observed, except that a globally estimated variogram model is used, since it would of course be impossible to estimate this from the small local neighbourhoods. Clearly each prediction requires a separate inversion of the local $\mathbf{C}$ matrix, whereas using the full data set only requires this inversion to be performed once for all predictions. However, the trade off is that inverting many small matrices is considerably easier and faster than inverting one large matrix.

The use of local search neighbourhoods involves more than just computational considerations. It really changes the whole complexion of the approach outlined above. There is no longer any real attempt to model global trend and second order effects, since now trend is estimated locally within the search window and no attempt is made to construct a global picture of first order effects. The emphasis is purely on local prediction, rather than description and explanation of the process $Y(\boldsymbol{s})$. This brings us to a slightly different, although equivalent, way of looking at the prediction strategy outlined above, which we take up in the next section.

5.5.7 Ordinary and universal kriging

The general strategy for spatial prediction outlined in the previous section is mathematically equivalent, but developed somewhat differently in approach and emphasis, to the techniques of *ordinary kriging* and *universal kriging*, which are commonly regarded as the natural extensions of the simple kriging method that we discussed in an earlier section. Both of these techniques are primarily focused on optimal local spatial prediction and as such are less concerned with specific identification and isolation of any global first order component of spatial variation, which is made explicit in the general model-based approach to prediction which we have described in the previous section. Instead, first order effects are simultaneously and implicitly estimated as part of the prediction process and are not naturally made explicit in the results. In the case where these techniques are applied globally to the full set of the n sample sites, estimates of global first order effects may be recovered indirectly from the computations. When the techniques are applied, as they often are, in moving local 'search neighbourhoods' of the prediction points, a 'trend' is estimated not only implicitly but also solely within the local search window and a global picture of first order effects cannot be sensibly recovered. This is not necessarily a disadvantage of the techniques—similar comments were made in relation to the use of local search windows in our previous section, and as has been said, ordinary and universal kriging are mathematically equivalent to different cases of the general approach given there—it merely reflects their emphasis on prediction as the primary objective rather than estimation of an explicitly parameterised model. In the final section of this chapter we describe these techniques and relate them to the approach of the previous section.

Ordinary kriging is mathematically equivalent to the approach in the previous section when it is assumed that the process to be predicted, $Y(\boldsymbol{s})$, has a constant mean value. In other words $\mu(\boldsymbol{s}) = \mu$. This corresponds to our previous trend model when $\boldsymbol{x}(\boldsymbol{s})$ is simply the scalar 1 and $\beta_1 = \mu$. However, rather than estimate μ by generalised least squares and then predict the zero mean residual process by simple kriging, adding back $\hat{\mu}$ at the final stage, ordinary kriging instead forms the prediction, $\hat{y}(\boldsymbol{s})$, of the original process $Y(\boldsymbol{s})$, at $\boldsymbol{s}$, in 'one step', by using a weighted linear combination of the observed values y_i at the sample sites $\boldsymbol{s}_i$.

The derivation of the appropriate weights mirrors the argument used in relation to simple kriging, and we refer the reader back to that discussion for more detail on some of the steps involved. We start with the weighted linear combination of random variables:

$$\hat{Y}(s) = \sum_{i=1}^{n} \omega_i(s) Y(s_i)$$

We then choose values for the weights, $\omega_i(s)$, so that the mean value of $\hat{Y}(s)$ is constrained to be μ, the same as that of the random variable $Y(s)$ that we are trying to predict. Since it has been assumed that the mean of $Y(s)$ and each of $Y(s_i)$ are all μ and the weights are just constants, it is clear that the mean of $\hat{Y}(s)$ will also be μ provided that the sum of the n weights, $\omega_i(s)$, is unity. Subject to this constraint we then wish, as before, to minimise the expected mean square error between values of $Y(s)$ and values of $\hat{Y}(s)$. Bearing in mind that the random variables $Y(s)$ and $Y(s_i)$ all have mean μ, we can show by a very similar argument to that used for simple kriging that this expected mean square error is:

$$E\left(\left(\hat{Y}(s) - Y(s)\right)^2\right) = \omega^T(s)\boldsymbol{C}\omega(s) + \sigma^2 - 2\omega^T(s)\boldsymbol{c}(s)$$

where as before $\boldsymbol{C}$ is the $(n \times n)$ matrix of covariances, $C(s_i, s_j)$ between all possible pairs of the n sample sites, and $\boldsymbol{c}(s)$ is an $(n \times 1)$ column vector of covariances, $C(s, s_i)$, between the prediction point s and each of the n sample sites.

We wish now to minimise this expression, subject to the sum of the weights being unity; that is, subject to the matrix constraint $\omega^T\mathbf{1} = 1$, where $\mathbf{1}$ is an $(n \times 1)$ column vector of 1s. To do this we use the standard approach of introducing a Lagrange multiplier, say $v(s)$, and consider the modified expression:

$$\omega^T(s)\boldsymbol{C}\omega(s) + \sigma^2 - 2\omega^T(s)\boldsymbol{c}(s) + 2(\omega^T\mathbf{1} - 1)v(s)$$

Differentiating with respect to both $v(s)$ and $\omega(s)$ in order to minimise leads to the two simultaneous equations:

$$\begin{aligned} \omega^T\mathbf{1} &= 1 \\ \boldsymbol{C}\omega(s) + \mathbf{1}v(s) &= \boldsymbol{c}(s) \end{aligned}$$

which can be re-expressed as a single equation by using an augmented matrix $\boldsymbol{C}_+$ and augmented vectors $\omega_+(s)$ and $\boldsymbol{c}_+(s)$, as:

$$\begin{array}{ccc} \boldsymbol{C}_+ & \omega_+(s) = & \boldsymbol{c}_+(s) \end{array}$$

$$\begin{pmatrix} C(s_1, s_1) & \cdots & C(s_1, s_n) & 1 \\ \vdots & \ddots & \vdots & \vdots \\ C(s_n, s_1) & \cdots & C(s_n, s_n) & 1 \\ 1 & \cdots & 1 & 0 \end{pmatrix} \begin{pmatrix} \omega_1(s) \\ \vdots \\ \omega_n(s) \\ v(s) \end{pmatrix} = \begin{pmatrix} C(s, s_1) \\ \vdots \\ C(s, s_n) \\ 1 \end{pmatrix}$$

This leads to the solution:

$$\omega_+(s) = C_+^{-1} c_+(s)$$

Furthermore, with the same notation the minimised expected mean square error corresponding to this choice of weights (the mean square prediction error, or kriging variance) is then:

$$\sigma_e^2 = \sigma^2 - c_+^T(s) C_+^{-1} c_+(s)$$

These are exactly the same results obtained earlier for simple kriging, except that they now involve the augmented matrix C_+ and augmented vectors $\omega_+(s)$ and $c_+(s)$.

To obtain our prediction, $\hat{y}(s)$, we solve the above equation for $\omega_+(s)$, extract from this the vector $\omega(s)$, and then $\hat{y}(s) = \omega^T(s)y$, where y is the vector of original observations y_i. Once we have obtained this prediction σ_e provides us with a standard error for prediction and hence a confidence interval as in the case of simple kriging.

This, then, is ordinary kriging. Clearly, we can repeat the process for any point s where a prediction is required. The prediction results, as we have said before, will be identical to those obtained using the approach of the previous section, in the case where $x(s) = 1$ and therefore $\mu(s) = \mu = \beta_1$. The ordinary kriging weights, $\omega(s)$, are of course different from those obtained using the earlier approach, $\lambda(s)$, since the former sum to unity and are applied to y_i rather than the residuals $\hat{u}_i = y_i - \hat{\mu}$. Ordinary kriging delivers the prediction in one step on the basis of a single, conceptually simple, matrix equation, avoiding explicit identification of the estimated global mean $\hat{\mu}$.

We attempt to convince the reader of these points by reference back to the simple numerical example we used earlier, with just five sample sites as shown in Figure 5.19. We assume, as we did for this example in the previous section, that $\mu(s) = \mu$, where μ is unknown, and use the same variogram model as before. We form the augmented matrix C_+ from the covariance matrix of the five sites, together with an extra row and column of 1s except for the bottom right, corner element, which is set to zero. The augmented vector $c_+(s)$ is formed from covariances between the prediction site s and each of the sample sites, with an extra row of 1.

Inverting C_+ and multiplying it by $c_+(s)$ gives the following ordinary kriging weights as the first five elements of the resulting augmented vector $\omega_+(s)$.

Sites	$\omega(s)$
s_1	0.251
s_2	0.094
s_3	0.363
s_4	0.151
s_5	0.141

Note that these ordinary kriging weights differ from the simple kriging weights obtained earlier and sum to unity. Our estimate $\hat{y}(\boldsymbol{s})$ is now formed from these weights and the original observations y_i as:

$$\begin{aligned}\hat{y}(\boldsymbol{s}) &= 0.251 \times 122.0 + 0.094 \times 183.0 + 0.363 \times 148.0 \\ &\quad + 0.151 \times 160.0 + 0.141 \times 176.0 \\ &\simeq 150.5\end{aligned}$$

a result, with rounding error, identical to that obtained in the previous section, where μ was estimated separately by generalised least squares. The kriging variance follows from our above expression as:

$$\begin{aligned}\sigma_e^2 &= 20.0 - 6.81 \\ &= 13.19\end{aligned}$$

again the same result as in the previous section.

Returning to our general discussion of ordinary kriging, in practice, as with our previous techniques, $C()$ would be unknown and have to be derived from a globally estimated variogram model. Note that this variogram may safely be estimated from the original observations y_i, since ordinary kriging assumes that $\mu(\boldsymbol{s})$ is constant over $\mathcal{R}$. Therefore, any trend does not have to be estimated and removed from the observations before a variogram can be estimated. This avoids any bias that may be introduced by estimating variogram models on the basis of residuals. Also, if required, the ordinary kriging equations can be re-written to use the variogram directly, rather than the covariogram, since under our assumptions the two are simply related by $C(h) = \sigma^2 - \gamma(h)$. Thus in terms of the variogram the ordinary kriging equations become:

$$\omega_+(\boldsymbol{s}) = \Gamma_+^{-1}\gamma_+(\boldsymbol{s})$$

and

$$\sigma_e^2 = \gamma_+^T(\boldsymbol{s})\Gamma_+^{-1}\gamma_+(\boldsymbol{s})$$

with an obvious extension to our earlier notation, where nothing changes except that covariances $C()$ are replaced with semi-variances $\gamma()$.

These issues become particularly significant when the technique is applied, as it often is, within moving local search neighbourhoods of each prediction point. For each point, $\boldsymbol{s}$, a local matrix Γ_+ and vector $\gamma_+(\boldsymbol{s})$ are formed from just the sample sites in the local search neighbourhood and the above equations are then solved with these to form the prediction and its standard error. This requires numerous inversions of Γ_+ as the search window moves with each new prediction point, but on each occasion only a small matrix is involved, rather than the $((n+1) \times (n+1))$ matrix that is involved when all n sample sites are used together. Note also that when ordinary kriging is used in local search neighbourhoods, $\mu(\boldsymbol{s})$ is only assumed constant within the search neighbourhood and not over the whole of $\mathcal{R}$. This sort of 'local constant mean' may often even be a viable assumption in some data sets which exhibit global trends, and obviate the need to resort to explicit trend estimation, particularly given our earlier comments that ordinary kriging makes no attempt to explicitly estimate global effects in the first place. A possible drawback is that a globally estimated variogram model must of course be used to derive the local matrix and vector elements, since it would be impossible to estimate this within local neighbourhoods. The estimate of this global variogram would be distorted if global trends were not explicitly estimated and removed before variogram modelling. However, it is only semi-variances at small distances which are used in applying ordinary kriging in local neighbourhoods and it could be argued that an estimate of a global variogram using data with a global trend would not be distorted at small distances. At this point the reader may feel that too many *ad hoc* modifications are being introduced into what was originally a fairly rigorous statistical method based on an explicit model of the spatial process concerned. We leave this judgement to the reader and simply remind them that ultimately spatial prediction will always be something of an 'art' involving judgement and intuition as well as statistical methods.

Leaving aside the preceding discussion of its use within local neighbourhoods, ordinary kriging basically assumes a constant mean value, $\mu(\mathbf{s}) = \mu$, for the process $Y(\mathbf{s})$ over $\mathcal{R}$. The natural extension to ordinary kriging to accommodate a global trend is referred to as *universal kriging*. Universal kriging is mathematically equivalent to the general case discussed in the previous section where a first order trend component, $\mu(\mathbf{s}) = \mathbf{x}^T\boldsymbol{\beta}$, is included. Recall that $\mathbf{x}(\mathbf{s})$ is a $(p \times 1)$ vector, $(x_1(\mathbf{s}), \ldots, x_p(\mathbf{s}))^T$, of values of explanatory variables at $\mathbf{s}$, usually involving the spatial coordinates of $\mathbf{s}$ and possibly functions of these such as their powers and cross products. Like ordinary kriging, universal kriging forms the prediction, $\hat{y}(\mathbf{s})$, directly in 'one step', by using a weighted linear combination of the observed values y_i at the sample sites $\mathbf{s}_i$. The trend is simultaneously and implicitly estimated as part of the prediction process and not made explicit in the results. We do not go into a detailed derivation of the method here. It largely follows the steps that we have already laid out for both simple and ordinary kriging. The universal kriging equations that result are a direct generalisation of those for ordinary kriging, except that now p Lagrange multipliers are involved and the matrix $\mathbf{C}$ and vector $\mathbf{c}(\mathbf{s})$ are augmented by p rows and columns as follows:

$$\underset{\mathbf{C}_+}{\begin{pmatrix} C(\mathbf{s}_1,\mathbf{s}_1) & \cdots & C(\mathbf{s}_1,\mathbf{s}_n) & x_1(\mathbf{s}_1) & \cdots & x_p(\mathbf{s}_1) \\ \vdots & \ddots & \vdots & \vdots & \ddots & \vdots \\ C(\mathbf{s}_n,\mathbf{s}_1) & \cdots & C(\mathbf{s}_n,\mathbf{s}_n) & x_1(\mathbf{s}_n) & \cdots & x_p(\mathbf{s}_n) \\ x_1(\mathbf{s}_1) & \cdots & x_1(\mathbf{s}_n) & 0 & \cdots & 0 \\ \vdots & \ddots & \vdots & \vdots & \ddots & \vdots \\ x_p(\mathbf{s}_1) & \cdots & x_p(\mathbf{s}_n) & 0 & \cdots & 0 \end{pmatrix}} \underset{\boldsymbol{\omega}_+(\mathbf{s})}{\begin{pmatrix} \omega_1(\mathbf{s}) \\ \vdots \\ \omega_n(\mathbf{s}) \\ \nu_1(\mathbf{s}) \\ \vdots \\ \nu_p(\mathbf{s}) \end{pmatrix}} = \underset{\mathbf{c}_+(\mathbf{s})}{\begin{pmatrix} C(\mathbf{s},\mathbf{s}_1) \\ \vdots \\ C(\mathbf{s},\mathbf{s}_n) \\ x_1(\mathbf{s}) \\ \vdots \\ x_p(\mathbf{s}) \end{pmatrix}}$$

This leads to the solution

$$\omega_+(\boldsymbol{s}) = \boldsymbol{C}_+^{-1}\boldsymbol{c}_+(\boldsymbol{s})$$

and mean square prediction error:

$$\sigma_e^2 = \sigma^2 - \boldsymbol{c}_+^T(\boldsymbol{s})\boldsymbol{C}_+^{-1}\boldsymbol{c}_+(\boldsymbol{s})$$

exactly as for ordinary kriging. Once $\omega_+(\boldsymbol{s})$ is obtained, the vector $\omega(\boldsymbol{s})$ is extracted and then the prediction is $\hat{y}(\boldsymbol{s}) = \omega^T(\boldsymbol{s})\boldsymbol{y}$, where $\boldsymbol{y}$ is the vector of original observations y_i. Again σ_e provides a standard error and confidence interval for the prediction.

The prediction results of applying universal kriging to the full set of n sample sites will, as we have said before, be identical to those obtained using the approach of the previous section where the trend $\boldsymbol{x}^T(\boldsymbol{s})\boldsymbol{\beta}$ is explicitly estimated by generalised least squares. However, the arguments for using universal kriging as opposed to generalised least squares are less convincing than they were in the equivalent comparison of ordinary kriging with the generalised least squares approach. Universal kriging calculates the trend implicitly and automatically, whereas one would really be interested in studying an explicit estimate of the trend and judging whether it makes sense in the light of what is known about the phenomenon under study. In this case it is generally better to proceed by the generalised least squares approach, estimating trend explicitly and making a conscious choice to remove it from the observed sample values, working from then on with the residuals. This point is reinforced by the fact that one would need in any case to make some estimate of trend, in order to obtain residuals for variogram modelling. The argument for applying universal kriging within local neighbourhoods is even less convincing. If it is really believed that there is a trend within the local neighbourhood, then why not just apply ordinary kriging within a smaller neighbourhood, within which its effect can be ignored?

Having spent some considerable time discussing various kriging methods, let us now attempt to apply some of these to our real data sets. We start by trying to derive a map of the distribution of the logarithm of nickel concentration for a sub-region of the larger study area over which samples were taken. This sub-region is defined by the rectangle whose south-west corner is given by map reference (612700, 546050) and whose north-east corner is (640500, 605000). We do this using ordinary kriging, within local neighbourhoods of each prediction point. The variogram model used was that derived earlier and illustrated in Figure 5.17. The local neighbourhoods used in the kriging were defined simply in terms of the ten sample observations nearest to each prediction point. The results in the form of a map of kriged values are shown in Figure 5.20, with accompanying variances also shown.

There is a broad area of relatively low estimates of nickel concentration to the west of the study region, rising to a local peak further north. Values are of the order of 2.2 to 4.6. The variances are of the order of 0.40–0.45, but extend

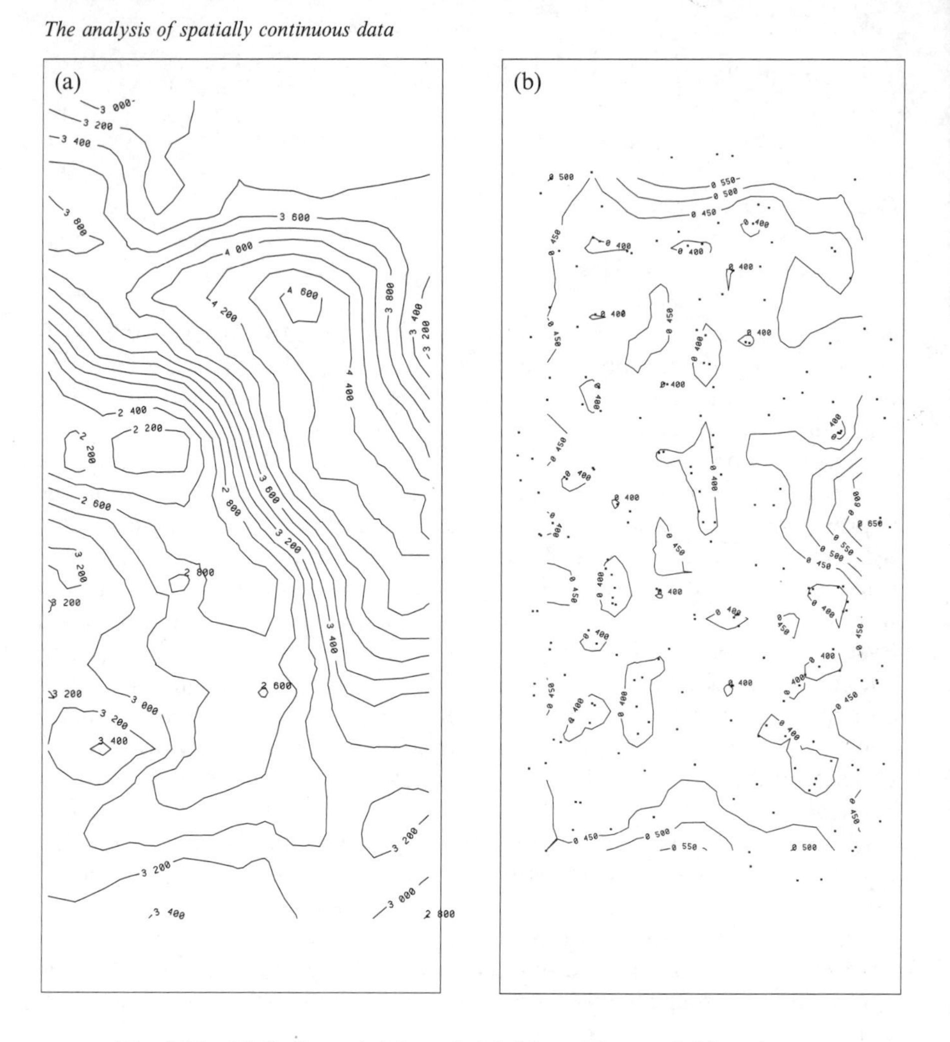

Fig. 5.20 (a) Ordinary kriging of nickel logarithms and (b) variances. The dots on (b) show the sample sites.

to 0.50 at the edge of the region being predicted. Comparison of the variance map with the distribution of sample sites (also shown on Figure 5.20b) shows how the variances are lower near sampled points. Examining the 'peak' of the surface, the kriged estimate is about 4.6. The variance here is about 0.4, giving a standard error of about 0.63. There is thus a 95 per cent chance that the true value is 4.6 ± 1.26, that is, within the range 3.34–5.86.

What results do we obtain from kriging other data sets? Universal kriging of the 1973 Venice groundwater data yields the results shown in Figure 5.21; the sample variogram is shown in Figure 5.22. A trend surface was first fitted to the data and the variogram constructed from the residuals. The variogram is very nearly linear up to the range of about 10 km. Further details of the exact

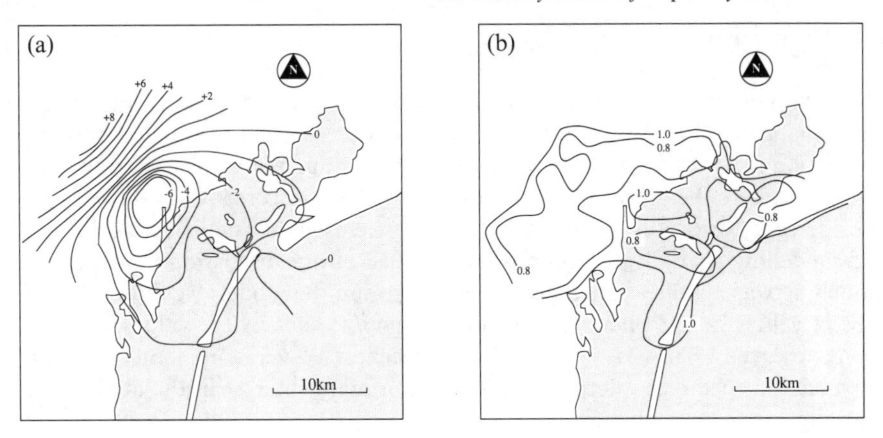

Fig. 5.21 (a) Universal kriging of groundwater and (b) standard errors

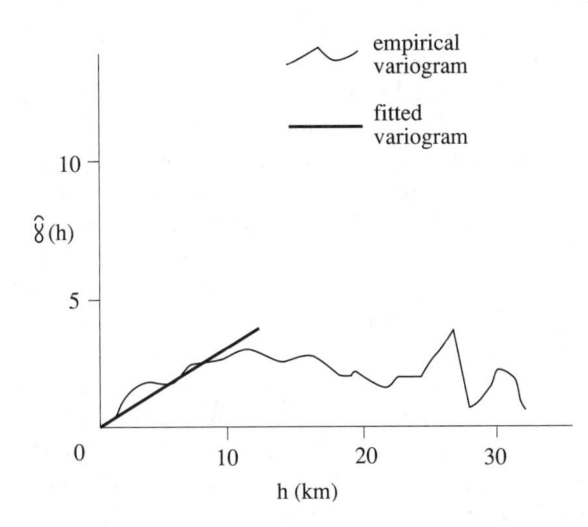

Fig. 5.22 Sample variogram for Venice groundwater

method used are given in the paper which originally reported these results, referenced at the end of this chapter.

The big cone of depression over the industrial area is highlighted on the kriged map, while the map of standard errors shows that the predictions become much less accurate where sampling is sparse (most obviously, at the edges of the study area). Such a map is of use in identifying locations where new measurements (in this case, boreholes) are required.

5.6 Summary

This chapter has turned out to be rather lengthy and at times quite mathematically complex. We would not wish our reader to become confused by detail and lose the main thread of our discussion. We will therefore try to summarise here the gist of the methods that we have covered. Overall, we have considered a basic 'core' set of tools that may be used in visualising, exploring, and modelling spatially continuous data. Much of our discussion in respect of modelling was actually concerned with how to use the models we developed to predict values of the phenomenon under study at sites where values had not been observed. This is not surprising, since, as we commented in the introduction to the chapter, prediction is a primary interest in the analysis of spatially continuous phenomena.

In the early sections of the chapter we examined simple tools for visualising spatially continuous data and for exploring spatial variation and spatial structure in such data. We used our usual approach of seeing whether we could 'decompose' such spatial variation into two components: first order variation or a trend component; and second order effects. In the first instance exploratory techniques such as spatial moving averages, TIN interpolation and kernel smoothing were suggested as being valuable. For the exploration of second order effects we introduced the ideas of the sample variogram and covariogram. We then turned our attention to the modelling of these components. Our first suggestion for modelling first order variation, referred to as 'trend surface analysis', involved the use of the standard ordinary least squares multiple regression model widely employed in non-spatial data analysis. We then introduced a modification to this, referred to as generalised least squares, in order to allow for second order effects. In order to be able to fit such a model we needed to specify more precisely the form of the second order effects in our process. This led us into a discussion of the modelling of variograms and covariograms. Using these ideas we were then able to develop ways to fit the generalised least squares model to our observed data.

Having obtained a plausible modelling framework for spatially continuous data, we then turned our attention to ways of using such models for prediction purposes. We could predict first order effects in an obvious way directly from our model, but could we also find a way to predict local fluctuations from the trend, given that these are assumed to be correlated in the model? This led us to a discussion of simple kriging, where a zero mean spatially correlated process is predicted at a site, by using an optimal weighted linear combination of neighbouring observed values of that process. We found that we were able to combine the idea of simple kriging with the earlier generalised least squares model of trend, to produce a generally applicable prediction strategy for spatially continuous data.

We went on to describe how this suggested strategy was mathematically equivalent, under different conditions, to two direct generalisations of simple kriging known as ordinary kriging and universal kriging. The first applied in the case where the first order component was a constant mean that did not vary geographically, and the second in the more general situation when a trend applied. We found the ordinary kriging method to lead to a prediction in 'one step' and somewhat easier to work with in practice than the general strategy; particularly when applied to moving local neighbourhoods around the prediction points. Universal kriging also provided a 'one step' prediction, this time in the trend case, where, in the language of the geologist the mean 'drifts' over $\mathcal{R}$. However, we were less convinced that this was really an advantage and might even be a drawback to using the technique. There is a strong argument for preferring the trend to be made explicit, as in the generalised least squares model, so that a judgement can be made as to how appropriate it is for the particular phenomenon under study.

In the next chapter, we go on to discuss some further, more specialised, methods for the analysis of spatially continuous data. We consider some extensions of kriging. We also introduce a selection of methods directed to the analysis of multivariate measurements. This relates to a situation which we have not discussed at all in this chapter, where values of several variables have been recorded at each site and interest lies in studying pattern in these values taken together, rather than each in isolation.

5.7 Further reading

The original Californian rainfall data come from:

Taylor, P.J. (1980) A pedagogic application of multiple regression analysis: precipitation in California, *Geography*, 65, 203–12.

The data were re-analysed in:

Jones, K. (1984) Graphical methods for exploring relationships, in Bahrenberg, G., Fischer, M. and Nijkamp, P. (eds) *Recent Developments in Spatial Data Analysis*, Gower, Aldershot, 215–27.

The Sudan rainfall data were kindly provided by Mike Hulme of the Climatic Research Unit, University of East Anglia. See:

Hulme, M. (1990) The changing rainfall resources of Sudan, *Transactions, Institute of British Geographers*, 15, 21–34.

As mentioned in the text, the temperature data for England and Wales were taken from the *Monthly Weather Report*. A similar set of data are analysed in the book by Upton and Fingleton, referenced below.

The case study on groundwater levels in the Venice region is taken from the following paper, which reports results using kriging:

Gambolati, G. and **Volpi, G.** (1979) Groundwater contour mapping in Venice by stochastic interpolators 1: Theory, 2: Results, *Water Resources Research*, 15, 281-90 and 291–97.

The radon data for Lancashire were originally made available for the study reported in:

Vincent, P.J. and **Gatrell, A.C.** (1991) The spatial distribution of radon gas in Lancashire (UK): a kriging study, in *Proceedings, Second European Conference on Geographical Information Systems*, EGIS Foundation, Utrecht, 1179–86.

Data on levels of PCBs in soils near Pontypool are reported in the following publication, but were provided in machine-readable form courtesy of Andrew Lovett, School of Environmental Sciences, University of East Anglia. See:

Polychlorinated Biphenyls, Dioxins and Furans in the Pontypool Environment: The Panteg Monitoring Project, Final Report to the Welsh Office, Environmental Risk Assessment Unit, School of Environmental Sciences, University of East Anglia.

The geochemical data for north Vancouver Island were generously provided by Graeme Bonham-Carter of the Geological Survey of Canada. For a good introduction to the use of such data in general see:

Howarth, R.J. (1983) (ed) *Statistics and Data Analysis in Geochemical Prospecting*, Elsevier Scientific Publishing, Amsterdam.

Climate data for South America were extracted from:

Schwerdtfeger, W. (1976) *Climates of Central and South America*, World Survey of Climatology, Volume 12, Elsevier, Oxford.

Issues concerning the visualisation of spatially continuous data are discussed in standard cartographic texts. See, for example:

Dent, B. (1990) *Cartography: Thematic Map Design*, Wm C. Brown Publishers, Dubuque, Iowa.

Robinson, A.H., Sale, R.D., Morrison, J.L. and **Muehrcke, P.C.** (1984) *Elements of Cartography*, Fifth edition, John Wiley, Chichester.

Both books contain discussions of the class interval selection problem, though see also:

Monmonier, M. (1993) *Mapping it Out: Expository Cartography for the Humanities and Social Sciences*, University of Chicago Press, Chicago.

and the classic paper:

Evans, I.S. (1977) The selection of class intervals, *Transactions, Institute of British Geographers*, 2, 98–124.

The literature on non-spatial multiple regression and related methods is huge; a classic general text is:

Draper, N. R. and **Smith, H.** (1981) *Applied Regression Analysis*, Second edition, Wiley, New York.

Discussions of some of the other methods considered in this chapter may be found in a variety of sources. In order of increasing difficulty, try:

Davis, J.C. (1986) *Statistics and Data Analysis in Geology*, Second edition, John Wiley, Chichester.

Upton, G.J.G. and **Fingleton, B.** (1985) *Spatial Data Analysis by Example: Volume 1: Point Pattern and Quantitative Data*, John Wiley, Chichester.

Ripley, B.D. (1981) *Spatial Statistics*, John Wiley, Chichester. See, in particular, Chapter 4, upon which we have based our discussion of spatial prediction.

Cressie, N. (1991) *Statistics for Spatial Data*, John Wiley, Chichester, especially Part I: Geostatistical Data.

On practical kriging you should consult the excellent book by:

Isaaks, E.H. and **Srivastava, R.M.** (1989) *An Introduction to Applied Geostatistics*, Oxford University Press, Oxford.

and for methodology see the following clear expository accounts:

Oliver, M.A., Webster, R. and **Gerrard, J.** (1989) Geostatistics in physical geography. Part 1: theory, *Transactions, Institute of British Geographers*, 14, 259–69.

Webster, R. and **Oliver, M.A.** (1990) *Statistical Methods in Soil and Land Resource Survey*, Oxford University Press, Oxford, Chapters 12–14.

The use of kriging in a hydrological context is summarised in:

Bras, R.L. and **Rodriguez-Iturbe, I.** (1985) *Random Functions and Hydrology*, Addison-Wesley, Reading, Massachusetts, Chapter 7.

We have used version 6 of ARC/INFO to derive the empirical results concerning the estimation of semi-variance, the fitting of a model to the variogram, and kriging. Other software, for example the UNIRAS software package, might also be used; there are also special-purpose kriging packages available.

5.8 Computer exercises

Here are suggested exercises that you can try on ideas discussed in this chapter using INFO-MAP and our example data sets. These exercises have been referenced at appropriate points in the chapter by numbered symbols in the margin.

Exercise 5.1

For an example of using proportional symbols in mapping, open the file 'PCBs near Pontypool' and load the data. Select six equal-interval classes using `Scaling` from the `Map` menu. Then choose the option `Symbol map` in the `Map` menu and display the resulting map. Using `Open` from the `Overlay` menu add the industrial site described in the text. What does this tell you about the distribution of PCB concentrations?

Next, try the 'trimmed equal' class interval scheme and create a new symbol map. How does the impression of concentration differ from the first map?

Now try taking logarithms of the PCB values. You will need to calculate a new variable using `Calculate` from the `Data` menu with the formula:

```
logpcb = log({1})
```

Map the new variable using 'standard deviates' as the scaling option. Compare this map with that you obtained previously.

You could try out similar ideas on some of the other spatially continuous data sets, for example 'Radon in Lancashire', 'South American Climate' or 'Groundwater in Venice'. Note that you can change the maximum symbol size used in such maps via `Symbol type` from within the `Map` menu. You specify an integer value between 1 (smallest) and 10 (largest and the default); this can be used to reduce the amount of symbol overlap.

Exercise 5.2

Again using the 'PCBs near Pontypool' file, first calculate a new variable containing the logarithms of PCB concentrations as described above. Then use `Contour map` in the `Analysis` menu and choose the 'contour line' option to create a contour map for logarithm of PCB concentration based upon the TIN method described in the text. Add the industrial site overlay if you wish. Experiment with numbers of classes and class interval schemes using `Scaling` within the `Map` menu and, continuing with the contour map option, try to produce the most informative result.

To superimpose the original observations onto the contour map, draw the contour map, then switch `Superimpose` in the `Map` menu to 'on' and follow this with a `Dot map` of the logarithms of the PCB values. Remember to switch `Superimpose` back to 'off' afterwards, otherwise subsequent maps will continue to be 'superimposed'.

Try similar ideas for contour mapping the 'Sudan rainfall' or 'Groundwater in Venice' data sets.

Exercise 5.3

For an example of deriving a kernel estimate of mean value open the file 'Groundwater in Venice' and load the data. Obtain an estimate of mean value of groundwater level in 1973. Do this by use of `Kernel map` from the `Analysis` menu and the option 'mean value'. You will need to specify a bandwidth, by marking the start and end points of a line on the map, the length of which will be used as the bandwidth. You mark the start point by a 'double click' at some convenient point on the map and then 'drag' the line (hold down the left button and move the mouse) to an end point which is fixed with another 'double click'. This will produce a 'raster' image of the kernel estimate of the mean value. Experiment with different bandwidths, noting the increasing smoothness of the resulting estimates as you increase bandwidths. Compare results with the 1977 figures to explore broad trends in the changing groundwater levels.

Try similar ideas for other data sets such as the logarithms of 'PCBs near Pontypool'.

Exercise 5.4

Here we consider a simple way to explore spatial dependence in the 'Vancouver Island geochemistry' data set. First calculate the logarithm of either zinc or copper concentrations. Assuming this new variable is variable number 6, we can then calculate a further variable which contains for each site, not the value of variable 6 at that site, but that at the nearest neighbouring site. The appropriate formula is:

```
nearby = {6}[near(1)]
```

The calculation will take some time, since there are a large number of observations in this data file. A scatter plot of 'nearby' against variable 6 gives some visual indication of the degree of spatial covariance. You might also calculate a correlation coefficient between them using the worksheet and the '`corr()`' function. Assuming 'nearby' is variable 7 then

```
work[1] = corr({6},{7})
```

will suffice. However refer again to the scatter plot of the logarithms against their neighbouring values. Notice three 'outlying' observations in this plot. `Dot map` the logarithms and use `Profile locations` with the 'ascending data values' option to locate these 'outlying' sites. In fact the recorded data values for all elements for these sites are somewhat suspect. You might try removing them from your analysis by giving them missing values. You would do this by using `Calculate` from the `Data` menu with the two formulae:

```
{6} = if({6}=0.0,miss(),{6})
{7} = if({7}=0.0,miss(),{7})
```

The resulting scatter plot of variable 6 against 7 and their correlation coefficient then gives a more accurate reflection of the true relationship.

You could repeat this exercise with 'second', 'third' etc. nearest neighbours, by simply changing the argument of the `near()` function in the earlier formula. This will give some idea of the persistence of this spatial covariance.

Exercise 5.5

Derive an estimated variogram for one of the geochemical elements in the 'Vancouver Island geochemistry' data set. Before doing so, check the frequency distribution of your chosen element to decide whether a transformation is advisable. Then `Dot map` the values (or the transformed values) and use `Variogram` from the `Analysis` menu to derive the variogram. Consider carefully the maximum distance over which to estimate the variogram, which you indicate by marking a line of that length on the screen. You do this in the same way as described for defining bandwidth for kernel estimation in a previous exercise. There is a large number of observations in this data set and the estimation will take some time. Choosing long distances will increase the computation time problem and little extra useful information will be obtained at distances in excess of 2–2.5 km. (Use `Grid references` to get some idea of distances in this file, the grid referencing is in metres, so 2 km is really a very small distance on the displayed map).

Try deriving variograms for other data sets such as the logarithmically transformed PCB values or 'Groundwater in Venice'. Try repeating the results given in the text for 'Radon in Lancashire', again first calculating the logarithm of the raw values.

Exercise 5.6

For an example of trend surface modelling open the file 'England & Wales temperature' and load the data. Dot map the August 1991 temperatures and observe the general pattern of spatial variation. Use `Calculate` from the `Data` menu with the following formula to fit a simple linear trend surface:

```
fitslin=regr({2},east(),north())
```

By using `Regression parameters` within the `Analysis` menu, check the fit of the regression, and the significance of each parameter. Use `Contour map` within the `Analysis` menu to produce a contour fill map of the fitted trend surface (the new variable 'fitslin').

Then use the formula:

```
fitsquad=regr({2},east(),north(),east()**2,north()**2,
              east()*north())
```

to fit a quadratic trend surface to the same variable. Look at the significance of the regression parameters and consider whether the quadratic surface represents a better model.

Derive the residuals for the linear trend surface by subtracting the fitted values from variable 2. Plot the residuals against the fitted values, to check for adequacy of the model. Check for second order effects in the residuals, by using plots of residuals against neighbouring residuals, as in Exercise 5.5, or by estimating a variogram (use a reasonably large upper limit for distance range). What are your conclusions about temperature variation in England and Wales? Are your conclusions unchanged in similar analyses of the August 1981 data?

Exercise 5.7

Following on from the previous exercise we might look at the effect of including 'elevation' as a covariate into the linear trend surface for August 1991 temperature in

England and Wales. First plot the residuals from the linear trend surface against 'elevation'. What do you observe?

Use the formula

```
fitselev = regr({2},east(),north(),{1})
```

followed by the `Regression parameters` option to formally examine the significance of including elevation into the linear trend surface. Note the value (–0.00535) of the regression coefficient for the 'elevation' term. Next correct the fitted trend surface to sea level by subtracting the appropriate elevation effect, using the formula:

```
corrfits = regr({2},east(),north(),{1})-(-0.00535)*{1}
```

Contour map the trend that is now corrected for elevation. How does this compare with the uncorrected linear trend contoured in the previous exercise—are there minor differences?

Exercise 5.8

In order to grasp the idea of a 'residual covariance structure' or second order effect, consider the 'Groundwater in Venice' data set. Open the file and load the data, and map the 1977 levels. Use the following formula to fit a simple linear trend surface:

```
fitslin = regr({2},east(),north())
```

By using `Regression parameters` within the `Analysis` menu, check the fit of the regression, and the significance of each parameter.

Derive the residuals for the linear trend surface by subtracting the fitted values from variable 2. Map these residuals. Do you observe any remaining spatial pattern? Now use the `Variogram` option with the `Analysis` menu to examine the nature of the residual covariance structure (you will need to use a large distance range for the variogram). Does this confirm the presence of a second order effect?

You might also try repeating the analysis given in the text of fitting a quadratic trend surface to logarithms of the PCB levels in the 'PCBs near Pontypool' data, then obtaining residuals from the regression and forming a variogram. How does the variogram compare with that for the logarithms of PCBs themselves, rather than residuals? Now compare the variogram with that for the residuals from a linear trend surface instead of a quadratic.

Exercise 5.9

In an earlier exercise you estimated variograms for different geochemical elements in the 'Vancouver Island geochemistry' data set. Now choose one of these elements (other than the nickel example which appears in the text). Consider appropriate transformations and re-estimate the variogram.

Whilst the variogram is displayed use of <TAB> will initialise the 'nugget', 'range' and 'sill' parameters of a spherical variogram model and superimpose this curve onto the displayed variogram. Use of <TAB> again will do the same for an exponential variogram model instead. A third use of <TAB> returns to the initial state with no model displayed. This 'tabbing' between models can be repeated as often as required. When either the spherical or exponential model is displayed, use of the up or down cursor keys increases or decreases the 'sill' parameter and redisplays the model. Similarly, use of left or right cursor keys increases or decreases the 'range' parameter, and use of < + > or < − > increases or decreases the 'nugget' parameter. Use of <INS> at any time will display the type of model and the current value of the 'sill', 'range' and 'nugget' parameters; use of any key will then remove this display and allow parameter values to be further adjusted.

By using this kind of 'interactive' modelling, try to determine a suitable model for the sample variogram you have derived; that is, the functional form and rough guesses of the 'range', 'sill' and 'nugget' effect.

Is there much difference between similar analyses of other geochemical elements?

You could try repeating this kind of variogram modelling for other data sets such as logarithmically transformed PCB values or 'Groundwater in Venice'.

Exercise 5.10

Open the 'Vancouver Island geochemistry' data set. Calculate a new variable containing the logarithms of nickel concentrations, by using `Calculate` from the `Data` menu. `Dot map` this variable. Use `Grid references` from the `Map` menu to identify the kriged region given in Figure 5.20; the grid references of this region are given in the text. Compare your displayed values with the kriged map. Do the kriging results appear to make sense? Can you identify why the standard errors are higher in some areas?

Next choose `Zoom/Restrict` within the `File` menu and select this subset of the study region to 'zoom'. You do this by marking one corner of the rectangular region by a 'double click' and then 'dragging' a window away from that corner, fixing the final position with another 'double click'. This defines the 'source' area for the zoom. You then similarly define a rectangular 'target' area (choose a moderate sized area roughly in the centre of the working screen). In this case use the 'raster' zoom option (there are no polygons). Once the zoom is completed, you may redraw the `Dot map` of the logarithms of nickel concentration to check that you have achieved what you required. It should now be easier to compare values with the kriged area in the text.

One quick comparison would be to create a contour map for the logarithms of the nickel data in your 'zoomed' area, based upon the TIN method described in the text. Use `Contour map` in the `Analysis` menu and choose the 'contour line' option. Superimpose the transformed data as a dot map. Experiment with numbers of classes and class interval schemes using `Scaling` within the `Map` menu and repeating the contour map, to try and produce the best result you can. How does this compare with the kriged predictions? Are there significant differences? In what way is the kriged map an improvement on the simple TIN interpolation?

Now draw a `Dot map` of the logarithms of nickel concentration and then use `Kriged map` from the analysis menu, to try to reproduce the kriging results given in Figure 5.20. Select the option 'kriged values' and specify a spherical variogram model, with nugget, range and sill parameters as suggested in the text, that is respectively, 0.298, 2130 (recall grid references here are in metres) and 0.827. Finally specify a radius for the local search neighbourhoods to be used in deriving the kriging estimates. Do this by drawing a line whose length will be used for the 'radius' of this neighbourhood, in the same way as described for defining bandwidth for kernel estimation in a previous exercise. In INFO-MAP, sample values will be used within this radius of any point when kriging that point, unless there are more than 15 of these, in which case the nearest 15 will be used, so choose your 'radius' to be small, about 3–4 km. The estimate is then computed on a 'raster grid' over the area. The computation can take some time. Compare your results with those given in the text. Try repeating the procedure but using the 'standard error' option rather than 'kriged values'.

You could then try kriging other elements within the same area, using variogram models derived in earlier exercises. You might also consider kriging for other data sets, such as the logarithmically transformed PCB values or 'Groundwater in Venice', again using variogram models derived in earlier exercises.

6

Further methods for spatially continuous data

Here, we consider various more specialised topics concerned with the analysis of 'spatially continuous' data. First, we outline two extensions to the basic kriging methods introduced in the last chapter: the techniques of 'block kriging' and 'co-kriging'. Second, we examine a selection of methods for the analysis of multivariate measurements. These apply when values of several variables have been recorded at each site and interest lies in studying pattern in these values taken together, rather than in each taken in isolation.

6.1 Introduction

In the previous chapter we were concerned with exploring and modelling spatial variation in a continuous measurement on the basis of sampled values at a number of specific point sites. One objective was to obtain good estimates of the value of the variable at points in our study region that had not been sampled. In this regard the most powerful technique that we introduced was that of kriging. We mentioned both ordinary and universal kriging, but a number of other, more specialised, variants of kriging exist. We start this chapter by discussing two of the most commonly used of these, *block kriging* and *co-kriging*.

The first of these relates to the situation, which arises particularly in mining geology, where an estimate is required not at a point site, but rather for the average value of the variable over some small area or *block* (for example, average ore grade in a section of a mine). One way to address this problem is simply to calculate many point estimates within the block concerned, using the techniques of the previous chapter, and then to average the resulting estimates. Though conceptually simple, this procedure may be computationally very intensive. Block kriging provides a technique to obtain mathematically equivalent results, but with significant reduction in computation, by constructing kriging estimates directly for the blocks concerned.

In all the kriging methods we considered in the previous chapter our estimates of unknown values of a variable of interest were based on the values of this *same* variable at a set of sampled sites. However, sometimes we may be in the fortunate position of also having measurements on other variables either at the same sites or different sites in our study region. For example, in exploration for a mineral such as copper, core samples at n sites may be taken and assayed. However, copper does not occur in isolation, so along with the percentage of copper we would probably also have readings on percentages of lead and zinc and other minerals at each of these sites. Other examples include data on a series of elements sampled in stream sediments, as in one of our case studies; or several properties measured at a set of soil pits; and so on. As well as each variable displaying in isolation some form of spatial dependence, we might expect there also to be some *cross-covariance* between them. We have already mentioned in Chapter 5 the inclusion of covariates into trend surface analysis in order to better understand first order variations in the mean value of the variable of primary interest; but we have not investigated the possibility of incorporating this sort of information into the covariance structure, or second order properties, in order to secure better local predictions in kriging. Co-kriging is a technique which allows for this.

We devote the second half of the chapter to a general discussion of a different kind of problem. The techniques mentioned above extend our discussion in the previous chapter of prediction of a spatially continuous variable. However, as discussed in Chapter 5, there are many cases where our interest may stop short of a desire to predict, and we may simply be concerned with understanding any observed pattern in spatially continuous data and the relative contribution of various possible factors to that pattern. Some of the techniques of the previous chapter allow us to explore such issues when interest lies in the pattern exhibited by a single variable. For example, we may use the ordinary or generalised least squares framework to assess how such pattern relates to spatial location and possibly to the values of other available covariates. However, we have not so far considered the case where we have several variables measured at each site and are interested in understanding pattern in more than one of these taken together. We may, for example, wish to characterise the regional geochemistry of a large area on the basis of measurements on several elements in stream sediments, or study climate variations on the basis of several measures of temperature, humidity and precipitation.

Such questions involve us in statistical methods for *multivariate analysis*, the analysis of patterns and relationships in the values of several variables or attributes treated simultaneously. This is a very wide field in its own right, and we can attempt little more in the second half of this chapter than to draw such methods to the attention of the reader and outline some of the techniques under this heading that are useful in spatial data analysis. Our treatment of these techniques will not include mathematical and computational detail; most of the methods we discuss are general purpose multivariate analysis techniques, which are better left to specialised texts on that subject. We reference a selection of these at the end of this chapter. Our reason for including a discussion of them here at all is because readers may not have come across them to the degree that they have encountered more elementary statistical techniques, such as regression, some familiarity with which we have assumed.

6.2 Extensions to kriging

The methods of ordinary and universal kriging discussed in the previous chapter represent only the most basic of a wide range of closely related spatial prediction techniques. For example, *robust kriging* is an extension designed to be particularly resistant to the effect of extreme or 'outlying' observations, whilst *disjunctive kriging* relates to obtaining a predictor which is a non-linear, rather than a linear, combination of neighbouring observations. Amongst such extensions we select two here for further consideration—those of block kriging and co-kriging.

6.2.1 Block kriging

We have already outlined the objective of block kriging in our introductory remarks to this chapter. To recap, the technique is concerned with the prediction of the average value of a spatially continuous variable over an area or block, say $\mathcal{A}$, rather than at a specific point location, $\mathbf{s}$, as was the case with the kriging methods presented in the previous chapter. This is a typical requirement in mining applications, where it is invariably blocks which are processed to extract minerals. As mining proceeds, large blocks are progressively selected for exploitation and it is therefore of more interest to predict average block yield than to make point predictions. A block may have a rich seam running through it, but may still be unprofitable to exploit if the rest of the block is of very low yield.

As mentioned previously, a conceptually simple way to obtain a block estimate is to predict for many point locations in the block and average the resulting values. However, since many hundreds of block estimates may be required in a typical application, this approach may prove computationally very intensive if more than a trivial number of points are used in each block. The number of computations would clearly be significantly reduced if block estimates could be constructed more directly.

The modifications required to the point location kriging methods given in Chapter 5 are straightforward and intuitive. We present them here only for the case of ordinary kriging. Modifications in other cases can be obtained by a straightforward analogy.

Recall that ordinary kriging applies to the case where the process $Y(\boldsymbol{s})$, with covariogram, $C()$, is assumed to have a constant mean value $\mu(\boldsymbol{s}) = \mu$, over $\mathcal{R}$ (or at least within the moving local neighbourhood within which predictions are calculated). Prediction of the process at a general point $\boldsymbol{s}$, is obtained from a weighted linear combination of observed values, y_i, at n sample sites, $\boldsymbol{s}_i$. The appropriate weights, $\boldsymbol{\omega}(\boldsymbol{s})$, are the first n elements of the augmented $((n+1) \times 1)$ vector, $\boldsymbol{\omega}_+(\boldsymbol{s})$, which is obtained as the solution to the system of equations:

$$\begin{array}{ccc} \boldsymbol{C}_+ & \boldsymbol{\omega}_+(\boldsymbol{s}) \quad = & \boldsymbol{c}_+(\boldsymbol{s}) \end{array}$$

$$\begin{pmatrix} C(\boldsymbol{s}_1, \boldsymbol{s}_1) & \cdots & C(\boldsymbol{s}_1, \boldsymbol{s}_n) & 1 \\ \vdots & \ddots & \vdots & \vdots \\ C(\boldsymbol{s}_n, \boldsymbol{s}_1) & \cdots & C(\boldsymbol{s}_n, \boldsymbol{s}_n) & 1 \\ 1 & \cdots & 1 & 0 \end{pmatrix} \begin{pmatrix} \omega_1(\boldsymbol{s}) \\ \vdots \\ \omega_n(\boldsymbol{s}) \\ \nu(\boldsymbol{s}) \end{pmatrix} = \begin{pmatrix} C(\boldsymbol{s}, \boldsymbol{s}_1) \\ \vdots \\ C(\boldsymbol{s}, \boldsymbol{s}_n) \\ 1 \end{pmatrix}$$

We refer the reader back to Chapter 5 for further details and clarification. To obtain our prediction, $\hat{y}(\boldsymbol{s})$, we solve the above equation to obtain $\boldsymbol{\omega}_+(\boldsymbol{s})$ as $\boldsymbol{C}_+^{-1}\boldsymbol{c}_+(\boldsymbol{s})$, extract from this the vector, $\boldsymbol{\omega}(\boldsymbol{s})$, and then calculate $\hat{y}(\boldsymbol{s})$ as $\boldsymbol{\omega}^T(\boldsymbol{s})\boldsymbol{y}$, where $\boldsymbol{y}$ is the $(n \times 1)$ vector of original observations, y_i, at the sample sites $\boldsymbol{s}_i$. Once we have obtained this prediction the mean square prediction error, or kriging variance, is given by:

$$\sigma_e^2 = C(\boldsymbol{s}, \boldsymbol{s}) - \boldsymbol{c}_+^T(\boldsymbol{s})\boldsymbol{C}_+^{-1}\boldsymbol{c}_+(\boldsymbol{s})$$

The only change to this whole process that is required for block prediction is to replace $\boldsymbol{c}_+(\boldsymbol{s})$, which includes the column vector of covariances between the location $\boldsymbol{s}$ and each of the n sample locations $\boldsymbol{s}_i$, by a corresponding vector, $\boldsymbol{c}_+(\mathcal{A})$, which replaces these with average covariances, $C(\mathcal{A}, \boldsymbol{s}_i)$, between the block $\mathcal{A}$ and each of the sample locations $\boldsymbol{s}_i$. The matrix $\boldsymbol{C}$, of covariances between the n sample sites is independent of the location at which a prediction is required and so remains unchanged. The term $C(\boldsymbol{s}, \boldsymbol{s})$ in the kriging variance is replaced by $C(\mathcal{A}, \mathcal{A})$, the average covariance in the block $\mathcal{A}$.

Strictly speaking we should compute such averages as:

$$C(\mathcal{A}, \boldsymbol{s}_i) = \frac{\int_{\mathcal{A}} C(\boldsymbol{s}, \boldsymbol{s}_i)\,\mathrm{d}\boldsymbol{s}}{A} \qquad i = 1, \ldots, n$$

$$C(\mathcal{A}, \mathcal{A}) = \frac{\int_{\mathcal{A}}\int_{\mathcal{A}} C(\boldsymbol{s}, \boldsymbol{s}')\,\mathrm{d}\boldsymbol{s}\mathrm{d}\boldsymbol{s}'}{A^2}$$

where A is the area of the block $\mathcal{A}$ and, as before, $C()$ represents the chosen covariogram model. In practice these integrals would be approximated by numerical quadrature. In the crudest case we would simply estimate the integrals by the averages of the values of the covariances involved over some appropriate grid of point locations within the block $\mathcal{A}$.

The advantage of using block kriging is that estimates are produced by solving only one set of kriging equations. The calculation of average covariances involves extra computation, but this is not as significant as that involved in forming a block estimate by averaging numerous point estimates. The ability to derive block predictions by simply replacing point covariances by block average covariances, in the same set of basic equations, is a particularly attractive feature of the kriging method, one that is not shared by most other interpolation methods.

A very similar situation to block kriging can occur when the original observations, y_i, themselves relate not to point sites $\mathbf{s}_i$, but to blocks $\mathcal{A}_i$. This is often referred to as prediction with *block support* as opposed to *point support* and is again typical in mining applications. Often, original observations are in the form of point support, such as core samples from a drill bit. However, as mining proceeds, data on average ore grade from blocks already mined becomes available and can be used to improve original predictions. Ordinary kriging with *block support* involves an obvious extension, where the covariance elements in the $\mathbf{C}$ matrix and $\mathbf{c}()$ vector, which involve $\mathbf{s}_i$, are replaced by corresponding average covariances over the blocks $\mathcal{A}_i$. For example, $C(\mathbf{s}_i, \mathbf{s}_j)$ becomes $C(\mathcal{A}_i, \mathcal{A}_j)$ and $C(\mathbf{s}, \mathbf{s}_i)$, becomes $C(\mathbf{s}, \mathcal{A}_i)$ (or $C(\mathcal{A}, \mathcal{A}_i)$ if a block $\mathcal{A}$ is being predicted). These average covariances are calculated by direct analogy with the expressions given above for $C(\mathbf{s}_i, \mathcal{A})$ and $C(\mathcal{A}, \mathcal{A})$.

So, ordinary kriging is really very flexible when it comes to block prediction or block support. Mixtures of point and block support can also be accommodated. In addition, the same idea can be extended to universal kriging and hence used when a global trend exists in the process to be predicted.

6.2.2 Co-kriging

In our introduction to this chapter we mentioned that in many studies concerned with the prediction of some primary variable of interest the available data will also contain values of one or more secondary variables recorded at the same sites. In some cases, values of these secondary variables may even be available at a more extensive set of sites than apply for the primary variable. For example, observations on the primary variable at some sites might be 'missing' because of lost samples, or unreliable test results, whereas they may exist at those sites for secondary variables. Alternatively, the primary variable may have been intentionally *under sampled* in the original data collection. For example, in mining it is common that all core samples be assayed for one particular mineral, but only those that yield promising results will be assayed for others. If the primary variable of interest, in some later prediction exercise, happens to be one of these other minerals, then the situation arises where observations are available on this variable only for a subset of the original core samples, whilst values for another mineral exist for all of them. Another possible kind of 'under sampling' arises where values of the same variable have been observed for a set of sites at a number of different time periods and in some of these periods certain site values are 'missing'. We had an example of this kind of situation with the Venice groundwater data set, discussed in the previous chapter. Groundwater levels were measured in both 1973 and 1977, but in 1977 observations were available for only 35 of the 40 sites at which levels were measured in 1973.

All of these cases can be represented mathematically as having observations on a primary variable, y_i, at n sample sites and additional observations on p possible covariates $(x_{i1}, \ldots, x_{ip})$ at a more extensive set of $n + m$ sites, the first n of which equate to those sites where the primary variable has also been recorded. In the simplest case $m = 0$ and values of both the primary variable and the covariates exist at all sites. The problem addressed in co-kriging is that of whether the covariate information can be used to improve the prediction of the value of the primary variable at a general point, $\boldsymbol{s}$, in $\mathcal{R}$, where it has not been sampled. Intuitively, this would seem a reasonable possibility. The argument is that if the secondary variables are correlated with the primary variable, then we can use observations at sites where they are both recorded to estimate this correlation. Via this correlation, values of the secondary variables at sites where the primary variable has not been observed provide useful information about the primary variable at those sites, which in turn can be used to help predict the value of the primary variable nearby.

This is what is done in co-kriging. We will develop the technique purely as an extension to ordinary kriging. Similar modifications of universal kriging are also possible, but we do not give details here. Also, for simplicity we will consider only the case $p = 1$ in any detail, that is where we have observations, x_i, on a single secondary variable. Extension to the case where there is more than one secondary variable can be obtained by analogy with the single variable case. The intuitive argument, given above, as to how we might use the information provided by secondary variables relies on the secondary and primary variables being correlated. We must start therefore, by developing a formal method of estimating and modelling this correlation. We do this by extending the idea of the covariogram and variogram for a single variable, introduced in Chapter 5, to a *cross-covariogram* and *cross-variogram* between two variables.

Formally, if $\{Y(\boldsymbol{s}),\ \boldsymbol{s} \in \mathcal{R}\}$ is the process relating to the primary variable, and $\{X(\boldsymbol{s}),\ \boldsymbol{s} \in \mathcal{R}\}$ that relating to the secondary variable, and both these processes are assumed stationary, then the cross-covariogram is defined as:

$$C_{YX}(\boldsymbol{h}) = E((Y(\boldsymbol{s}+\boldsymbol{h}) - \mu_Y)(X(\boldsymbol{s}) - \mu_X))$$

where $\boldsymbol{h}$ is an arbitrary vector separation in $\mathcal{R}$. We can similarly extend the idea of a variogram to a cross-variogram defined as:

$$2\gamma_{YX}(\boldsymbol{h}) = E((Y(\boldsymbol{s}+\boldsymbol{h}) - Y(\boldsymbol{s}))(X(\boldsymbol{s}+\boldsymbol{h}) - X(\boldsymbol{s})))$$

Notice that in general the cross-covariogram need not be symmetric, that is

$$C_{YX}(\boldsymbol{h}) \neq C_{XY}(\boldsymbol{h})$$

whereas the cross-variogram is always symmetric, so that

$$\gamma_{YX}(\boldsymbol{h}) = \gamma_{XY}(\boldsymbol{h}).$$

However, in practice the cross-covariogram is most often modelled as a symmetric function and this model is usually derived from an estimated cross-variogram model, by the relationship:

$$C_{YX}(\boldsymbol{h}) = \gamma_{XY}(\infty) - \gamma_{XY}(\boldsymbol{h})$$

This is similar to the way that we previously estimated a variogram model and converted this to a covariogram model when required, through the relationship:

$$C(\boldsymbol{h}) = \gamma(\infty) - \gamma(\boldsymbol{h})$$

The natural sample estimator of the cross-variogram, given n pairs of observations (y_i, x_i) at sample sites $\boldsymbol{s}_i$, is:

$$2\hat{\gamma}_{YX}(\boldsymbol{h}) = \frac{1}{n(\boldsymbol{h})} \sum_{\boldsymbol{s}_i - \boldsymbol{s}_j = \boldsymbol{h}} (y_i - y_j)(x_i - x_j)$$

where the summation is over all pairs of observed data points with a vector separation of $\boldsymbol{h}$, and $n(\boldsymbol{h})$ is the number of these pairs. As $\boldsymbol{h}$ is varied a set of values is obtained, so giving a sample cross-variogram. In practice, as in the estimation of the variogram for a single variable, there will rarely be enough observations with an exact vector separation of $\boldsymbol{h}$, and so a series of intervals are used and the estimator calculated on the basis of separations which lie in these intervals. In general this will involve specifying a tolerance on both distance and direction. Under the assumption of isotropy the cross-variogram is estimated over all directions, which implies replacing $\boldsymbol{h}$ in the above expression simply with its length h.

As with single variable variograms, before an estimated cross-variogram can be used in conjunction with co-kriging methods it is first modelled with a smooth parametric function in order to provide a continuous description of the cross-covariance structure at any spatial separation that may be required in the prediction process. The same families of models are used as in the case of the variogram and the reader is referred back to the discussion of variogram modelling in Chapter 5 for details. An extra consideration that arises in modelling cross-variograms is a requirement that the cross-covariogram between any two variables be mathematically consistent with the individual variograms for each of the variables taken in isolation. We do not give details here of the minimum necessary mathematical restrictions that this imposes on the two individual variograms and the cross-covariogram taken together. We simply note that it is sufficient that the cross-covariogram model is constructed only from basic families of variogram models that appear in both of the individual variograms. The parameters do not have to be the same, but the model types do. Recall that in Chapter 5 we mentioned that single variable variogram models could be constructed from the basic model families, exponential, spherical, Gaussian etc., by 'mixing' them, that is by using a model which is a linear combination of other models where all the weights involved are positive. We are now saying, for example, that if one of the individual variograms is a mixture of spherical and exponential, and the other is a mixture purely of exponentials, then the cross-covariogram can only be a mixture of exponentials. A further requirement is that the square of the weight that a particular family is given in the cross-variogram has to be strictly less than the product of the weights that family is given in the two individual variograms.

So far we have said nothing about co-kriging as such, we have simply described how to obtain a formal description of the cross-covariance between the values of primary and secondary variables. We assume from here on that these methods, along with those in Chapter 5, have been used to derive appropriate variogram and cross-variogram models, and that these have been converted, via the appropriate relationships, into covariogram models $C_Y()$, $C_X()$ and cross-covariogram model $C_{YX}()$. We now turn our attention to how to use this information to improve the prediction of the primary variable.

Recall from the previous section on block kriging, and in more detail from Chapter 5, that ordinary kriging applies to the case where the process $Y(\boldsymbol{s})$, to be predicted, is assumed to have a constant mean value $\mu(\boldsymbol{s}) = \mu$, over $\mathcal{R}$ (or at least within the moving local neighbourhood within which predictions are calculated), and that optimal prediction, at a general point $\boldsymbol{s}$, is based on a weighted linear combination of the observed values, y_i, at n sample sites, $\boldsymbol{s}_i$. In the co-kriging extension to ordinary kriging we assume that both the primary and secondary processes have constant mean values, μ_Y and μ_X, respectively. We then extend the previous idea to construct a prediction, $\hat{y}(\boldsymbol{s})$, of the primary process in terms of a linear combination of both values of that process, y_i $(i = 1, \ldots, n)$, at the n sites where these have been recorded, and also those of the secondary process, x_j, $(j = 1, \ldots, n+m)$, recorded at the same n sites and at an additional m other sites. In order to derive the appropriate optimal weights, we proceed by analogy with the derivation of the ordinary kriging weights. That is, we start by considering an appropriate weighted linear combination of random variables:

$$\hat{Y}(\boldsymbol{s}) = \sum_{i=1}^{n} \omega_{yi}(\boldsymbol{s})\,Y(\boldsymbol{s}_i) + \sum_{j=1}^{n+m} \omega_{xj}(\boldsymbol{s})X(\boldsymbol{s}_j)$$

We then choose values for the weights, $\omega_{yi}(\boldsymbol{s})$ and $\omega_{xj}(\boldsymbol{s})$, so that the mean value of $\hat{Y}(\boldsymbol{s})$ is constrained to be μ_Y, the same as that of the random variable $Y(\boldsymbol{s})$ that we are trying to predict; and also so that the expected mean square error between values of $Y(\mathbf{s})$ and values of $\hat{Y}(\mathbf{s})$ is minimised. It is assumed that the mean of $Y(\mathbf{s})$ and of each $Y(\mathbf{s}_i)$, are all μ_Y and that μ_X is the mean of all the $X(\mathbf{s}_j)$. One way in which we can obtain the required mean value μ_Y for $\hat{Y}(\mathbf{s})$, is to ensure that $\sum_{i=1}^{n} \omega_{yi}(\mathbf{s}) = 1$ and $\sum_{j=1}^{n+m} \omega_{xj}(\mathbf{s}) = 0$. Other possibilities exist, but these are the most frequently used constraints. Subject to these constraints we then wish, as with ordinary kriging, to minimise the expected mean square error between values of $Y(\mathbf{s})$ and values of $\hat{Y}(\mathbf{s})$. The relevant mathematics follows closely that for the case of ordinary kriging, except that now two Lagrange multipliers, $v_1(\mathbf{s})$ and $v_2(\mathbf{s})$, are involved, rather than one. The result is that the optimal co-kriging weights are found to satisfy a similar system of equations as before. That is:

$$\boldsymbol{C}_+\boldsymbol{\omega}_+(\mathbf{s}) = \boldsymbol{c}_+(\mathbf{s})$$

where now the augmented matrix $\boldsymbol{C}_+$ is:

$$\boldsymbol{C}_+ = \begin{pmatrix} C_Y(\mathbf{s}_1,\mathbf{s}_1) & \cdots & C_Y(\mathbf{s}_1,\mathbf{s}_n) & C_{YX}(\mathbf{s}_1,\mathbf{s}_1) & \cdots & C_{YX}(\mathbf{s}_1,\mathbf{s}_{n+m}) & 1 & 0 \\ \vdots & \ddots & \vdots & \vdots & \ddots & \vdots & \vdots & \vdots \\ C_Y(\mathbf{s}_n,\mathbf{s}_1) & \cdots & C_Y(\mathbf{s}_n,\mathbf{s}_n) & C_{YX}(\mathbf{s}_n,\mathbf{s}_1) & \cdots & C_{YX}(\mathbf{s}_n,\mathbf{s}_n) & 1 & 0 \\ C_{XY}(\mathbf{s}_1,\mathbf{s}_1) & \cdots & C_{XY}(\mathbf{s}_1,\mathbf{s}_n) & C_X(\mathbf{s}_1,\mathbf{s}_1) & \cdots & C_X(\mathbf{s}_1,\mathbf{s}_{n+m}) & 0 & 1 \\ \vdots & \ddots & \vdots & \vdots & \ddots & \vdots & \vdots & \vdots \\ C_{XY}(\mathbf{s}_{n+m},\mathbf{s}_1) & \cdots & C_{XY}(\mathbf{s}_{n+m},\mathbf{s}_n) & C_X(\mathbf{s}_{n+m},\mathbf{s}_1) & \cdots & C_X(\mathbf{s}_{n+m},\mathbf{s}_{n+m}) & 0 & 1 \\ 1 & \cdots & 1 & 0 & \cdots & 0 & 0 & 0 \\ 0 & \cdots & 0 & 1 & \cdots & 1 & 0 & 0 \end{pmatrix}$$

and the augmented vectors $\boldsymbol{\omega}_+(\mathbf{s})$ and $\boldsymbol{c}_+(\mathbf{s})$ are:

$$\boldsymbol{\omega}_+(\mathbf{s}) = \begin{pmatrix} \omega_{y1}(\mathbf{s}) \\ \vdots \\ \omega_{yn}(\mathbf{s}) \\ \omega_{x1}(\mathbf{s}) \\ \vdots \\ \omega_{xn+m}(\mathbf{s}) \\ v_1(\mathbf{s}) \\ v_2(\mathbf{s}) \end{pmatrix} \qquad \boldsymbol{c}_+(\mathbf{s}) = \begin{pmatrix} C_Y(\mathbf{s},\mathbf{s}_1) \\ \vdots \\ C_Y(\mathbf{s},\mathbf{s}_n) \\ C_{YX}(\mathbf{s},\mathbf{s}_1) \\ \vdots \\ C_{YX}(\mathbf{s},\mathbf{s}_{n+m}) \\ 1 \\ 0 \end{pmatrix}$$

The co-kriging weights $\boldsymbol{\omega}_y(\mathbf{s})$ and $\boldsymbol{\omega}_x(\mathbf{s})$ are derived respectively from the first n and the next $n+m$ elements of the vector, $\boldsymbol{\omega}_+(\mathbf{s})$, which in turn is given by

$\boldsymbol{C}_+^{-1}\boldsymbol{c}_+(\boldsymbol{s})$; then the prediction $\hat{y}(\boldsymbol{s})$ is formed as $\boldsymbol{\omega}_y{}^T(\boldsymbol{s})\boldsymbol{y} + \boldsymbol{\omega}_x{}^T(\boldsymbol{s})\boldsymbol{x}$, where $\boldsymbol{y}$ is the $(n \times 1)$ vector of observations, y_i, on the primary variable and $\boldsymbol{x}$ is the $((n+m) \times 1)$ vector of observations, x_j, on the secondary variable. Once we have obtained this prediction the mean square prediction error, or kriging variance, is given by:

$$\sigma_e^2 = C_Y(\boldsymbol{s}, \boldsymbol{s}) - \boldsymbol{c}_+^T(\boldsymbol{s})\boldsymbol{C}_+^{-1}\boldsymbol{c}_+(\boldsymbol{s})$$

an analogous expression to that obtained for ordinary kriging.

The co-kriging equations given above are for point prediction and have assumed only one secondary variable; an extension to $p > 1$ is straightforward. Co-kriging may also be extended to predict block values, following the approach described in the previous section of this chapter. It can also be modified, if required, to be used in conjunction with universal kriging, when a trend exists in the mean value of the primary variable. Note in connection with the latter that in Chapter 5 we have already mentioned the inclusion of covariates into a trend model in order to better understand global variations in the mean value of the variable of primary interest. However, this is a somewhat separate issue to that with which co-kriging is concerned. Co-kriging allows us in addition to exploit the covariate information in predicting local fluctuations from the mean.

In conclusion to this section we should note that there are certain situations where co-kriging will not improve prediction estimates obtained from straightforward kriging. This applies particularly when $m = 0$ (that is, values of both the primary and secondary variables exist at all sites), and when the individual variograms and cross-variograms are all proportional to the same basic family of variogram model. Hence if all variogram models are 'similar' in shape and the primary variable is not appreciably under sampled, there would be little advantage to employing the more complicated co-kriging over other more simple kriging methods.

We do not report any empirical results applying co-kriging to our data sets. We have indicated above what the relevance of this technique might be in a mining context, and the same principles could be extended to the geochemical data for north Vancouver Island, discussed in Chapter 5. Recall that we include data for nearly 1000 sites, at each of which five elements are sampled. Suppose we also had data on another element (perhaps arsenic or fluorine, for instance), but that this element had been sampled at a much smaller subset of locations. Suppose further that it was variation in this other element, the primary variable, that we were particularly interested in, maybe because of its possible link with animal or human ill-health. We could then use information on spatial variation in one or more of the more widely sampled secondary variables (zinc, lead, etc.) to give predictions about the primary variable at locations where it had not been sampled. As software for co-kriging becomes more widely available we expect to see it used more extensively in a range of environmental science applications.

6.3 Multivariate methods

In the remainder of this chapter we intend to concentrate on a rather different set of considerations from those with which we have been preoccupied so far. We refer to the case where we have several variables measured at each site and are interested in understanding pattern in more than one of these values taken together. This is a different problem from either studying the pattern in any one of them taken in isolation, or predicting the value of one, 'given' values of the others, which has been our concern throughout Chapter 5 and in the early part of this chapter.

For example, we may wish to characterise the regional geochemistry of a large area on the basis of measurements on several elements in stream sediments, using some or all of these to derive a regional geochemical map. In practice geochemists often do this by using combinations of three elements. This aids in the search for ore bodies, or in determining the 'health' of soils or vegetation. As we suggested in our description of the Vancouver Island geochemical data in Chapter 5, particular elements tend to be associated, and these multi-element associations often prove useful in mineral exploration. Another example is provided by the climate data for South America. Here we have various measures on precipitation and temperature at each site and wish to find appropriate combinations of these that help us to summarise and understand climatic variation over the region; possibly, we may wish to use this information in order to derive broad climatic regions in the continent.

Such questions involve us in consideration of general statistical methods for multivariate analysis—the analysis of relationships in the values of several variables or attributes treated simultaneously. Here we give only a brief introduction to some of these techniques that may be useful in spatial data analysis. We also depart somewhat in that discussion from our main theme in the book. Whereas we have mostly been concerned with forms of analysis that are explicitly spatial in nature, spatial forms of multivariate analysis are less well developed. Where spatial extensions do exist, they tend to be relatively specialised. We assume that our reader will be less familiar with multivariate techniques than with other forms of general statistical methods and so introduce methods from first principles, rather than concentrating purely on the few specialised spatial multivariate techniques that do exist.

Our discussion will therefore focus mainly on relatively straightforward, standard, non-spatial multivariate techniques. In practice, at least in the first instance, it is these that are most commonly used in spatial analysis when multivariate measures need to be dealt with. The general approach is to use them for purposes of data reduction and exploration in the multidimensional attribute space, with the objective of identifying a small number of 'interesting' sub-dimensions (combinations of attributes) which may then be examined from a spatial perspective, either exploring for spatial patterns and relationships, or for use in spatial classification or discrimination.

Because the techniques we will be discussing are general purpose statistical methods, which are well described in many standard texts on multivariate analysis, we restrict ourselves to an outline description of what the techniques do and to what situations they apply, rather than give mathematical or computational details. A range of well documented software packages, such as SPSS, SAS, GENSTAT, or even the simple to use MINITAB, to name but a few, are readily available to perform all of the standard analyses we describe here, as well as numerous other more sophisticated multivariate techniques. The reader who wishes to follow up further mathematical details, or more specialised techniques, can refer, as a starting point, to one of the texts we list at the end of this chapter.

There will also be little in our discussion of multivariate methods that is restricted to spatially continuous data in particular. We take the opportunity to introduce multivariate techniques generally in this chapter, because they are useful in connection with such data and this is where we encounter them first. However, we shall also find them to be applicable to various questions which arise in the analysis of area data which we consider in Part D of the book. It will be convenient then to refer the reader back to their introduction in this chapter.

Throughout this section we shall assume that we have observations, $(y_{i1}, \ldots, y_{ip})$, on p attributes at each of n sites, $\boldsymbol{s}_i$. As is conventional in multivariate analysis, we shall refer to these collectively as the $(n \times p)$ matrix $\boldsymbol{Y}$, each row of which is the $(1 \times p)$ vector, $\boldsymbol{y}_i^T$, of observations on the p attributes at the ith site. We shall also find it useful to be able to refer to each of the attributes (variables) in a general sense, rather than values of them at a specific site, and we will use the $(1 \times p)$ vector, $\boldsymbol{y}^T$, to denote the p variables $(y_1, \ldots, y_p)$.

One useful way to view the matrix $\boldsymbol{Y}$, is as a set of n points in a p-dimensional space, where $(y_{i1}, \ldots, y_{ip})$ are the p coordinates of the ith observation in this space. We shall refer to this space as the *data space* or the *attribute space*, to distinguish it from the location of the observations in *geographical space*, for which we continue to use our usual notation where $\boldsymbol{s}_i = (s_{i1}, s_{i2})^T$ is the location of the ith site.

In general we shall assume that $(y_{i1}, \ldots, y_{ip})$ are observations on variables $(y_1, \ldots, y_p)$ measured on a quantitative scale; that is, integer counts, or continuous measurements on some interval. Whilst some of the methods we describe may be applied when categorical or qualitative variables (measured on nominal or ordinal scales) are involved, the reader is cautioned against blindly applying methods that we discuss to data of this type. We will not cover special techniques for categorical or qualitative data and emphasise here that many methods appropriate for quantitative variables are simply not applicable to this kind of data. In the cases where they are, modifications are usually required to handle categorical or qualitative measures. The reader is referred to some of the general multivariate texts listed at the end of this chapter for details of methods specifically designed to handle categorical or qualitative data.

Another issue concerned with measurement is the degree to which the quantitative scales of the p observations at any site are comparable. It is typical to have mixtures of attributes, some of which may be measured in millimetres, others in metres, others relating to ratios or percentages, constrained to lie in either a $(0, 1)$, or a $(0, 100)$ range. The results of some multivariate techniques are unaffected by such widely differing units of measurement. This applies when the method concerned inherently 'standardises' values so that all variables are effectively treated equally. However, this is certainly not true of all, or even the majority of, multivariate techniques. In most cases the results and their interpretation can be significantly affected by the relative quantitative scales of the data. The question then arises as to whether the data should be *standardised* before use; that is, whether each data value should first be *mean centred* by subtracting from it the mean of its corresponding variable, and then this mean centred result further scaled by dividing by the standard deviation of the same variable so that all values are then measured on the same standard quantitative scale. The degree to which this is appropriate in any particular application is largely a question of whether one indeed wishes to 'treat all variables equally'. In most cases this will be sensible, particularly in preliminary exploratory analyses, but in special cases one may wish to retain the relative 'weightings' between variables imposed by their original scales of measurement. Even then it is almost always sensible to mean centre the observations before use, if only to reduce numerical problems in computation. A pragmatic approach when in doubt as to whether to standardise is to perform the analysis with and without standardisation and see whether and what substantive differences in interpretation are obtained. Throughout our discussion here, except where explicitly stated otherwise, we will tend to assume that $\boldsymbol{Y}$ refers to mean centred observations. In most of the techniques we discuss $\boldsymbol{Y}$ could in general be replaced by $\boldsymbol{Z}$, a matrix of standardised observations (often referred to as 'z scores').

6.3.1 Principal components

We start our brief review of multivariate methods with perhaps the most commonly used of these techniques, *principal components analysis.* We mentioned earlier that a useful way to view the data matrix, $\boldsymbol{Y}$, is as a scatter of n points in a p-dimensional data space, where $(y_{i1}, \ldots, y_{ip})$ are the p coordinates of the ith observation in this space. Given that a primary interest in multivariate analysis is to try to understand any pattern or structure in the n observations over the p attributes or variables $(y_1, \ldots, y_p)$, a natural approach is to look for directions in the data space in which the n points are most 'spread out'. In other words, we wish to rotate the original coordinate system $(y_1, \ldots, y_p)$, of the data space to a new one, $(u_1, \ldots, u_p)$, where u_1 is measured along the direction in which the observations have the most 'separation', u_2 is measured along the direction in which they are next most 'separated' and so on. Put simply, we wish to find a more informative frame of reference within which to view the n points in the data space.

To put this idea into practice, we need to specify more precisely what is involved mathematically. Any 'direction' in the data space consists of a linear combination of the original variables, so 'rotating' the original coordinate system involves specifying a linear transformation of our original variables to new ones:

$$\begin{aligned} u_1 &= a_{11}y_1 + a_{12}y_2 + \cdots + a_{1p}y_p \\ u_2 &= a_{21}y_1 + a_{22}y_2 + \cdots + a_{2p}y_p \\ \vdots &= \vdots \\ u_p &= a_{p1}y_1 + a_{p2}y_2 + \cdots + a_{pp}y_p \end{aligned}$$

where the a_{jk} are constants. Hence the original coordinates of the ith data point, y_{ij}, $j = 1, \ldots, p$, become in the rotated coordinate system:

$$u_{ij} = a_{j1}y_{i1} + a_{j2}y_{i2} + \cdots + a_{jp}y_{ip}$$

In principal components analysis we restrict ourselves to 'rigid' rotations of the original axes. The new coordinate axes must be 'at right angles' to each other, or *orthogonal*, as were the original ones. This implies that the $(p \times p)$ matrix, $\boldsymbol{A}$, of coefficients, a_{jk}, should be an orthogonal matrix. That is, its row vectors, $\boldsymbol{a}_j^T$, are of unit length, so that $\boldsymbol{a}_j^T\boldsymbol{a}_j = 1, \quad j = 1, \ldots, p$, and in addition they are pairwise orthogonal, so that $\boldsymbol{a}_j^T\boldsymbol{a}_k^T = 0, \quad j \neq k$. Finally, we need to be explicit about what we mean by points being 'spread out' along a direction. In principal components analysis we decide to measure this by the variance of the observations in that direction.

So, to summarise, we wish to choose the vector of weights $\boldsymbol{a}_1$ in such a way as to make the variance of the transformed observations u_{i1} as large as possible, then to choose $\boldsymbol{a}_2$ so that u_{i2} have the next largest variance, subject to $\boldsymbol{a}_2$ being orthogonal to $\boldsymbol{a}_1$, and so on through to the choice of $\boldsymbol{a}_p$. It turns out that choice of the appropriate vectors, $\boldsymbol{a}_j$, corresponds to the solution of a common mathematical problem, that of finding the *eigenvectors* of a matrix. More specifically, $\boldsymbol{a}_j$ are the eigenvectors, ranked in decreasing order of their eigenvalues, of the $(p \times p)$ matrix $\boldsymbol{S} = \boldsymbol{Y}^T\boldsymbol{Y}/(n-1)$, the matrix of sample covariances between the p original variables (recall that we are taking $\boldsymbol{Y}$ to be mean centred). The corresponding *eigenvalues*, λ_j, are conveniently the variances of the observations in each of the new directions. Readers who are unfamiliar with the ideas of eigenvectors and eigenvalues, or who wish to follow up details, should consult one of the standard multivariate texts listed at the end of this chapter.

The jth new variable u_j is conventionally referred to as the jth *principal component*, whilst u_{ij} is referred to as the *score* of the ith observation on the jth principal component. The relationship between the jth principal component and the kth original variable can be evaluated through the covariance between them, which is given by $a_{jk}\sqrt{\lambda_j/s_{kk}}$ (where s_{kk} is the kth diagonal element of $\boldsymbol{S}$, the estimated variance of the kth original variable). This is often called the *loading* of the kth original variable on the jth principal component. Using these loadings it is possible to interpret what each of the principal components is measuring in terms of the original variables. For example, the first principal component may simply, as is typical, reflect the 'average' magnitude of all p original variables. Others may represent a contrast between the values of one set of the original variables and another.

Since each component is by definition the 'best' linear summary of variance left in the data after the previous components are accounted for, the first r components (where r is usually much smaller than p) may explain most of the variance in the data. The degree to which this is the case may be assessed directly from the relative magnitude of the eigenvalue, λ_j, associated with each principal component, since this is the variance of the component. Rotation of axes cannot change the overall original variance in the data, so since the principal components are orthogonal this must be $\sum_{j=1}^{p} \lambda_j$. The proportion 'explained' by the first r principal components is thus simply $\sum_{j=1}^{r} \lambda_j / \sum_{j=1}^{p} \lambda_j$. On the basis of this we may choose to retain only the first r new variables to study further, knowing that we are not losing much information by doing so. In particular, plots of the principal component scores on pairs of the first few principal component axes can reveal useful information about structure in the original observations.

Note in passing that principal components analysis is not invariant to standardisation of the original variables; an analysis performed using $\boldsymbol{Y}$ will not yield the same components as an analysis using standardised values $\boldsymbol{Z}$. In general, unless there are good reasons to the contrary, standardisation of the initial variables is recommended before a principal components analysis is performed. In this case the eigenvalues and eigenvectors that form the basis of the analysis are derived from the matrix $\boldsymbol{R} = \boldsymbol{Z}^T\boldsymbol{Z}/(n-1)$, which is simply the correlation matrix between the p original variables.

Principal components analysis, then, identifies a small number of linear combinations of attributes which account for a large proportion of the variability of observations in attribute space. The first few principal components are a *projection* of the original high-dimensional attribute space onto a subspace of lower dimension which retains the most important 'separation' of observations in terms of attribute values. The principal component loadings allow us to interpret the subspace in terms of the original variables.

These ideas can be illustrated with reference to one of our case studies, that concerned with climate variation in South America. The first four components, based upon the correlation matrix of all 16 original variables in the data set, account for 88.1 per cent of the total variation. In order, the eigenvalues account for 41.4%, 24.8%, 13.6% and 8.3% respectively. We do not reproduce the full matrix of loadings here. Suffice it to say that the following variables load highly on the first component: mean annual temperature; minimum January temperature; mean July temperature; minimum July temperature; and temperature range. All have positive loadings, suggesting that this component describes the main characteristic of latitudinal variation in temperature, from the hot tropical climates of the north to the extreme cold of the far southern

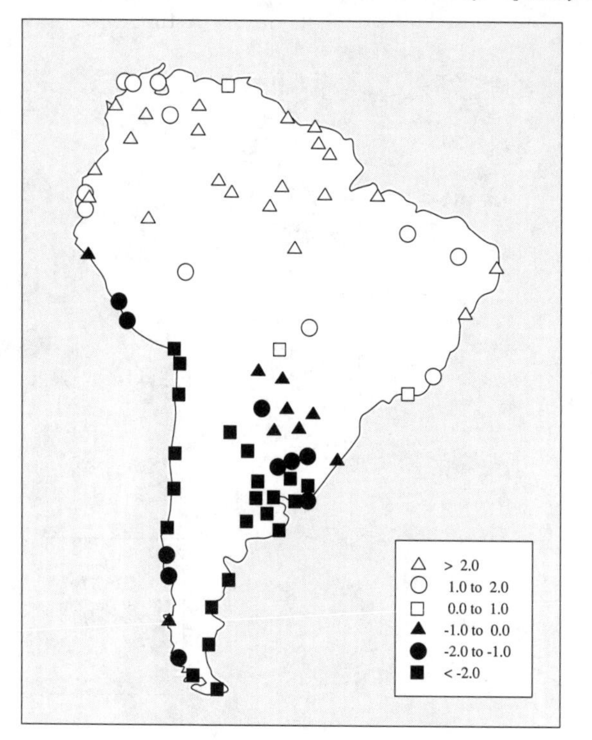

Fig. 6.1 First component scores for South American climate

zones. This may be confirmed by a map of the scores on this first component (Figure 6.1). The Amazon region gives high scores, while to the south, both along the west and east coasts, there is a preponderance of negative scores.

Variables with high positive loadings on the second component are: mean January temperature, maximum January temperature, and maximum July temperature. Those loading negatively include: number of days with precipitation over 1 mm, and mean number of days with greater than 1 mm of precipitation in July. In general, this component separates dry zones from wetter ones. There is a cluster of positive scores along the north west coast of Venezuela, where conditions are quite arid, but also high positive scores along the north west coast of Chile (the Atacama Desert). Negative scores are found further south, where temperatures are cold, but also on the west coast of Ecuador and Peru, where rainfall is often heavy (Figure 6.2).

Component three is characterised by high negative loadings of the precipitation ratio (July/January) and the rain days ratio (also July/January). This component reflects the seasonality in rainfall, with positive scores for locations where summer precipitation is relatively high and negative scores where the ratio of winter to summer rainfall is extremely high. July precipitation and maximum January temperature are the only variables with

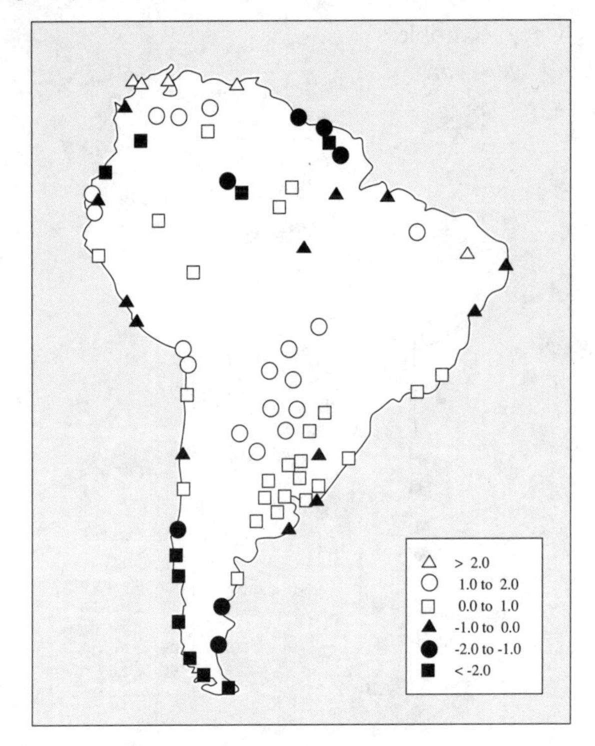

Fig. 6.2 Second component scores for South American climate

moderately high positive loadings on component four. The distribution of component scores is more fragmented here, with tropical parts of Venezuela scoring high and the drier coastal areas on the Caribbean having negative scores. Again, the Pampas region has positive scores.

As well as giving us an insight into climate variation, one of the main uses of principal components analysis as applied to such data is to derive a set of uncorrelated, synthetic variables that may subsequently be used in climate classification. We shall see shortly how to perform such classifications and what results are obtained when we use the components derived here in such tasks.

Before leaving the subject of principal components analysis we should emphasise one point about the use of the technique in a spatial context. There is no guarantee that the directions which maximally separate observations in attribute space (as identified by the principal components) will *necessarily* be those that correspond to the configuration of observations in geographical space. Indeed, in general, this is unlikely. Therefore, principal components analysis will not necessarily be of use if the objective of the analysis is to determine which combinations of attributes demonstrate the most significant spatial pattern. Including the spatial coordinates (s_1, s_2), of the observations as extra variables into the principal components analysis is not particularly helpful in resolving this problem either, since the 'geography' is then treated on an equal basis to any other attribute. Transformation of the attribute space is confounded with that of the geographical space in the analysis, preventing any interpretation of the results in terms of a meaningful and readily understood geographical structure.

6.3.2 Factor analysis

Factor analysis is superficially somewhat similar to principal components analysis, and the two techniques can be (and often are) confused. This is unfortunate, as factor analysis is both conceptually and mathematically very different from principal components analysis. It is based on an assumption that the observed correlations between the attributes $(y_1, \ldots, y_p)$ are mainly the result of some *a priori* underlying regularity or structure in the data, rather than one that is defined purely on the basis of a mathematical criterion, such as maximising the variance or 'separation' of observations, as in principal components analysis. More specifically, an *a priori* model is proposed whereby each of the observed variables is assumed indirectly and partially to measure a fixed number, r, of pre-defined characteristics or *latent factors* which cannot be directly measured themselves.

The analysis attempts to identify how many of these factors are significant in the data, in what relative order, and how each of these relate to the observed attributes $(y_1, \ldots, y_p)$. The most commonly quoted applications arise in psychological or educational testing, where $(y_1, \ldots, y_p)$ are scores on a series of tests, each of which is assumed to reflect, to a differing extent, certain latent factors; for example general intelligence, verbal facility, arithmetical ability and so on. Such factors cannot be measured directly, but are partially measured by each of the observable test scores. In a spatial context one might have a similar situation in applications in geology, where 'controls' such as palaeotemperature, deformation of bedrock and rock permeability may be considered important in determining the occurrence of lead or zinc deposits, but may not be themselves directly observable. Methods are sought to identify such controls indirectly from a set of observations on related chemical, mineralogical and rock-deformation variables.

In factor analysis the parts of an observed variable that relate to the *latent structure* or shared factors are usually called *common*, and the remaining part that is idiosyncratic to that variable is referred to as *unique* or *specific*. Under this assumption, the unique part of a variable cannot contribute to relationships between the observed variables. It follows that the observed correlations between $y_1, \ldots, y_p$ must result from the common factors shared by the variables. The idea is that these assumed common factors will account for all the observed correlations in the data, and will also be smaller in number than the original variables. Factor analysis can thus be thought of as a technique by which a minimum number of hypothetical variables are specified in such a way that after controlling for these hypothetical variables all the remaining correlations between the original observed variables become zero. The analysis assumes the existence of residual 'uninteresting' unique components in each of the original variables, not related to the common factors and not contributing to the correlation between the original variables.

So we represent the original variables $(y_1, \ldots, y_p)$ as linear combinations of the r hypothesised common factors $(u_1, \ldots, u_r)$ plus a residual component unique to each of the original variables:

$$
\begin{aligned}
y_1 &= a_{11}u_1 + a_{12}u_2 + \cdots + a_{1r}u_r + \epsilon_1 \\
y_2 &= a_{21}u_1 + a_{22}u_2 + \cdots + a_{2r}u_r + \epsilon_2 \\
\vdots &= \vdots \\
y_p &= a_{p1}u_1 + a_{p2}u_2 + \cdots + a_{pr}u_r + \epsilon_p
\end{aligned}
$$

and choose the constants, a_{jk}, to account as much as possible for correlations between the original p variables.

At first sight it looks as if we could get a *factor solution* like this by 'inverting' the relationships between the first r principal components and the original variables, given in the previous section. However, the fundamental difference between principal components and factor analysis is that the former is simply concerned with accounting for *variances* of each of the original variables by rotating axes, whilst in factor analysis the emphasis is on explaining *correlations* in the original variables in terms of a model which proposes a certain number of common factors. If the first r principal components were employed as a 'solution' for an r-factor model, then it would certainly not in general be a very good 'solution', since it focuses on maximising variance and much of the correlation in the original variables would be left 'unexplained' by the first r principal components.

In fact, the mathematical details of obtaining good solutions to factor analysis are more complex than those of principal components analysis and we do not present them here. Readers interested in details can follow them up in the references listed at the end of the chapter. One general problem with factor solutions is that they are not unique. One factor solution can be transformed into another without violating its basic mathematical properties in terms of explaining correlations. Such transformations are those that rotate the original factor solution (either an oblique or an orthogonal rotation). Having obtained an initial factor solution it is therefore typical to rotate this to obtain a set of factors more meaningful in theoretical terms when the loadings, a_{jk}, on the original variables for each factor are considered. There are several rotational methods that can be employed, as well as several ways of obtaining the original factor solution. Such techniques are widely available and relatively easily accessible in the kind of statistical packages mentioned earlier.

Once obtained, the results of a factor analysis may be used in a similar way to the results of principal components analysis. That is, factors may be interpreted in terms of their loadings for the original variables and plots of the factor scores for each of the original observations on pairs of the factor axes can reveal useful information about structure in the original observations. The r factors may be considered a useful reduction in the dimensionality of the original data, in a similar way that principal components can be regarded, although of course with a different interpretation. However, it should be emphasised that unlike principal components analysis, which does not depend on an imposed model, the interpretation of the results of a factor analysis and its usefulness in a study are largely a function of whether the *a priori* assumptions concerning the number and conceptual validity of the proposed common factors are in fact justified for the data under study.

6.3.3 Principal coordinates and multidimensional scaling

At one level both principal components analysis and factor analysis can be thought of as techniques which represent observations from an original p-dimensional data space in a subspace with a smaller number of dimensions, within which certain characteristics of the original observations are highlighted. *Metric* and *non-metric scaling* techniques, which are often referred to under the general heading of *multidimensional scaling*, are also concerned with establishing 'interesting' graphical representations of a set of multivariate data in a small number of dimensions. However, rather than start from the $(n \times p)$ data matrix, used by both principal components and factor analysis, they start from an $(n \times n)$ matrix, $\boldsymbol{D}$, consisting of elements which represent some form of *dissimilarity* between every pair of observations in the original sample. We will have more to say about how such dissimilarities may be specified shortly. The idea of the analysis is to construct from this matrix a low-dimensional graphical representation or 'view' of the observations, usually in two or three dimensions, in which each of them is represented by a point, and the Euclidean distance between any two points in this space matches as closely as possible the original dissimilarity specified between the two corresponding observations. Such a 'model' would then hopefully give a good visual impression of any pattern in the original dissimilarities and highlight relationships between groups of observations.

Our original $(n \times p)$ data space is of course exactly such a 'model', but it is p-dimensional and thus in general one with too large a dimension to be able to be easily viewed and interpreted. Also, it is a 'model' which is appropriate only when dissimilarity between observations is assumed to equate with p-dimensional Euclidean distance. Multidimensional scaling in general allows us to seek a representation in a lower dimensional space than the original p dimensions *and* to use a more general definition of dissimilarity than p-dimensional Euclidean distance. It is for this reason that we may be prepared to 'give up' our original $(n \times p)$ representation of the data in favour of an $(n \times n)$ starting point.

Recall our earlier comments about using principal component scores as a way of projecting the original p-dimensional attribute space onto an

r-dimensional subspace of the first few principal components. We might argue that since the original $(n \times p)$ data space is a perfectly reasonable graphical representation of the observations when dissimilarity is measured in terms of Euclidean distance, then the r-dimensional subspace projection formed from scores on the first few principal components should represent a suitable multidimensional scaling solution when the dissimilarity matrix $\boldsymbol{D}$ is specified in terms of Euclidean distance. This is in fact exactly the case. When dissimilarity is specified as p-dimensional Euclidean distance, multidimensional scaling simply reproduces the principal components scores, and in that case the analysis is usually termed *metric* scaling or alternatively *principal coordinates analysis*.

The different nomenclature for a technique which produces essentially the same projection as principal components analysis is useful, since it reminds one that the mathematics of principal coordinates analysis is different to that of principal components analysis. In particular, although we do not give details here, multidimensional scaling in general involves the spectral decomposition of the $(n \times n)$ dissimilarity matrix, whilst principal components analysis is concerned with the spectral decomposition of the $(p \times p)$ covariance matrix (or correlation matrix if standardised variables are used). Now whilst it is usually the case that the number of variables p is less than the number of observations n, there are applications where the reverse is the case. Principal coordinates analysis, being a special case of multidimensional scaling, then provides a more numerically efficient way of arriving at what is essentially a principal components projection, than principal components analysis itself. This is one benefit of principal coordinates analysis, although it should be noted that such an analysis can only reproduce the principal components scores. The loadings of the original variables on the principal components cannot be derived from principal coordinates analysis, since the analysis starts only with a dissimilarity matrix of Euclidean distances between observations and there is no notion of variables as such.

Although this last point might be felt to be a drawback of the technique, looked at differently it relates precisely to a second benefit of principal coordinates analysis. This is that the method can be used to find a useful low-dimensional graphical projection of a dissimilarity matrix between observations, when data are originally presented only in the form of dissimilarities and we do not have the typical structure of n observations on p variables which we have hitherto considered. For example, in a spatial context, starting with a matrix of distances between the major towns and cities in the UK, principal coordinates analysis allows us to construct a two-dimensional visualisation or 'map' of their relative locations which, to a good approximation, reproduces the given distances between them. Notice the judicious use of the word 'relative' here—the resulting 'map' may well not look immediately like the UK that we are used to; it could be a mirror image or alternatively upside down!

These benefits become more pronounced when we allow dissimilarities to be defined in ways other than just Euclidean distance; as a result, principal coordinates analysis then becomes more the general multidimensional scaling.

The mathematics of the technique remains much the same, but we are now free to define dissimilarities between observations in any way we chose. We might replace Euclidean distance with some form of weighted Euclidean distance, where the weights reflect the differing importance of the original variables. Many other possible measures of dissimilarity could be used; for example, the so called *city block* metric where $d_{ij} = \sum_{k=1}^{p} \left| y_{ik} - y_{jk} \right|$. In addition, we can define various dissimilarity measures appropriate to the case where some of the original variables are measured on nominal or ordinal, rather than quantitative, scales. Finally, our data may be presented directly in the form of observed dissimilarities, and not as observations on variables at all. Multidimensional scaling thus allows us, for example, to construct a two-dimensional graphical configuration of a set of locations on the basis of a given matrix of the travelling times between them, or even the cognitive 'distance' between them, as obtained from individuals' judgements of their separation. In this way, we obtain 'maps' of different kinds of space, maps which perhaps give a truer representation of spatial relationships than traditional ones.

To summarise, the object of multidimensional scaling techniques is to uncover a space of minimal dimension such that when the observations are plotted in this space the distance between them reproduces to a good approximation the observed overall dissimilarity structure between them. The researcher can therefore get an intuitive feel for the basic structure of such data. We have not discussed the specific mathematics of the technique here and refer the reader to the standard texts listed at the end of the chapter for details.

6.3.4 Procrustes analysis

The multivariate techniques which we have discussed all result in graphical configurations of the original observations in an r-dimensional space. Typically, in the course of a multivariate analysis a number of such configurations may be obtained, arising from the use of different criteria. It then becomes of interest to compare these various configurations to see if there are overall similarities. Recall, for example, that techniques like multi-dimensional scaling are concerned only with the *relative* locations of observations; different configurations may therefore be better compared if first reflected or rotated.

Procrustes analysis is concerned with exactly this problem, the comparison of two or more multidimensional graphical representations, or configurations, of n objects. In the simplest case, comparing two configurations $\boldsymbol{X}$ and $\boldsymbol{Y}$ of equal dimension $(n \times r)$, a Procrustes analysis performs the 'best' combination of translation, rotation, reflection or dilation of $\boldsymbol{Y}$, in order to most closely match the $\boldsymbol{X}$ configuration of the same n points, the criterion used being one of minimising Euclidean distances between corresponding points of the two configurations. The analysis provides summary measures of the 'fit' between the two configurations as well as a graphical indication of their match after the optimal scaling and rotation has been performed. It is typical to 'view' this as a plot, showing the location of each of the observations in one of the original configurations with arrows from each pointing to their corresponding location in the 'best' match of the other, determined from the analysis. If all arrows in this plot were very short we would conclude that the configurations were broadly 'similar'. Intersecting, or long, arrows indicate poor fit for the observations concerned. Further discussion and mathematical details of the technique may be found in the standard multivariate texts listed at the end of the chapter.

We mention the technique here because of a potentially useful spatial application. This is where we have a map of locations given as $\boldsymbol{X}$, an $(n \times 2)$ spatial configuration, with $\boldsymbol{Y}$ representing two attributes that have been measured at each location. Procrustes analysis would then rotate and scale the configuration of the n observations in the bivariate attribute space to most closely align with the map, giving useful insight into whether the relative positioning of the observations in the attribute space in any way corresponds to their relative spatial locations on the map.

In the more general case, where one has $p > 2$ attributes measured over a map, the problem is one of comparing the $(n \times p)$ attribute configuration $\boldsymbol{Y}$ with the $(n \times 2)$ spatial configuration $\boldsymbol{X}$. It is perhaps useful to visualise this in terms of the case where $p = 3$. One has a map of the observations and a separate configuration of the same observations as points in three-dimensional attribute space. We seek a view of the attribute 'cube', from which the arrangement of points looks as much as possible like their arrangement on the map, given that we can stretch or reflect the view of the cube to get a good match.

The conventional general solution to comparing configurations with unequal dimensions is *augmented Procrustes analysis*. Here, the smaller dimension configuration is augmented with columns of zeros and then a conventional Procrustes analysis is performed. In the spatial example this would amount to approximating the spatial configuration in the higher dimensional attribute space before performing the Procrustes analysis. All distances used in the analysis are p-dimensional distances; so we are asking to what extent the map accounts for the *entire* attribute configuration. This may not be desirable in the spatial case. The requirement may be, rather, to look for any two-dimensional projection or plane in the attribute space which, when shifted, rotated, reflected or scaled, fits well with the configuration of observations on the map.

An alternative approach, referred to as the *reducing Procrustes method*, achieves just this result. In contrast to the augmentation method, reducing Procrustes first approximates the attribute configuration in the lower-dimensional spatial configuration and then performs the standard Procrustes analysis, effectively asking whether the attribute has any planar component that is correlated with the map. Such an analysis is implemented through an iterative adjustment to the standard Procrustes algorithm, details of which are discussed in the references given at the end of this chapter.

The results of applying reducing Procrustes analysis to some of the variables in the South American climate data are shown in Figure 6.3. The three

Fig. 6.3 Procrustes analysis of temperature variables for South America

variables used here are average annual temperature and mean temperature in January and July. Clearly, other variables could have been chosen, for example, a mixture of precipitation and temperature measures, although some care has to be taken in interpreting the results of this technique when the variables used are on very different measurement scales. The arrows from each of the sites indicate where these sites would be placed in the best fitting two-dimensional configuration of the three attributes. The analysis is clearly identifying the general north–south temperature gradient, but it is interesting to note where individual sites are placed on this scale. For example, there is a suggestion that on the basis of the three temperature variables used, the four most southern sites, should really be further 'south' than they are; similarly, there are some central sites that are more 'like' some of their more northerly counterparts.

6.3.5 Cluster analysis

In some problems the object of the analysis is to see whether individuals (such as the climate stations above) can be formed into a set of groups on the basis of the measured variables. The number of groups might not be specified in advance; the investigator seeks a system such that the individuals within a group resemble each other (according to the values taken by the variables) more than do individuals in different groups. The techniques of principal components analysis, factor analysis or multidimensional scaling may be useful in suggesting groupings of observations, as discussed above, but only indirectly, in that they do not explicitly deliver a division of the original observations into groups. Cluster analysis (also referred to as *numerical taxonomy*) addresses this problem more directly, attempting to split the individuals into a number of groups on the basis of the observed variables.

Several different clustering methods have been proposed and these are discussed fully in the references given at the end of this chapter. The starting point for all the techniques is the same as that used in multidimensional scaling, namely calculation of an $(n \times n)$ matrix, $\boldsymbol{D}$, of dissimilarities between every pair of observations in the original sample. Euclidean distance is often used but there are many other possible dissimilarity measures, as we have discussed previously. We might, for example, choose to perform a principal components analysis of the original variables first and then calculate a dissimilarity matrix for a subsequent cluster analysis on the basis of the Euclidean distance between observations in the subspace defined by the first few principal components. Whichever dissimilarity matrix is defined, cluster analysis proceeds to break observations into groups on the basis of the chosen dissimilarity measure.

Hierarchical clustering techniques start with all observations in separate groups and then at each stage successively join the most similar observations or groups of observations, according to the optimisation of some criterion at each step. Alternatively, they may start with all observations in one group and successively split observations or groups of observations, on a similar basis. One of the simplest such methods, *single linkage* clustering (also called *nearest neighbour* clustering), starts with all observations in separate groups and then at each step finds the pair of observations which have the smallest nearest neighbour distance of all pairs whose two members are in different groups. The two groups to which these belong are then fused. One can impose some overall 'stopping rule' on the process in terms of the size of nearest neighbour distance that permits fusion, or a required final number of groups.

The *minimum spanning tree* can also be used to achieve single linkage clustering. A spanning tree is a set of links joining all of the observations, such that firstly, every observation is connected to every other observation by some subset of such links and secondly, no 'closed circuits' occur in the links. If the length of a link is defined by the dissimilarity between the two observations at its endpoints, then the length of a spanning tree is the total length of all the links in the tree. The minimum spanning tree is the spanning tree of shortest length among all spanning trees that could be constructed. Such a minimum spanning tree can be constructed by a simple iterative algorithm. Start with any observation and join it to the 'most similar' other observation (as defined by the dissimilarity measure being used). Next, find that currently unconnected observation that has the smallest dissimilarity with *any* of the observations already connected and join it to that observation. Repeat this process until all nodes are connected. If at any stage a tie is encountered one of the possibilities is chosen arbitrarily. Quite apart from its use in clustering, as described below, the minimum spanning tree is itself a useful exploratory device. It is a convenient method of highlighting 'close' neighbours on the basis of the chosen dissimilarity measure. We might derive a minimum spanning tree on the basis of a dissimilarity measure defined in respect of p attributes measured at a set of spatial locations, and then superimpose this spanning tree onto a map. This would give a good visual indication of whether observations 'close' in attribute space are also 'close' in geographical space, and, if not, which particular observations are 'out of line'.

Using the minimum spanning tree in simple single linkage clustering, one starts with all observations in one group and, at each step, successively splits observations into further groups on the basis of removing the largest link remaining in the minimal spanning tree. Thus, at each step, one extra group of observations is formed. If two links have the same length both are removed and two extra groups formed. Successive clusters are the same as those produced by the single linkage process described earlier. Again, one can impose some overall stopping rule on the process in terms of relative size of link that can be removed, or a required final number of groups.

Ex 6.2

Single linkage is only one possible criterion for hierarchical clustering. Starting with all observations in different groups one could fuse those two groups that have the smallest average dissimilarity, or those that would lead to the smallest increase in the within-group sum of squares, or some other such criterion. Different criteria lead to clusters with different properties; some, like single linkage, produce clusters which are invariant to a monotonic transformation of the dissimilarity matrix; others do not. Some, like single linkage clustering, lead to 'elongated' clusters, the so called *chaining* effect, because an observation can join a group on account of its similarity with just *one* member of the group. Other criteria, like those based on average dissimilarity or within-group sums of squares, produce more 'spherical' or homogeneous clusters.

Such hierarchical clustering methods can be contrasted with *optimisation clustering* techniques. In the former, although a criterion is optimised at each step, there is no guarantee that, if one ends up with k groups, this is the partition of the observations which would optimise this same criterion over of all possible partitions of the observations into k groups. Optimisation clustering techniques, on the other hand, do attempt to split observations into a pre-specified number of groups such that the specified criterion is optimised globally over all possible splits. The disadvantage is that one has to specify a number of groups *a priori*. Additionally, many of the algorithms suggested do not guarantee a global optimisation, but one conditional on a particular initial configuration of k groups. One thus needs to experiment both with different numbers of pre-specified groups and with different starting configurations, until a 'satisfactory' solution is obtained.

One of the simplest optimisation methods is *k-means* clustering. Given a pre-specified value of k and an initial starting split of observations into k groups, this method reallocates observations between the groups so that the within-group variance of the dissimilarity measure is minimised. Global optimisation is usually not practically possible, but various algorithms exist for finding an acceptable local optimum given a reasonable starting configuration, and for determining such a starting configuration simply. As with hierarchical techniques, many alternative optimisation techniques have been suggested.

Faced with the range of different clustering methods that are available, it is usually a good idea to use several methods in any particular application. If these different methods tend to agree broadly on group membership for a particular choice of the number of groups, then this is probably an indication that this grouping is reasonably 'robust' and not just an artefact of the particular method being used. Clustering methods will always produce groups; it is the responsibility of the analyst to determine whether these have any substantive meaning.

Let us now illustrate the application of these ideas to one of our data sets, that on climate in South America. There is a long tradition of attempts to derive 'objective' classifications of world climate, beginning with Köppen in the early part of this century, continuing with Thornthwaite in the 1940s, and in the 1960s seeing the application of the kinds of multivariate methods we have reviewed above. The early classifications relied not so much on observed climate data, as on 'external' factors such as vegetation. Subsequent numerical taxonomic approaches have taken a broad range of climate variables and used these to derive regional classifications. However, because of the relative lack of data on humidity, many classifications are forced to rely on annual and seasonal data on temperature and precipitation. Earlier, we used such variables in a principal components analysis for our South American climate example. We now attempt to derive a classification of South American climates using the principal components that were derived there.

Suppose we use k-means clustering. To begin with, suppose we specify that we are seeking to group sites into four classes and use all four principal components derived earlier. Since these principal components were themselves derived from standardised original variables, we do not standardise them again when using them for clustering. In other words we allow them to have a differential weighting in the clustering which reflects their relative importance in explaining variance in the original standardised observations. We obtain the results shown in Figure 6.4.

Climate type 1 comprises sites in the far north of the continent, together with those sites in the Atacama desert region. This is broadly a dry, arid, regional type, but we might feel it a little odd to mix very warm areas north of the equator with much colder zones further south. Class 2 corresponds with the warm temperate Pampas region, while class 3 comprises southern sites, both on the west and east coast; we might expect to see these in separate classes. Class 4 is the extensive tropical belt.

These results seem reasonably sensible. However, at least two regional classes are not as homogeneous as we might expect. It is probably worth attempting a six-class solution; the results are shown in Figure 6.5. This is more

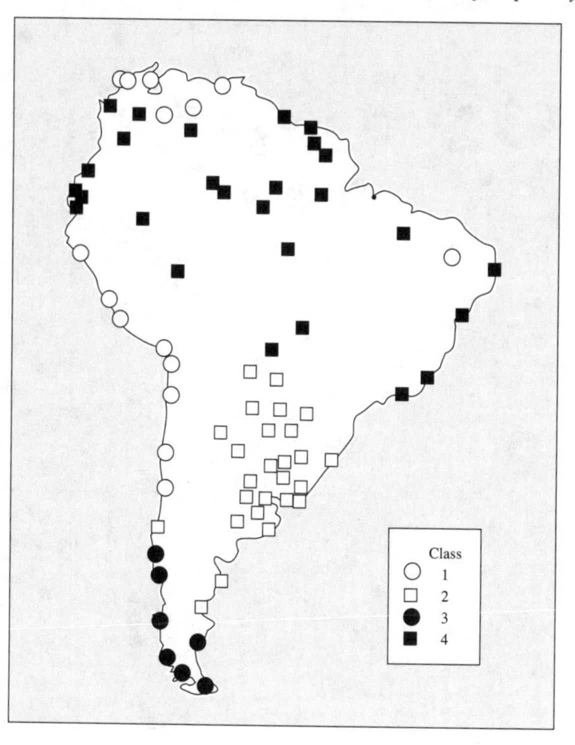

Fig. 6.4 South American climate, k-means cluster solution ($k=4$)

satisfactory. Class 1 comprises only four sites, but these are all in west Venezuela, where there is a high ratio of winter to summer rainfall. Regional type 2 is again the distinctive Pampas region. The broad tropical zone in the four-class solution is now broken up into two zones, one (3) is the tropical zone to the south of the Amazon where it tends to be drier in the winter, the other (5) is a classically tropical zone of high rainfall. Type 4 comprises the sites that are dry, cold desert areas, while type 6 is a very distinctive region, where the four sites are in the south-west of Chile; a region with high rainfall but moderate temperatures that encourage forest growth. Clearly, the six-class solution paints a richer picture of climate variation than does the four-class solution.

We should emphasise again that the results we get are dependent wholly on the selection of variables, and of course on the particular set of sites chosen. As mentioned in Chapter 5, we omitted sites above 200 m elevation. It is also apparent on any of the maps we have shown that the distribution of sites is spatially uneven, reflecting, in part, the distribution of population. Thus there are relatively few climate stations in much of the Amazon region, but quite dense coverage in the Pampas region and along the west coast. Results of multivariate classification using geographical data, as with any other spatial

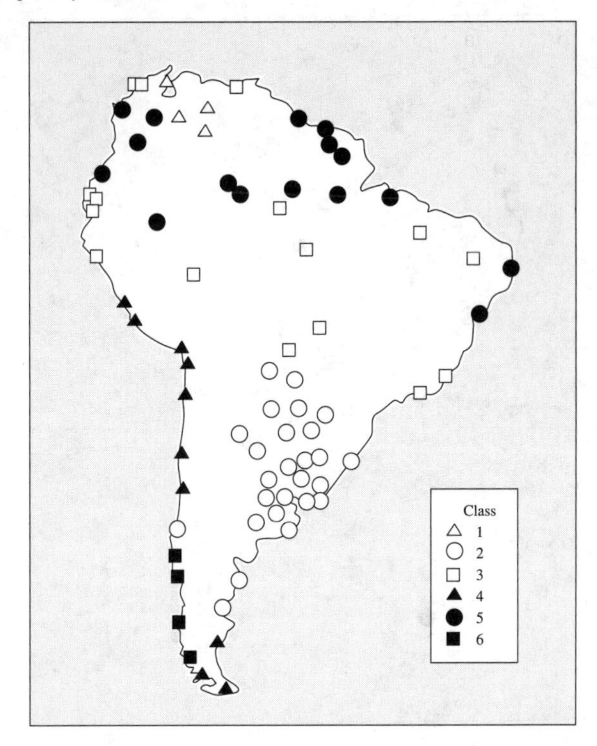

Fig. 6.5 South American climate, k-means cluster solution ($k = 6$)

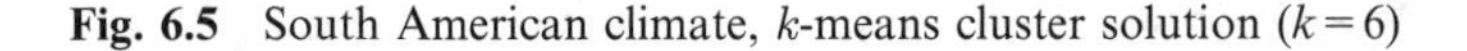

analysis, depend on appropriate data input, both that relating to location and attributes.

Before leaving our consideration of clustering, we note that in some spatial applications there may be a requirement to constrain clustering in such a way that locations within a cluster are also geographically 'close'; in other words, to ensure that any clusters uncovered by the analysis form coherent spatial regions. If clustering is carried out solely on attribute values, there is of course no guarantee that this will be the case. In cases where this is a requirement there are various ways in which such *contiguity constraints* may be introduced into the analysis. The simplest is to include the spatial coordinates of locations into the analysis as extra variables and therefore allow them to influence the dissimilarities between observations. However, this does not insist on spatial contiguity within clusters, it merely encourages it. Modifications which impose stronger and more sophisticated contiguity requirements are possible. Of course, the imposition of a contiguity constraint in some contexts is nonsensical. Given that we have hot desert climates in different locations on the earth's surface we will hardly find it illuminating to require our climate regions to be contiguous! Contiguity constraints only really make much sense when applied to area data and we will return to a discussion of this issue in Chapter 8.

As with the other multivariate methods we have discussed, we have only given an overview of clustering techniques and not enough mathematical details to use them without further reference. Again, we direct readers to the standard texts listed at the end of this chapter.

6.3.6 Canonical variates

The multivariate methods we have discussed so far have been concerned with exploring multivariate data for structure, rather than attempting to investigate and highlight differences in a pre-defined, known structure in the data. An alternative situation is when the observations are known *a priori* to fall into different groups and our objective is to distinguish statistically between such groups of observations, given the set of p variables $(y_1, \ldots, y_p)$ which have been observed.

We may be interested simply in a description of group differences, or in going further and using such descriptions to classify new observations into one of the groups in such a way that the probability of incorrect classification is small. For example, in remote-sensing applications, which we discuss in Chapter 8, there is a need to classify land cover in images of unknown regions, by relating the reflectance values to those obtained from images of areas with known land cover. This latter area is generally referred to as *discriminant analysis*. Here we simply look at a common method for describing and understanding group differences, that of *canonical variates*; we do not discuss the more general problem of discrimination.

The objective of *canonical variate analysis* is to weight and linearly combine the original variables, y_i, into a new set of variables, u_i, so that the pre-defined groups of observations are forced to be as statistically distinct as possible when measured on these new variables or *canonical variates*. These new variables are again linear combinations of the original variables, familiar from our earlier discussion of principal components analysis:

$$\begin{aligned} u_1 &= a_{11}y_1 + a_{12}y_2 + \cdots + a_{1p}y_p \\ u_2 &= a_{21}y_1 + a_{22}y_2 + \cdots + a_{2p}y_p \\ \vdots &= \vdots \\ u_r &= a_{r1}y_1 + a_{r2}y_2 + \cdots + a_{rp}y_p \end{aligned}$$

where again the a_{jk} are constants. Hence the original coordinates of the ith data point, y_{ij}, $j = 1, \ldots, p$, become in the canonical variate coordinate system:

$$u_{ij} = a_{j1}y_{i1} + a_{j2}y_{i2} + \cdots + a_{jp}y_{ip}$$

If the observations are initially in k groups then in general, r, the number of canonical variates obtained, will be the smaller of $k-1$ or p. The criterion used to select the weights a_{jk} is that successive canonical variates should be uncorrelated and that each should maximise the remaining variance between groups and minimise variance within groups, after the previous canonical variates have been accounted for. The appropriate decomposition of between-group variance and corresponding choice of successive vectors of canonical variate weights, $\boldsymbol{a}_j$, again results from a set of eigenvalues and corresponding eigenvectors. In this case they are not those of the covariance matrix (or correlation matrix if standardised variables are used) as in principal components analysis, but rather those of the product of the inverse of the 'within-groups' covariance matrix and the 'between-groups' covariance matrix (or corresponding correlation matrices if standardised variables are being used). Further explanation and mathematical details of the derivation of the canonical variate weights a_{jk}, may be found in the standard multivariate texts referenced at the end of this chapter.

Once obtained, the canonical variates can be used in analogous ways to those used in principal components or factor analysis, to understand the structure of the main differences between the groups. In particular, canonical variate loadings may be interpreted to understand their relationship to the original variables, and plots of the canonical variate scores of the observations on pairs of the first few canonical variates can prove very useful in visualising and further interpreting group differences.

6.3.7 Canonical correlation

Canonical variates are useful when a pre-defined categorical grouping exists in the data. *Canonical correlation analysis* applies where one has two groups of continuous variables: $(y_1, \ldots, y_p)$ considered jointly to be 'response' variables and $(x_1, \ldots, x_q)$ considered jointly to be 'explanatory' variables. One wishes to uncover any important 'joint' relationships between these two sets of variables. This can be thought of as a generalisation of the situation dealt with in the previous section where the *a priori* categorical grouping is replaced with a continuous 'configuration' of observations defined by their values on the set of 'explanatory' variables $(x_1, \ldots, x_q)$.

The basic approach is to derive successive pairs of linear combinations, u_j and v_j, one from each of the two sets of variables, 'explanatory' and 'response' respectively, in such a way that each successive pair is maximally correlated after the previous pairs have been accounted for. These pairs of *canonical variables* are chosen to be uncorrelated pairs of variables. If $q < p$, then the analysis will determine $r = q$ pairs of such canonical variates, u_j and v_j, $j = 1, \ldots, r$, u_j being a linear combination of $(x_1, \ldots, x_q)$ and v_j a linear combination of $(y_1, \ldots, y_p)$. Each successive pair, $j = 1, \ldots, r$, therefore represents that pair of linear combinations of the two sets of variables $(x_1, \ldots, x_q)$ and $(y_1, \ldots, y_p)$ which are 'next most correlated' with each other, given the previous pairs. Further explanation and mathematical details may be found in the standard multivariate texts referenced at the end of this chapter.

Once obtained, the canonical variables can be used to reduce the dimensionality of the data by subsequently using scores on the first few (u_j, v_j) pairs, rather than values on the original variables. They can also be used to gain insight into the structure of the data by interpreting the linear combinations that make up each of the canonical variates in terms of their loadings on the original variables.

One particular application in the spatial context is in seeking spatial structure in p multivariate attributes. Then we may take $q = 2$ and $(x_1, x_2) = \boldsymbol{s}^T = (s_1, s_2)$ (the spatial coordinates of the observations), with $\boldsymbol{y}^T = (y_1, \ldots, y_p)$ being p attributes at these locations. Canonical variate analysis then gives rise to just two pairs of canonical variates, $u_j = \boldsymbol{a}_j^T \boldsymbol{s}$ and $v_j = \boldsymbol{b}_j^T \boldsymbol{y}$, $j = 1, 2$. The analysis determines those two linear combinations of attributes which are most correlated with any two linear combinations of spatial coordinates and identifies which combinations of coordinates these are.

This is useful information, but interpretation is not straightforward. The 'meaning' of the attribute canonical variates, v_j, can be inferred in terms of the original variables by inspection of standardised forms of the weight vectors $\boldsymbol{b}_j$ (at least in respect of 'major contributions', which is usually all that is required). However, the question that remains is what spatial direction each of these is correlated with. The spatial canonical variates, u_j, to which the v_j correspond, will in general represent projections of the original spatial locations which are difficult to visualise or interpret. The problem is that a standard canonical correlation analysis is successful in achieving a reduction in the attribute space to the two 'most spatially correlated' combinations of attributes, but suffers from the problem that the spatial directions to which these relate are an undesirable distortion of space which is difficult to interpret.

One suggestion to alleviate this problem is to consider how the information provided by the canonical correlation analysis concerning the most spatially appropriate combinations of attributes can be visualised directly in relation to the original untransformed geographical space. This requires some fairly straightforward manipulation of the output from the standard analysis. If $\boldsymbol{A}$ is the matrix with rows $\boldsymbol{a}_j^T$ and $\boldsymbol{B}$ the matrix with rows $\boldsymbol{b}_j^T$ for $j = 1, 2$, then the set of relationships defining the canonical variates $\boldsymbol{u} = (u_1, u_2)^T$ and $\boldsymbol{v} = (v_1, v_2)^T$ can be written $\boldsymbol{u} = \boldsymbol{A}\boldsymbol{s}$, $\boldsymbol{v} = \boldsymbol{B}\boldsymbol{y}$, which can then be thought of differently as $\boldsymbol{s}$ and $\boldsymbol{w} = \boldsymbol{A}^{-1}\boldsymbol{B}\boldsymbol{y}$, where the first is now the original untransformed spatial coordinates.

We then argue that $\boldsymbol{w}$, so defined, is that two-dimensional projection of observations in the original attribute space which 'best' corresponds to their configuration $\boldsymbol{s}$, in the original geographical space. 'Best' refers to the criterion of maximum correlation as identified via the canonical correlation analysis.

This two-dimensional attribute configuration, $\boldsymbol{w}$, can be interpreted by looking at the contribution of each of the original attributes in the linear combination $\boldsymbol{w}$. In addition the 'fit' of the projection can usefully be visualised in a similar way to that suggested in relation to Procrustes analysis earlier; that is, by superimposing it directly onto the spatial configuration of the observations, using an arrow diagram as described earlier. However, we first require a centring and overall scaling of $\boldsymbol{w}$ to account for possible differences in measurement scales in attribute and geographical space. The centroid of observations in the attribute projection is shifted to correspond to their centroid in geographical space and the attribute coordinates are then scaled by an overall constant to force the total Euclidean distance of all observations from their centroid to correspond in both the spatial and the attribute configurations. Further details of this kind of technique may be found in one of the references we give at the end of this chapter.

6.4 Summary

In this chapter we have considered some more specialised topics concerned with the analysis of spatially continuous data. These have fallen into two distinct groups.

Firstly, we have looked at various extensions to the basic kriging methods introduced in Chapter 5; in particular the techniques of 'block kriging' and 'co-kriging'. Such techniques were a direct continuation of the methods of that earlier chapter, explicitly spatial in nature and oriented particularly towards problems involving spatially continuous data.

Secondly, we have considered a selection of methods which, although useful, were only partially concerned with spatially continuous data and in many cases not explicitly designed for the analysis of spatial data. These were directed to the analysis of multivariate measurements, where values of several variables or attributes were recorded at each location and interest was in studying pattern in these values taken together, rather than in each taken in isolation. As we saw, some methods involved finding combinations of the attributes which represented particularly significant components of the overall variation and helped us to more easily understand and interpret its main features (principal components, factor analysis, multidimensional scaling, Procrustes analysis). We also considered techniques for classifying sites on the basis of the multivariate attributes (cluster analysis); or using them to understand the differences between pre-defined groups of sites (canonical variates), or between sites differentiated by a pre-defined set of 'explanatory' covariates (canonical correlation).

The multivariate techniques that we discussed were mostly general purpose data analysis techniques, not designed specifically for spatial data and not particularly restricted to spatially continuous data. We included a discussion of them in this chapter because they are useful in addressing certain kinds of problem that arise in the analysis of spatially continuous data, and this is where we encounter use of these methods for the first time. We will also find multivariate methods to be of use in the next part of the book, where our concern is with a different kind of spatial data, that associated with spatial zones or areal units.

6.5 Further reading

On block kriging and co-kriging you should consult:

Isaaks, E.H. and **Srivistava, R.M.** (1989) *An Introduction to Applied Geostatistics*, Oxford University Press, Oxford, Chapters 13 and 17.

Webster, R. and **Oliver, M.A.** (1990) *Statistical Methods in Soil and Land Resource Survey*, Oxford University Press, Oxford.

Cressie, N. (1991) *Statistics for Spatial Data*, Wiley, Chichester. Part I: Geostatistical Data.

There is a very wide range of texts on general multivariate statistical methods. A very good place to start is:

Krzanowski, W.J. (1988) *Principles of Multivariate Analysis: a user's perspective*, Oxford University Press, Oxford.

Other multivariate texts are extensively referenced there.

For texts which are more specifically oriented towards multivariate analyses in geography, geology and environmental science see:

Reyment, R. and **Joreskog, K.G.** (1993) *Applied Factor Analysis in the Natural Sciences*, Cambridge University Press, Cambridge.

Johnston, R.J. (1977) *Multivariate Statistical Analysis in Geography*, Longman, Harlow.

Mather, P.M. (1980) *Multivariate Statistical Analysis in Physical Geography*, Wiley, Chichester.

but also see Webster and Oliver, cited above.

On climate classification using statistical methods a useful overview is given in:

Balling, R. (1984) Classification in climatology, in Gaile, G.L. and Willmott, C.J. (eds.) *Spatial Statistics and Models*, D. Reidel, Dordrecht, 81–108.

Multivariate analysis of geological and geochemical data is discussed in some detail in the book by Reyment and Joreskog quoted above.

The application of multidimensional scaling to data on travel time, cognitive distance, and so on, is discussed in:

Gatrell, A.C. (1983) *Distance and Space: A Geographical Perspective*, Oxford University Press, Oxford.

The application of both reducing Procrustes analysis and canonical variables for identifying spatial structure in multivariate data is more fully discussed in:

Bailey, T.C. and **Hinde, J.P.** (1993) Applications of canonical correlation and Procrustes analysis in exploratory multivariate spatial data analysis, in *Proceedings of the 4th European Conference on Geographical Information Systems*, Vol 1, EGIS Foundation, Faculty of Geographical Sciences, Utrecht, The Netherlands, 1993, 606–616.

6.6 Computer exercises

Here are suggested exercises that you can try on ideas discussed in this chapter using INFO-MAP and our example data sets. These exercises have been referenced at appropriate points in the chapter by numbered symbols in the margin.

Exercise 6.1

In order to get some experience of using principal components analysis, open the 'South American climate' file and load the data. The component scores discussed in the text are included as part of the data set. Map the first two sets of scores using proportional circles, to confirm the findings in Figures 6.1 and 6.2. Then go on to examine maps of the distribution of scores on components 3 and 4. What information do these give about climate variation?

These component scores involved all 16 of the original climate variables. It is of interest to experiment with subsets of these variables. For example, suppose we focus on just overall annual variations in precipitation and temperature and ignore seasonal effects. We could calculate principal components and scores based on just the variables 'mean annual temperature', 'annual precipitation', 'days > 1 mm', and 'temperature range'. The following formula would calculate a new variable containing scores on the first principal component relating to these four variables:

```
score1=prco(1,0,{1},{8},{11},{14})
```

Remember that to calculate new variables you use `Calculate` from the `Data` menu. Here the first parameter in the `prco()` function is the number of the component whose scores you wish to obtain, the second specifies whether or not (0 or 1) to standardise the data, and the remaining parameters are the variables to be analysed. The resulting loadings may be viewed from within the `Analyse` menu, specifying `PCA diagnostics`. Use this to find the loadings and the percentage of variance explained by the first component, whose scores you have calculated. Make a note of these results (or save them to the clipboard). Next repeat the process for the second principal component using the formula:

```
score2=prco(2,0,{1},{8},{11},{14})
```

again noting the loadings and percentage variance. Repeat again for the third component.

Now map the scores on the second component you have just derived, and plot these scores against the scores on the third principal component that you have derived. Do you notice any 'outlying' sites? Use `Profile locations` in the `Analysis` menu combined with 'descending data values' or 'ascending data values' and the information you noted about the principal component loadings to explore why such sites may be climatic anomalies.

You could experiment with other selections of the original variables in principal components analysis in similar ways. There are also various possibilities for using principal components analysis to explore certain features of the 'Vancouver Island geochemistry' data set, which was described in Chapter 5.

Exercise 6.2

You can try out ideas of hierarchical clustering on the South American climate data. Open the 'South American climate' file and load the data. Calculate a new variable containing eight clusters derived from single linkage clustering on standardised 'mean annual temperature', 'annual precipitation', 'days > 1 mm', and 'temperature range', by using the formula:

```
slclusters=slink(8,0,{1},{8},{11},{14})
```

Here the first parameter in the `slink()` function is the number of classes you wish to obtain and the second specifies whether you do or do not (0 or 1) wish to standardise the data before classifying. These parameters are then followed by the list of variables to be used in the classification.

By using `Scaling` in the `Map` menu set the class interval type to 'discrete' and then `Dot map` the new variable. What do you observe? What is the likelihood of clustering techniques like single linkage producing a cluster consisting of just a single site, and perhaps placing most of the observations into one large class?

Contrast this solution with that obtained by using only the two precipitation variables 'annual precipitation' and 'days > 1mm' for clustering.

Experiment with single linkage clustering with different subsets of variables and different number of clusters, in this data set.

Exercise 6.3

Try repeating the *k*-means clustering of the South American climate data described in the text, but using only the first two, or the first three, principal component scores, rather than all four. For example, you could calculate a new variable containing six *k*-means clusters on the first three principal components with a formula:

```
kclusters=kmean(6,1,,{17},{18},{19})
```

Here the first parameter in the `kmean()` function is the number of classes you wish to obtain, the second specifies whether you do or do not (0 or 1) wish to standardise the data before classifying, and the third represents an initial set of clusters (specify ,, as above, if a default is to be used). Note we suggest here that you do not standardise the data, since you are using principal components and they have themselves already been derived from standardised variables. You could try standardising if you wished, to see whether there is a substantial difference in the solutions. These parameters are then followed by the list of variables to be used in the classification.

`Dot map` the resulting variable of clusters, remembering to set the class interval type to 'discrete' before you do so. Do your results differ much from those presented in the text (Figure 6.5)? How does this depend upon which of four principal components you choose and how many clusters you choose? How do your clusters compare with those obtained from single linkage clustering in the previous exercise? You could also compare your classification with that of Köppen, a copy of which you will find in a good atlas (alternatively, consult, for example, Blouet, B.W. and Blouet, O.M. (1993) *Latin America and the Caribbean: A Systematic and Regional Survey*, John Wiley, Chichester, p. 15).

You could also try applying *k*-means clustering directly to subsets of the original variables, rather than the principal components. In this case you should use the option in the `kmean()` function to standardise before clustering, since some of the original variables are on very different scales of measurement.

One of the problems encountered following a cluster analysis is in interpreting the clusters so formed. This may be quite complex, but an obvious place to start is to look at how the mean values of the original variables differ between the clusters. Suppose you wish to know the mean value of 'annual temperature' (variable 1) in each cluster, following a cluster analysis which has determined six clusters and whose results are in, say, variable 26. You may do this quite simply by using `Calculate/display worksheet` from the `Analysis` menu and calculating the six means into six worksheet items with the following single formula:

```
i=1,6:work[i]=mean(if({26}=i,{1},miss()))
```

Experiment with this idea, calculating the means of different original variables across one of the cluster solutions you have determined. You could also modify the formula slightly to obtain the standard deviation of a chosen variable across the clusters instead of the mean. This might give some idea of how homogeneous different clusters were in respect of that variable. Various other possible statistics can also be tried across clusters, such as median, range and so on.

Use of `Profile Locations` from the `Analysis` menu combined with the ability to select groups of locations and total or average their data items (as described in one of the exercises in Chapter 2) can also be of use in interpreting clusters.

Part D

The Analysis of Area Data

7

Introductory methods for area data

Here and in the following chapter we consider the analysis of data associated with spatial zones or areas. The areas may form a regular lattice, as with remotely sensed images, or be a set of irregular areas, such as postal or administrative districts. Unlike the case considered in Part C, attributes associated with these areas cannot necessarily be assumed to vary continuously over space. As far as our analysis is concerned, the given areas are considered to be the only spatial 'locations' at which the attributes can be measured. Our objective is to model any spatial pattern in the values over the fixed areas and to determine possible explanations for this. Such methods are relevant to studies concerning demography, health, political or economic activity, and many similar areas.

7.1 Introduction

In Part C we were concerned with spatial variation in an attribute which varied continuously over the study area and which had been sampled at a number of fixed point locations. We now turn our attention to the situation where the attribute of interest, at least as far as our analysis is concerned, does not vary continuously, but has values only within a fixed set of areas or zones covering the study area. These fixed sites may either constitute a regular lattice (for example, plots used in agricultural field trials, or 'pixels' in remote sensing) or they may consist of irregular areal units (such as census tracts or health districts).

Our interest is not particularly in prediction of the value of the attribute at points in the study region at which it has not been observed, because it does not have a value 'between areas' and usually we have observed its value in all possible areas. Rather, we are interested in the detection and possible explanation of spatial patterns or trends in the area values. With spatially continuous data many of the models that we used described spatial variation purely in terms of location. Explanation, in terms of other possible covariates, did sometimes arise, but often only as part of a model oriented towards prediction. For example, the primary purpose of kriging models is one of optimal spatial interpolation, not explaining pattern or trend. With area data we will be more interested in 'explaining' spatial variation in the variable of interest in terms of covariates measured over the same set of areas, as well as in terms of the spatial arrangement of those areas. Such analyses are relevant in any context involving the spatial relationship between two or more measurements over a set of irregular or regular areas, such as the relationship between disease rates and various socio-economic or environmental factors.

Mathematically, whereas with spatially continuous data we were dealing with a process $\{Y(\boldsymbol{s}),\ \boldsymbol{s} \in \mathcal{R}\}$, indexed by any location in $\mathcal{R}$, we are now dealing with the case $\{Y(\mathcal{A}_i),\ \mathcal{A}_i \in \mathcal{A}_1, \ldots, \mathcal{A}_n\}$, a set of random variables indexed only by fixed sub-regions $(\mathcal{A}_1, \ldots, \mathcal{A}_n)$ of $\mathcal{R}$ with $\mathcal{A}_1 \bigcup \ldots \bigcup \mathcal{A}_n = \mathcal{R}$. For simplicity we shall generally refer to the random variables $Y(\mathcal{A}_i)$ as Y_i, the observed values of which we shall denote y_i. Our observations are now not at a sample of the possible locations at which attribute values could occur, but rather constitute one possible realisation of values of the attribute at all locations. We have to conceive of this as a sample from the 'super population' of all possible realisations of the process over these areas that might ever occur.

Of course, treatment of an attribute within areas in this way might simply arise by virtue of the way that data have been collected. If, for example, we are dealing with measures of some environmental pollutant averaged over a set of areas, then the attribute can theoretically vary continuously over space. Either we have only area aggregates and not specific site measurements available to us, or we have for some other reason chosen to analyse the data on an area basis.

To summarise then, our objective in this and the subsequent chapter is to infer whether there a spatial trend or pattern in attribute values which are recorded over a set of areas, $\mathcal{A}_i$. We may also seek to explain this in terms of the relative spatial arrangement of the $\mathcal{A}_i$ and the values of possible covariates, $\boldsymbol{x}_i^T = (x_{i1}, \ldots, x_{ip})$, recorded for each $\mathcal{A}_i$. We have the usual concepts of large scale global trend or first order variation in our response variable, Y_i and local or second order variation as well. These correspond, as before, to modelling first order variation in the mean value μ_i of Y_i and second order variation or spatial dependence between Y_i and Y_j; that is, $COV(Y_i, Y_j)$.

As usual, we first discuss a range of case studies or examples of data that lend themselves to the kinds of methods described in this and the subsequent chapter. Next, we consider the visualisation of such area data before examining a number of exploratory methods. We then move on to discuss more formal statistical models.

7.2 Case studies

Throughout this and the following chapter we shall use a selection of area data sets to illustrate various forms of analyses. We have provided copies of these data sets on disk. They include:

- Child mortality in Auckland, New Zealand
- Socio-economic data for Chinese provinces
- Census data for Enumeration Districts in the London Borough of Barnet
- Voting data in the 1992 US Presidential election
- Mortality rates and socio-economic measures in English Health Districts
- Prevalence of human blood group A in Ireland
- Emissions of nitrogen and ammonia in Europe
- LANDSAT TM data for part of High Peak region, England.

We begin by describing all of these data sets in more detail. Full references to the sources of the data are given at the end of this chapter.

Our first case study relates to the spatial distribution of child mortality among 167 census districts in Auckland, New Zealand (Figure 8.1), as recorded over a nine year period (1977–85). The region involved covers approximately 5000 square kilometres and the data include the number of deaths in children under five years old in each small area during the nine year period, together with the population of children under five as recorded in the 1981 census. We would like to understand any possible spatial patterns in child mortality over the region, perhaps identifying areas with particularly high levels. However, the numbers of deaths are quite small and there is likely to be a large amount of random variation in the rates from area to area. Can we perhaps smooth out some of these random variations and obtain more stable estimates of the risk?

The second data set concerns socio-economic measures for civil divisions in China (Figure 7.3). One of the aims of economic geography is to describe and understand patterns of inequality between different regions. Within the People's Republic of China such inequalities should not, in theory, exist, though since the mid-1970s the push towards higher levels of economic growth has meant that historical inequalities have persisted and been acknowledged as inevitable. Industrialisation has led, in general, to concentrations of the better paid and more educated in urban areas, with a poorer population in more rural areas. We would like to be able to visualise and explore such spatial variations. In doing so, we should acknowledge that such variations cannot sensibly be described by any one variable in isolation. We need to recognise the multivariate nature of development processes and use appropriate statistical methods to characterise this. What are the 'components' of development? Can we analyse these and perhaps obtain some kind of socio-economic regionalisation of the country, in much the same way as we discussed arriving at a climatic classification of South America in Chapter 6? The data set comprises 18 socio-economic variables recorded in 1982, for the 29 major civil divisions in China. The variables include measures of industrial and

Table 7.1 Description of socio-economic variables for Chinese provinces

1. Value of gross industrial output as % of total value of gross agricultural and industrial output
2. Tractor ploughed area as % of total cultivated area
3. Power irrigated area as % of total irrigated area
4. Electricity consumed in rural areas, in kWh, per member of commune labour force
5. Consumption of chemical fertiliser in kilograms per hectare of sown area
6. Grain yield in *jin* per *mu*
7. Gross industrial output in Rmb per inhabitant
8. Gross output of coal industry in Rmb per inhabitant
9. Hospital beds per 10 000 population
10. Doctors per 10 000 population
11. University graduates per 10 000 population
12. Students in secondary specialised schools per 10 000 population
13. Urban population as % of total population
14. Fertility rate
15. Value of retail sales of consumer goods in Rmb per inhabitant
16. Net income in Rmb per person in rural communes
17. Expenditure of peasants on consumer goods in Rmb per person
18. Male deaths from cancer 1973–75, per 100 000 males

agricultural structure and economic performance, income, education and health. A full list is given in Table 7.1.

Our third case study also involves multivariate measures, in this case those arising from recorded census counts. Classifications and regionalisations of such data are often used in the developed world, frequently for marketing purposes but more generally in planning contexts. Census data exist in many countries at quite small geographical scales. In Britain, for example, such data are available for *enumeration districts* (EDs), areas that comprise, on average, maybe only 200 households. For any of the approximately 110 000 EDs used in the 1991 Census, several thousand counts are recorded, allowing rich descriptions of population and household structure, housing quality, employment, and so on. Clearly, such data are of enormous value both to the public sector, concerned with areas such as the provision of health care and education, and to the private sector, where commercial companies want to 'target' or identify particular areas where their products or services might find a market. We have included here a data set that may be used to illustrate these ideas. The district of Barnet on the outskirts of north-west London was divided into 619 EDs for the 1981 Census (Figure 8.4). We have assembled a small set of variables for each ED. These include the count of the total number of households, together with nine other census counts; these are listed in Table 7.2. 'Private households' is simply the number of households (other than communal establishments) in an area; the next two variables are the numbers of such households that are either owned by the occupier or are rented from the Local Authority (Council). The next two variables are the numbers of households without a car, and the numbers with two or more. 'Dependent

Table 7.2 Description of census counts for London borough of Barnet

1. Private households
2. Owner occupied
3. Local authority
4. No car
5. Two cars
6. Dependent children
7. Single parents
8. Pensioners
9. Lone female pensioners
10. Migrants

children' is the number of households with 'dependent' children, aged under 15. 'Single parents' is the number of households containing at least one single parent family. 'Pensioners' is a count of households comprising one or more pensioners, while 'lone female pensioners' is that subset of households comprising a woman aged at least 75, living alone. 'Migrants' is a count of households containing members who were living at a different address one year before the census was taken.

Such counts can of course be analysed directly, but it will almost always be sensible to calculate new variables as proportions or percentages of the total number of households, or some other suitable denominator. Otherwise, we shall simply be analysing mostly spatial variations in the size (in terms of numbers of households) of EDs. In most cases we can use total numbers of households as a denominator. However, when looking at single parents it may be more sensible to use the number of households with dependent children, whilst with elderly women living alone we could use the total number of households containing pensioners as a 'base' variable.

Such data may be used in a variety of ways. Note first that the counts in some EDs will be very small. This means that estimates of rates and proportions in such areas may be unreliable. Can we perhaps use the spatial (proximity) information in the data to 'smooth' out such variations and obtain more reliable maps of rates? Alternatively, we may use the multivariate nature of the data to construct socio-economic 'scores' for the EDs and to group together those EDs that seem to be similar in composition. Such a strategy is widely used in marketing, but also in health research. In applications such as these, researchers often want simple, 'shorthand', ways of describing the character of an area. As a result, both academic researchers and commercial companies have invested considerable time and money in devising multivariate classifications and 'scoring' areas in particular ways. Examples of such systems include ACORN and 'Super Profiles'. The use of these systems forms part of the more general applied area of research known as *geodemographics*.

Census data of this description also raise another interesting area of spatial analysis. Suppose we wish to make estimates of a census variable for areas other than the EDs for which the data are published? For example, many businesses organise sales programmes not around EDs or simple aggregations of these, but rather based on postcode sectors. The boundaries of these areas cut across those of EDs in arbitrary ways. This poses a general problem with the analysis of area data, that of *cross-areal interpolation*. In other words, can we use data recorded within one set of zones to estimate data for an alternative set; can we interpolate ED data to match postcode sector boundaries? We shall discuss this kind of problem briefly in this chapter.

Our next three case studies differ somewhat from those outlined so far. We still have a set of variables for areal units, but now there is a primary variable of interest, or 'response', the spatial variations in which we wish to 'explain', at least partly, in terms of the values of the other variables. At one level this may appear to be a straightforward regression problem of the type first encountered in Chapter 1 and again in more detail in Chapter 5. However, in this chapter we shall discuss how to incorporate into such analyses information on the spatial configuration of areal units.

The first of this group of case studies concerns explaining geographical variations in the vote for President Clinton in the 1992 US election. Data are provided on the percentage support for Clinton and estimates of the 1992 voting age population, in each of the 48 'coterminous' states (Figure 7.1). (Alaska and Hawaii are excluded, as is Washington DC). One obvious candidate variable to explain the voting pattern is explicitly geographical: the proximity of states to Clinton's home state of Arkansas. However, we include eight variables, all recorded in either 1989 or 1990, as possible alternative explanations for variations in support for Clinton. They involve measures of population composition, poverty, unemployment, education, and business failures.

The next study is concerned with variations in mortality rates over 190 English District Health Authorities. The data relate to the year 1989 and readers should be aware that the number and boundaries of Health Authorities have changed since then. Three standardised mortality ratios are included, calculated over the five years up to the end of 1989, for: carcinoma of the lung (males aged 35–64); carcinoma of the breast (females aged 50–64) and acute myocardial infarction or heart attacks (males 35–64). In each case the numerator is the observed number of deaths in the District in the five year period and the denominator is the expected number of deaths during the same period, on the basis of the national rate corrected for the age–sex structure of the District population. In addition, there are two crude measures concerned with health care spending, and one concerned with usage of services. The first is the expenditure on health promotion and education in each District per 1000 resident population; the second is the total hospital expenditure for all acute specialities in each District per capita of the age–sex standardised resident population; the third is the average length of stay of patients in general medical specialities. Finally, three variables derived directly or indirectly from the 1981 Census are also included. The first is simply the percentage of owner occupied households, the other two are measures of 'deprivation'. One of these is known as the 'Jarman underprivileged areas score', an index based on a weighted average of the percentage of elderly living alone, children under five years of age, one parent families, unskilled socio-economic groups, unemployed, overcrowding, migration and ethnic minorities. The other is an alternative but similar index devised by the Department of the Environment. Both of these indices purport to measure aspects of 'social deprivation' and are derived from multivariate analyses of the sort we referred to earlier when discussing census data. Such measures should not be used uncritically; however, they have been commonly employed in health planning research within the UK.

Our next data set concerns the distribution of the adult population in Eire that has blood group A; a data set that has been analysed in other texts on spatial analysis. The proportion of adults with blood group A was measured in 1958 in samples drawn from the populations of each of the 26 Irish counties; in total, some 55 000 people were involved. We may wish to describe spatial variations in this proportion over the set of counties. The background to this problem lies in the history of settlement of the country, with Anglo-Normans settling there from the 12th century onwards. Settlement was primarily in the east of the country and it is counties in this area that have the highest proportions with blood group A. Can we explain variation in blood group in terms of settlement history? To do this we have one crude surrogate variable for Anglo-Norman settlement, the number of place names per unit area that contain the suffix 'town'. A further covariate is a categorical variable that denotes whether or not a county is within or beyond the 'Pale', that part of the country that was settled and dominated by the Anglo-Normans (which, by the way, gives rise to the colloquial expression 'beyond the pale'). To what extent do these simple variables explain variation in blood group?

Finally, we include two studies that relate to data collected on a regular lattice. The first of these involves emissions, in kilotonnes, of nitrogen and ammonia in Europe. These data take the form of estimated emissions in 1985 for a set of grid squares (each 150 × 150 kilometres) covering the continent. The grid is not 'complete', since observations for many of the cells (mostly those over sea areas) are unavailable. Those observations that are included are derived from official figures submitted by European countries participating in the European Monitoring and Evaluation Programme. As well as estimates of land-based emissions, these published data include estimates of emissions from international maritime trade, which are of particular relevance to the nitrogen oxides emissions. The major source of the ammonia emissions is agricultural activity. Such emissions will be of significance in adding to the nitrogen inputs in soils. We cannot emphasise too strongly that these data are estimates, and that they are 'error-laden'. Sources of error include the following. First, the data are in gridded (raster) form but will have been derived from estimates based on point, line and area-based sources of pollution. Second, not all European countries have supplied data, and for those that have, the quality is likely to be spatially variable; for example, the accuracy of nitrogen oxide data in western Europe is thought to be within 10% of the true figure, while for eastern Europe it may be around 20%. For ammonia emissions the errors are uncertain but may be as high as 50%. That the data are error-laden is of obvious concern, but the material seems to be the best currently available. Given the error, it might be sensible to use the data to paint a broad regional picture of acid emissions. We can do this using some of the 'smoothing' techniques discussed in this chapter.

The final case study again involves a regular grid of areas, in this case the 'pixels' typical in remote sensing. We touched on this kind of data briefly in Chapter 2 in the section on GIS. We noted there that such data are available from sensors mounted on board satellites and involve reflectance values available for different bands of the electromagnetic spectrum. In the case of the LANDSAT 5 'Thematic Mapper' (TM) sensor, data are available for seven bands, in the visible and infrared regions of the spectrum. A typical LANDSAT scene covers a swath approximately 185 kilometres wide, with a resolution of 30 × 30 metres; that is to say each data value is for a 'pixel' of 30 square metres. We have included a very tiny section of data for one small part of the earth's surface; a region, in fact, covering just one square kilometre! (Since it is divided into 30 × 30 pixels each pixel actually represents 33.33 square metres.) The region in question is part of the Peak District in England, known as the High Peak. For those familiar with the UK's Ordnance Survey grid, our small area covers the grid square whose south-west corner is given by: (417000, 387000) and north-east corner is: (418000, 388000). The area covers a mix of land uses. In the far south-west of the square is part of Ladybower Reservoir, while the south-central part of the square is occupied by a patch of mixed deciduous and coniferous woodland known as Grimbocar Wood. The land rises from an elevation of 700 metres in the south-west to over 1200 metres in the north. Much of the area to the north is rough grazing and moorland.

The variables included for each of the 900 pixels are the seven bands of TM data. These cover the following wavelengths, where the units are in millionths of a metre. Band 1: 0.45–0.52 represents the 'blue' region; band 2: 0.52–0.60 the 'green' region; and band 3: 0.63–0.69 the 'red' region. These first three bands represent the visible region of the spectrum. Band 4: 0.76–0.90 covers wavelengths in the near-infrared region, band 5: 1.55–1.75 the mid-infrared region, and band 7: 2.08–2.35 again the mid-infrared region. Finally, band 6: 10.4–12.5 covers the thermal infrared region. Pixel values for each band can take on values between 0 and 255, where 0 represents no reflectance and 255 denotes maximum reflectance. In practice, the range of values in any band will be quite limited. For example, in our small subset, values in band 1 vary from only 64 to 90, while in band 4 they show more variation (from 16 to 106).

The various bands, singly or in combination, perform different functions. This is because different earth and water features reflect varying amounts of energy at particular wavelengths. As a result, two features that are not distinguishable in one band may show up as very different in another. Band 1 is used mostly for studying water depth and quality, but is also used in understanding plant stress. Band 2 measures the green reflectance peak of vegetation, while band 3 aids in the identification of plant species. Band 4 also measures the vigour of vegetation. Bands 5 and 7 are used to measure the water content of vegetation. In terms of combinations, a normal colour composite of blue, green and red (bands 1–3) is useful in looking at variations in water sediment content. In the mapping of vegetation type and built-up areas, combinations of bands 2, 3 and 4, or 3, 4 and 7, or 3, 4 and 5 are often used. Regardless of what combination is selected, typically a coloured 'map' is produced by assigning each band to a different colour 'gun' on a computer monitor; in terms of the combinations mentioned above the colour guns blue, green and red would be used, in that order.

In practice, remotely sensed data require considerable processing before they are usable for spatial data analysis. The data we include have already undergone such processing. We do not go into the details of these important operations, since they are somewhat peripheral to our subject. Moreover, no-one who wishes to do serious remote sensing will be analysing in detail a LANDSAT scene that covers as small an area as the one we include. We offer the data here because they can be used to illustrate some techniques of interest and are of relevance in more realistic remote sensing applications. We discuss such techniques briefly in Chapter 8. The reader who wishes to get to grips with the subject in more detail can consult one of the texts we refer to at the end of that chapter.

7.3 Visualising area data

In Chapter 3, the natural way of visualising point patterns was a dot map, and in Chapter 5, the equivalent tool for continuously varying data was a proportional symbol map; or alternatively, a contour map based on some appropriate interpolation technique. With area data various possibilities exist for visualising the spatial variation of an attribute of interest.

Firstly, we may again use proportional symbols, as described in Chapter 5. We superimpose onto a base map of the areas $\mathcal{A}_i$, symbols whose size is proportional to the attribute value in that area. Colour scaling and symbol size can be used together if required to allow visualisation of the values of two attributes; the symbol size is proportional to one attribute and the symbol colour is based on the other. By 'size', we usually mean the area of a symbol such as a circle, perhaps the height of a symbol such as a rectangle (the base of which would normally be of fixed length). As we saw in Chapter 5 we could also use other kinds of symbols.

An alternative and perhaps more commonly used form of display for area data is the *choropleth* map. This is a map where each of the areas $\mathcal{A}_i$ is coloured (or shaded) according to a discrete scale based on the value of the attribute of interest within that area. The number of classes and the corresponding class intervals can be based on several different criteria, and we refer the reader back to the discussion of class interval and type selection in Chapter 5, in connection with proportional symbol maps. The discussion there on choice of shading colour is also of relevance in choropleth mapping.

Such issues merit careful thought and it is important to appreciate that although they are very widely used, choropleth maps can be very misleading. Alternative choices of class intervals for the colours or shading can lead to widely different visual impressions of trend and spatial pattern in the resulting maps. A good text on cartographic methods is invaluable reading here. We may illustrate this with one of our case studies, that involving the vote for President Clinton in the 1992 election (Figure 7.1).

The range of data values is from 25.9% in Utah to 53.9% in Clinton's home state of Arkansas. A map using six classes based on the standard deviations of the variable (Figure 7.1a), as described in Chapter 5, gives a reasonably sensible view of the data, highlighting the lower values in the broadly conservative west-central states (such as Idaho, Nebraska and North Dakota). Using six equal classes, or 'trimmed' classes, gives broadly similar results; however, if we choose instead six quantiles (that is, sextiles) we obtain rather a misleading message (Figure 7.1b), since Arizona (with 37% voting for Clinton) is forced into the same class as Utah.

There are also other problems associated with the use of choropleth maps. Firstly, physically large areas tend to dominate the display, in a way which may be quite inappropriate for the type of data being mapped. For example, in mapping socio-economic data, large and sparsely populated rural areas may dominate the choropleth map because of the visual 'intrusiveness' of the large areal units. Yet the real interest may be in physically smaller zones, probably in the more densely populated urban areas. After all, the socio-economic data we are wanting to map are usually properties of the people living in areas, not of the areas themselves. A good example of the problem might be Canada where, if we wished to map, say, voting data on a traditional choropleth map we would see little of interest in the major Ontario cities, since they would be swamped by the need to also show the vast (but rather uninteresting from the viewpoint of electoral geography!) areas of Yukon and the Northern Territories.

Secondly, when the attribute of interest has arisen from the aggregation of individual data to the areas, it must be appreciated that these areas may have been designed rather arbitrarily on the basis of administrative convenience or ease of enumeration. As a consequence, any pattern that is observed across the zones may be as much a function of the zone boundaries chosen, as it is of the underlying spatial distribution of attribute values. As we noted in Chapter 1, this has become known as the *modifiable area unit problem*. It can be a particularly significant problem in the analysis of demographic and socio-economic data, where the enumeration areas have rarely been arrived at on any basis which relates to the data under study. By contrast, area data relating to the physical environment, such as geological strata, soil types, and so on, may be recorded for 'natural' areas which better relate to the data at hand.

How may we 'solve' such problems? Appropriate selection of class intervals and colour choice has attracted considerable attention and, in effect, there is no 'solution' other than awareness of its existence and impact. One suggestion is to try and avoid the problem by use of continuous shading or gradation of

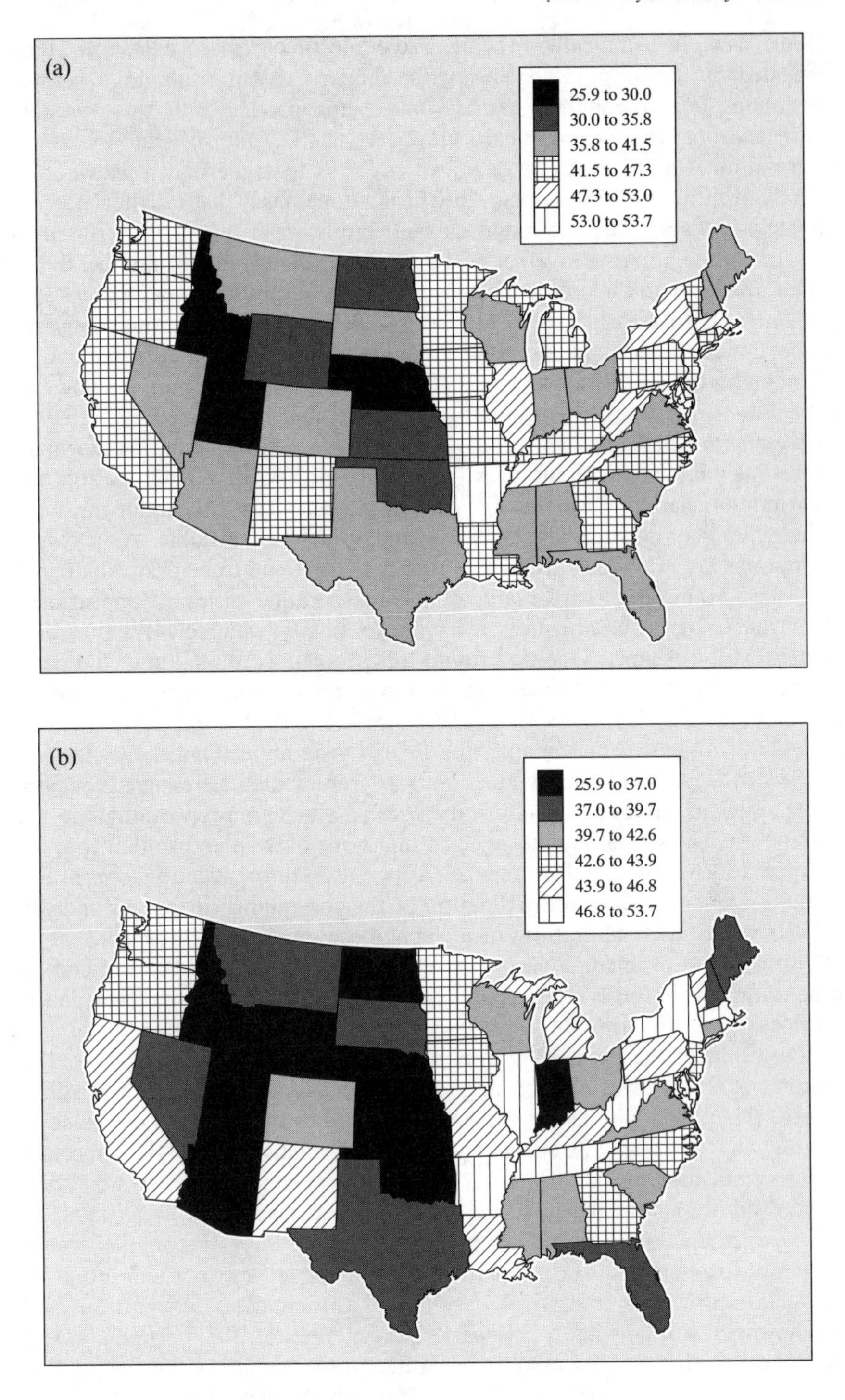

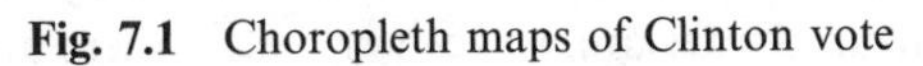

Fig. 7.1 Choropleth maps of Clinton vote

colour. This is technically feasible, since plotting devices can draw lines separated by any desired spacing, while modern colour technology permits continuous colour shading. While some maps produced in this way are aesthetically appealing many cartographers feel they are difficult to extract information from. Indeed, we might go as far as to argue that if we want an entirely accurate choropleth map, in which a zone has a shading that matches precisely its data value, we might as well simply write the value in the areal unit, or forego a map altogether and give the reader a table of numbers!

One approach to the problem of the dominance of physically large areas is to geometrically transform each of the zones A_i in such a way as to make its area proportional to the corresponding attribute value, whilst at the same time maintaining the spatial contiguity of the zones. The resulting map is known as a *density equalised* map or *cartogram*, an idea we first encountered in Chapter 3. As we mentioned there, algorithms to perform such transformations form an interesting mathematical study in their own right. With regular lattices the algorithm is relatively straightforward. For irregular zones the algorithms can be complex, computationally intensive and somewhat unstable. As a result, cartograms are not widely used or, if they are, they tend to be drawn by hand! Published examples of cartograms often have a rather inelegant appearance; this is due to the requirement of preserving contiguity, which gives rise to some rather tortuous shapes. One way round this problem is to relax the contiguity constraint and to endeavour to see, but not to insist, that adjacent areal units remain adjacent when the sizes of the zones are altered in proportion to the variable of interest. For example, the British geographer Daniel Dorling has devised an algorithm whereby areal units are represented as regular geometric shapes, such as circles or hexagons, the sizes of which are proportional to, say, population. The shapes are arranged so that none overlap and so that the vast majority touch those to which they are adjacent on the conventional map. The method is, then, a good approximation to the continuous cartogram and can provide some fine visualisations of social and economic data. Figure 7.2 shows the distribution of unemployment in 1988 across 852 areas in Britain. Darker zones have higher levels of unemployment. Notice generally high levels in the north-east, and the 'ring' of lower unemployment around London.

A solution to the problem of modifiable areal units is difficult. The ideal solution is to avoid using area aggregated data altogether if at all possible; indeed, this philosophy lay at the heart of some of the methods discussed in Chapter 4. We saw how, in some applications of practical spatial data analysis, such as epidemiology and crime, we could perform valuable analyses on point data, without aggregating this to a set of inherently arbitrary areal units. Of course, such an approach will not be viable in many cases and one has to live with the areal units for which data are available. If so, it is important to appreciate that the statistical results of any analysis of pattern and relationships will inevitably be partly a function of the particular areal configuration that is being used. In general, data should always be analysed on the basis of the smallest areal units for which they are available and aggregation to arbitrary larger areas avoided, unless there are good reasons to

Fig. 7.2 Density equalised map of unemployment in Britain, 1988

suspect that these areas make some *a priori* sense in relation to the data being studied. It is also a good idea to check any inferences drawn from the data by using different areal configurations of the same data, if at all possible. Later in this chapter we discuss some techniques for interpolating irregular area data to fine regular lattices which may also be of value in helping to alleviate this sort of problem.

Returning to our general theme of visualisation methods, so far we have mostly discussed the visualisation of a single variable. With cartograms, and with graduated or proportional symbols, we can also map not just one variable but two (and occasionally more). With graduated symbols, as we have seen, the symbol size might represent one variable and the shading another. With a cartogram, the transformed area represents one variable, while another can be mapped onto the transformed base map. Some cartographers have sought to map two variables onto the same conventional choropleth map. This can be done by drawing up a bivariate classification of data on, for example, education and income, so that any areal unit is assigned to a 'bin' that represents an income class and an education category (such as the percentage of high school graduates). Colours are assigned to each bivariate class, and this is where problems of interpretation arise, since if each variable is classified into five classes there will be 25 bins and 25 colours for the reader to interpret. Some researchers have argued for logical colour schemes, with extreme bivariate values given pure colours and those representing less unusual values being shaded with varying strengths of grey. Others have extended the idea of 'unclassed' choropleth maps to these two-variable maps, devising bivariate choropleth maps with continuous colours. Again, the issue of interpretability remains.

We have already had a preliminary look at the US voting data, noting the highest percentage in Clinton's home state, other high values in New York, Maryland and Illinois. What of other data sets?

We do not show a map here, but the data on the distribution of blood group A in Ireland has a rather limited range (from 23.9% to 35.9%). Most of the class interval schemes we might choose confirm that values are higher in the three counties south of Dublin (Cavan, Wexford and Wicklow), with low percentages in the south-west (Cork and Kerry). The data set for Chinese provinces, which we illustrate later in the chapter, comprises 18 variables that may be visualised as choropleth or graduated symbol maps. Many of the variables reveal high values in the provinces of Beijing, Tianjin and Shanghai. For example, those variables measuring industrial output, the use of mechanisation in agriculture, and the availability of doctors all show a strong concentration in these urban areas. The variable measuring output from coal production picks out the coal fields of Shanxi and Ningxia, while the fertility index highlights a very different pattern, with low values in urban areas and higher rates in the rural west and southern provinces. In view of our earlier comments about cartograms, it is interesting to note that while the major urban centres are not hard to visualise they are rather overwhelmed by the large provinces of Tibet and Xinjiang. It is therefore of interest to see that, in the original paper from which these data are taken, John Cole has mapped an aggregate 'modernisation' index (about which we shall have more to say in the next chapter) as a cartogram.

7.4 Exploring area data

In this section we follow the pattern we have established in previous parts of the book and start our discussion of analysis methods for area data by considering various exploratory techniques which can be used informally to investigate hypotheses of interest or suggest possible models. As usual some of these are more concerned with investigating first order effects in the observed process whilst others address the possibility of spatial dependence or second order effects.

One issue which we are forced to discuss immediately, since it is involved in most methods concerned with area data, is that of how to measure the 'proximity' of observations when they relate to possibly irregularly shaped areal units. We then discuss various simple approaches to assessing how the mean value, μ_i, of the attribute of interest varies across the study region. We conclude the section by looking at techniques for exploring spatial dependence in the data, through the 'correlogram'. These are equivalent to the techniques used in Chapter 5 for examining second order properties of spatially continuous data via the 'co-variogram' and 'variogram'.

It should be mentioned here that with certain kinds of area data, such as remotely sensed images or environmental measures collected over small zones, it may be useful in exploratory analysis to assume that the attribute value is located at a specific point in each of the areas $\mathcal{A}_i$. We can then use the kind of exploratory methods discussed in relation to spatially continuous data in Chapter 5, to investigate trends and relationships in the data. This *ad hoc* approach may be useful in some cases, but we do not repeat such methods here, referring the reader back to the earlier chapter. Here we focus on methods which are more specifically designed for area data. Many of these use various kinds of spatial proximity measures for areas which we now introduce.

7.4.1 Proximity measures with area data

In the case of a continuously varying attribute over a study area it is natural to use distance between point locations as the basis for measuring spatial proximity. In the case of area data we need to consider how to define spatial proximity measures between each of the areas $\mathcal{A}_i$. At the crudest level we could obviously use the distances between the geographical centres or 'centroids' of each of the areas; but in doing so we would be disregarding some aspects of the spatial nature of these areas. We require a mechanism which allows us to define spatial proximity in a more general way.

The general tool which we shall use is the idea of an $(n \times n)$ spatial proximity matrix $\boldsymbol{W}$, each of whose elements, w_{ij}, represents a measure of the spatial proximity of areas $\mathcal{A}_i$ and $\mathcal{A}_j$. As a rule, the choice of w_{ij} will depend upon the sort of data that one is dealing with and the particular mechanisms through which one expects spatial dependence to arise. Some possible criteria might be:

$$w_{ij} = \begin{cases} 1 & \text{centroid of } \mathcal{A}_j \text{ is one of the } k \text{ nearest centroids to that of } \mathcal{A}_i \\ 0 & \text{otherwise} \end{cases}$$

$$w_{ij} = \begin{cases} 1 & \text{centroid of } \mathcal{A}_j \text{ is within some specified distance of that of } \mathcal{A}_i \\ 0 & \text{otherwise} \end{cases}$$

$$w_{ij} = \begin{cases} d_{ij}^{\gamma} & \text{if inter-centroid distance } d_{ij} < \delta \quad (\delta > 0; \gamma < 0) \\ 0 & \text{otherwise} \end{cases}$$

$$w_{ij} = \begin{cases} 1 & \mathcal{A}_j \text{ shares a common boundary with } \mathcal{A}_i \\ 0 & \text{otherwise} \end{cases}$$

$$w_{ij} = \frac{l_{ij}}{l_i} \quad \begin{array}{l} \text{where } l_{ij} \text{ is the length of common boundary between } \mathcal{A}_i \text{ and } \mathcal{A}_j \\ \text{and } l_i \text{ is the perimeter of } \mathcal{A}_i \end{array}$$

Hybrid measures based on these various criteria can also be used; for example, combinations of length of shared boundary and distance between centroids. We could also use proximity criteria which incorporate various alternative measures of spatial separation. These might include travel time between zone centroids, for example. We might even consider using spatial interaction data of the type we discuss later in Chapter 9 to develop measures of spatial proximity; for example, flows of people, goods, or traffic. Alternatively, properties of the kind of tessellations discussed in Chapter 5 could be used, where these are applied to the centroids of the areas. For regular lattices it is often appropriate to use simple proximity measures such as:

$$w_{ij} = \begin{cases} 1 & A_j \text{ shares an edge with } A_i \\ 0 & \text{otherwise} \end{cases}$$

Note that in general there is no reason why $\boldsymbol{W}$ need be symmetric.

It is sometimes necessary to specify proximity measures of different orders, often referred to as *spatial lags*. For example we might require a series of proximity matrices $\boldsymbol{W}^{(1)}, \ldots, \boldsymbol{W}^{(k)}$ where $\boldsymbol{W}^{(1)}$ represents spatial proximity of the areas at spatial lag 1 (within some distance band, or 'first nearest neighbours' etc.) then $\boldsymbol{W}^{(2)}$ represents spatial proximity at spatial lag 2 (within next distance band or 'second nearest neighbours' etc.) and so on.

7.4.2 Spatial moving averages

Having looked at how we might define different measures of spatial proximity for area data, we now consider methods which explore how the mean value μ_i of the attribute of interest varies across the study region. As in the case of spatially continuous data, a very simple way to estimate global variations and trends in the values of an attribute over the areas is to estimate μ_i by an average (usually weighted) of the values in 'neighbouring' areas. The spatial proximity matrix $\boldsymbol{W}$ provides a flexible method of defining a suitable set of weights for 'neighbouring' areas and the smoothed estimate is then:

$$\hat{\mu}_i = \frac{\sum_{j=1}^{n} w_{ij} y_j}{\sum_{j=1}^{n} w_{ij}}$$

The denominator is clearly unnecessary if $\boldsymbol{W}$ has been *standardised* to have row sums of unity, which is often the case.

We may illustrate these ideas using the China data. We calculate a spatial moving average based on whether polygons do, or do not, share a common boundary (the fourth of the proximity schemes listed earlier). We obtain a result (Figure 7.3) for the variable measuring gross industrial output as a percentage of the total value of gross agricultural and industrial output. This

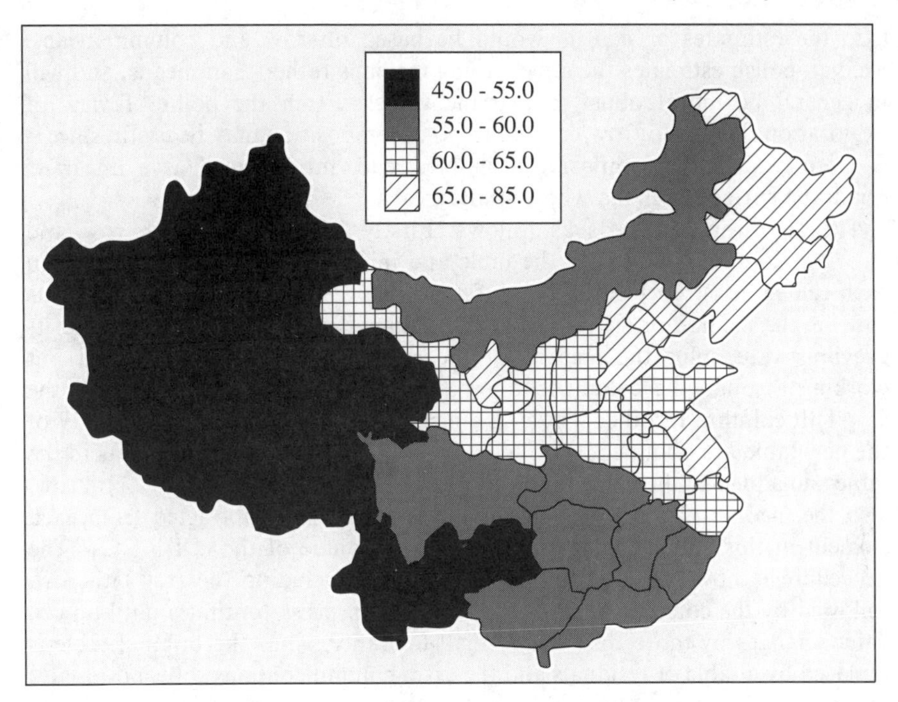

Fig. 7.3 Spatial moving average of gross industrial output in China

yields a smoother picture of spatial variation than a map of the raw data and serves to highlight broad regional trends. Of course, selecting an alternative proximity weighting scheme, or using fewer or more neighbours, will generate a different picture.

7.4.3 Median polish

When our data, y_i, are on a regular grid a simple technique which may help to identify broad spatial trends in μ_i and which is more resistant to extreme values or outliers in the data is *median polishing*. Since our data y_i are in the form of a regular $(r \times s)$ lattice of values, we will change our notation here temporarily and denote them as y_{ij}, in which case the μ_i in which we are interested also become temporarily μ_{ij}. The idea is to treat y_{ij} as if they were cell entries in a two-way table and then to obtain an additive decomposition of each cell entry into:

$$y_{ij} = \mu + \mu_i + \mu_j + \epsilon_{ij}$$

where μ is some fixed overall effect, μ_i and μ_j represent fixed row and column effects and ϵ_{ij} is a random error. Then the overall mean μ_{ij} is just $\mu + \mu_i + \mu_j$. Now this 'model' could be fitted by an ordinary analysis of variance, in which case the estimates of μ, μ_i, μ_j would be based on row and column means. Median polish estimates the effects using medians rather than means, and will in general be more robust to extreme values. From the point of view of exploration this sort of row/column trend decomposition may be useful since it is more flexible than imposing a simple trend 'model' such as a linear or quadratic surface over the whole study area.

The algorithm proceeds as follows. Firstly an extra $(r+1)$th row and $(s+1)$th column is added to the table and initialised with zero values. Then each cell value is replaced by the difference between y_{ij} and the median of its row i.e. the median of $(y_{i1}, \ldots, y_{is})$. The value of $y_{i,s+1}$ is then set equal to its previous value plus the median of the row. This process is repeated but working through the columns rather than the rows and including now the $(s+1)$th column. In other words, the medians of the columns in the body of the new table are obtained, subtracted from the corresponding elements in the table and added to the value in the corresponding element of the $(r+1)$th row; also the median of the $(s+1)$th column is found and subtracted from each element in this column and also added to the value of the cell $y_{i+1,s+1}$. The procedure is now repeated for the rows, again including the $(r+1)$th row, followed by the columns again and so on. This process continues until no cell value changes by more than some small tolerance. Thus the original table is replaced by a table of residuals and the extra column contains robust estimates of the μ_i, the extra row similarly for the μ_j, with the $(r+1, s+1)$ cell containing an estimate of μ. The estimated or 'fitted' value for each cell mean $\hat{\mu}_{ij}$ is then just the sum of these estimates $\hat{\mu} + \hat{\mu}_i + \hat{\mu}_j$.

We may illustrate this technique using the data on emissions of nitrogen oxide in Europe. As discussed previously, this data set contains many 'missing values', and the observations that are included are known to be subject to recording errors. Their values are also very highly variable. Use of a 'robust' technique like median polish to reveal broad trends would therefore seem a sensible way to proceed, at least in the first instance. Since the technique uses only medians and therefore only the ranking of the data, it is invariant to any transformation that preserves the original ordering of the data. It is also able to cope easily with missing values; these cells are simply omitted when the median of either a column or row which contains them is required.

A map of the raw nitrogen oxide observations is shown in Figure 7.4a. Non-shaded grid squares correspond to 'missing values'. These are mostly over sea areas and the broad outline of the European continent is just about distinguishable, the vertical orientation of the grid being roughly in a north-easterly direction in this case. There are some very highly variable and extreme values running in a band across central and eastern Europe. Figure 7.4b shows the result of median polish. The grey scales used have been deliberately restricted in this map so that they are directly comparable with those in the map of the raw values; there is in fact more detail in the median polish than implied here. However, the image illustrates well how in this case the median polish has dramatically suppressed the very extreme values in the raw data. The value of the polished image is that it represents a 'robust' estimate of the global

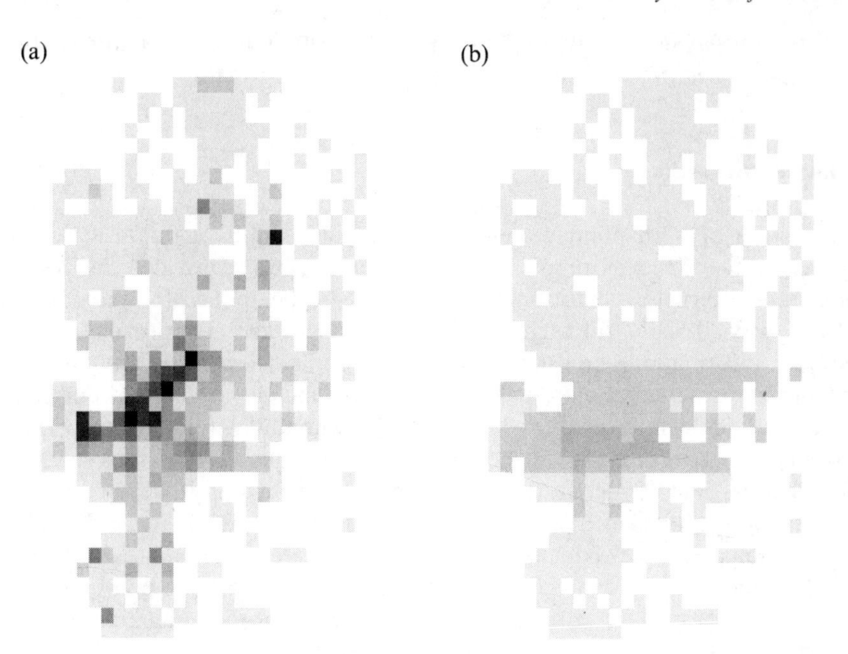

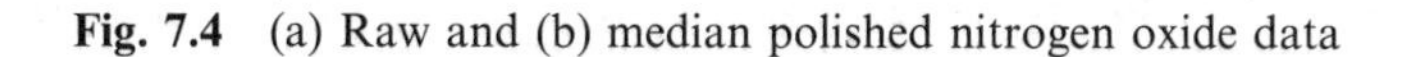

Fig. 7.4 (a) Raw and (b) median polished nitrogen oxide data

trend in the original data. Attention would then focus on the residuals for subsequent, more detailed, analysis. Note however that the median polish seems to suffer from 'banding' effects. These arise from the additive 'row effect', 'column effect' model which underlies the algorithm and which would be a very dubious model for this data, given that what trend there is in the original values is in no way oriented to the grid. One problem with median polish is that it attempts to decompose trend according to the directions of the grid, which often has no relationship to the spatial orientation of the trend. This is fine if trends are essentially circular 'hot spots', but not if they are elongated in a direction which is neither vertical nor horizontal in relation to the grid. Another problem with median polish is that one is unable in any way to control the degree of smoothing applied. We will come on to this point when we discuss applying kernel methods to area data in the next section.

Median polish can be adapted for systems of areal units other than regular lattices, by assigning each area to the closest cell of some suitably chosen grid which has been overlaid on the areas. Note that there is no reason why such a grid needs to be equally spaced in either direction. It should be chosen so that each cell of the grid corresponds to a single area as far as is possible. In general there may be two areas assigned to some grid cells and none to others. This does not affect the median polish algorithm; repeat observations in a grid cell are simply included as extra values for the computation of the median and grid cells with no associated areas are ignored. If a whole row or column of grid cells has no observations then it too is ignored.

7.4.4 Kernel estimation

In the case of both point patterns and spatially continuous data a flexible approach to the exploration of spatial pattern was provided by kernel estimation. Kernel estimation techniques can also be applied to area data, in various ways. Ideally, for such data one would wish kernel approaches to use as much information as possible about the specific geometry of the areas $\mathcal{A}_i$ with which observations are associated. Unfortunately, we do not currently have methods which enable the required computations to be performed efficiently enough to make this possible, since they involve numerous repeated integrations of a kernel function over possibly irregularly shaped regions. Consequently, most kernel approaches currently used on area data avoid using information about the geometry of the $\mathcal{A}_i$ and instead usually assume that each of the observations y_i can be associated with some appropriate point location $\boldsymbol{s}_i$. This might for example be the 'centroid' of the corresponding area $\mathcal{A}_i$, or a relevant major 'centre of population' in that area.

As a result, there is currently little difference between the application of kernel estimation to area data and that for spatially continuous data. For example, where the observation y_i in area $\mathcal{A}_i$ represents some 'average' measure over that area, such as a standardised mortality rate, or a per capita economic measure, then the kernel approach which we outlined for spatially continuous data in Chapter 5 can be applied directly to estimate the mean $\mu(\boldsymbol{s})$ of this measure at a general point, $\boldsymbol{s}$, in $\mathcal{R}$. Normally a fine grid of such points would be used. As before the relevant estimate is:

$$\hat{\mu}_\tau(\boldsymbol{s}) = \frac{\sum_{i=1}^{n} k\left(\frac{(\boldsymbol{s}-\boldsymbol{s}_i)}{\tau}\right) y_i}{\sum_{i=1}^{n} k\left(\frac{(\boldsymbol{s}-\boldsymbol{s}_i)}{\tau}\right)}$$

where $k()$ is a standardised probability density function (the *kernel*) symmetric about the origin, and $\tau > 0$ (the *bandwidth*) determines the amount of smoothing, being the radius of a disc centred on $\boldsymbol{s}$ within which points $\boldsymbol{s}_i$ will contribute 'significantly' to the estimate, $\hat{\mu}_\tau(\boldsymbol{s})$, of the mean $\mu(\boldsymbol{s})$ of y_i. How this estimate is to be interpreted will depend upon the particular data in question. We acknowledge that in the case of area data a sensible interpretation may often only be possible if $\boldsymbol{s}$ is taken to represent not a point, but rather a small area around $\boldsymbol{s}$.

A suitable choice for $k()$ is the quartic kernel discussed previously in Chapters 3 and 5. The reader is referred back to those discussions for more details and comments concerning choice of bandwidth τ.

When observations y_i in areas $\mathcal{A}_i$ represent totals such as census counts then the above approach is not applicable. An 'average value' at $\boldsymbol{s}$ of such a count is a meaningless concept and we need instead to think in terms of an estimate of the 'density' (count per unit area) at $\boldsymbol{s}$, say $\lambda(\boldsymbol{s})$. An obvious estimate is effectively the numerator of the above kernel estimate:

$$\hat{\lambda}_\tau(\boldsymbol{s}) = \sum_{i=1}^{n} \frac{1}{\tau^2} k\left(\frac{(\boldsymbol{s} - \boldsymbol{s}_i)}{\tau}\right) y_i$$

on the basis that we are now estimating 'total per unit area'. We came across this sort of idea in Chapter 4, in relation to the analysis of cases of disease where the 'at risk' population was heterogeneous. The density estimate can then be used to form a count estimate in a small region around $\boldsymbol{s}$ by multiplying by the area of such a region, or for larger regions perhaps by integrating $\hat{\lambda}_\tau(\boldsymbol{s})$ over the region.

This is essentially the kind of technique which Ian Bracken and David Martin have used to break down count data published for one fixed set of irregular areal units to a fine grid which can then be re-aggregated to other sets of areas and thus used in conjunction with information on an alternative areal base. Alternatively, data from both sets of areas can be interpolated to the grid and subsequent analyses carried out at the grid level.

Their technique has proved especially useful for interpolating census data from original enumeration districts to a finer grid which better captures the true distribution of population in a region. The centroids, $\boldsymbol{s}_i$, used in this case are those that are reported by OPCS (the census agency in Britain) for each enumeration district. These are not geometric centroids but reflect the location of the major population centre within that enumeration district. This implies that the kernel estimated population data may well reflect a truer picture of the way in which population is distributed over a study area, than would be given by assuming that population values are uniformly distributed within the enumeration districts.

The kernel approach Bracken and Martin use is in fact a form of *adaptive kernel estimation* where the fixed bandwidth τ above, is replaced by a varying bandwidth $\tau(\boldsymbol{s}_i)$, over $\mathcal{R}$. We commented on this method earlier in Chapter 3, in relation to kernel estimation of the intensity of point patterns. In their case, the bandwidth is made to depend on the local intensity of 'population centres' in the vicinity of $\boldsymbol{s}_i$, in a way specially developed to be appropriate to such population interpolation. In addition, since they require the total population over $\mathcal{R}$ to be reproduced exactly as the sum of the interpolated populations in all the small grid squares, the grid population produced from the density estimates is constrained to add to the original total population. This amounts to multiplying the kernel estimate by a normalising factor. In practice they also incorporate *a priori* decisions about certain grid squares which do not have any population. For example, those on high ground, or in the sea!

The results of this kind of analysis are illustrated in Figure 7.5, which shows total population from the 1991 Census, interpolated to grid cells of

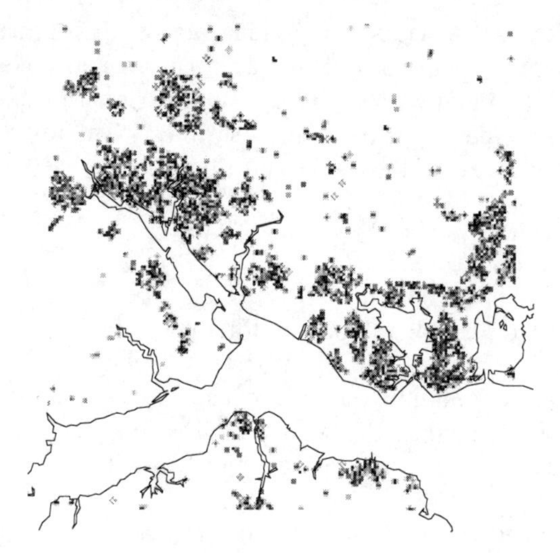

Fig. 7.5 Population interpolated to a fine grid in an area of S. England

200 m × 200 m, over a 39 km × 39 km region in the South of England. The area covers Southampton, Portsmouth and the northern Isle of Wight, with the coastline indicated. The kernel estimated surface shows the true distribution of population within this area considerably more clearly than would be possible by working with the published counts for each enumeration district in the area.

This technique is an application of kernel estimation to the general problem of transferring data from one set of areal units to another, or *cross-areal interpolation* as it is often known. In passing, it is perhaps of interest to comment briefly on other ways in which this may be done.

- Assign data to 'nearest' centroid. Data are interpolated from one set of areas to the other on the basis of assigning all the data associated with an area in one set to that area in the other set which is closest in terms of having smallest centroid–centroid distance.
- Interpolate data from one set of areas to the other on the basis of assigning all the data associated with an area in one set to that area in the other set in which its centroid lies. This is often referred to as *point in polygon* interpolation.
- Interpolate data from one set of areas to the other as a weighted average of data in the areas in the first set which the area in the second set overlaps. The weights are based on the proportion of the overlapping area. This is often referred to as *area weighting*.

Other more sophisticated and statistically based schemes have also been developed, for which we give some references at the end of the chapter.

Returning to kernel estimation, we commented earlier on its use where the observation y_i in area $\mathcal{A}_i$ represented some kind of 'average' measure over that area, and we derived an estimate of $\mu(\boldsymbol{s})$, using the kernel estimation methods from Chapter 5. Having now discussed kernel estimation of count data, it is worth remarking that if the former 'average' measure was in fact a rate and the raw counts from which the rate is derived are available, then rather than produce a kernel estimate of the 'average' rates y_i, as described earlier, it would be better to use instead a ratio of kernel estimates of the two counts. If the numerator count corresponding to the rate y_i is n_{i1} and the denominator n_{i2} then an appropriate estimate might be:

$$\hat{\mu}_\tau(\boldsymbol{s}) = \frac{\sum_{i=1}^{n} k\left(\frac{(\boldsymbol{s} - \boldsymbol{s}_i)}{\tau}\right) n_{i1}}{\sum_{i=1}^{n} k\left(\frac{(\boldsymbol{s} - \boldsymbol{s}_i)}{\tau}\right) n_{i2}}$$

where we have assumed that the same bandwidth is used in both the numerator and the denominator. It may sometimes be preferable to use different bandwidths, in which case slight modifications apply as discussed in Chapter 4.

Ex 7.6

7.4.5 Spatial correlation and the correlogram

The exploratory methods discussed so far in this chapter have been concerned with first order variations in attribute values, in other words with estimating the way in which the mean or expected value of the process varies in the study region. We now turn our attention to methods designed more explicitly to explore the spatial dependence of deviations in attribute values from their mean; that is, the second order properties.

In our discussion of spatially continuous data in Chapter 5 we used the variogram and covariogram to summarise and analyse covariance structure in the observed process. There is no reason why we cannot apply the same ideas to area data if we are prepared to make the assumption that attribute values are located at some point within each area, such as the centroid, so that we can calculate the Euclidean distance between them. However, we really need measures which incorporate the more general forms of spatial proximity that we discussed earlier.

Various techniques have been suggested in this regard. Those most commonly used tend to focus on estimating spatial correlation rather than covariance. Note that strictly we should probably refer to this as *spatial autocorrelation*, rather than just spatial correlation, since it involves the correlation between values of the *same* variable at different spatial locations. However, as in previous chapters, we will drop the prefix 'auto' in our subsequent references. The two measures that have been most widely used are:

Moran's I—for a spatial proximity matrix $\boldsymbol{W}$ spatial correlation in attribute values y_i is estimated as:

$$I = \frac{n \sum_{i=1}^{n} \sum_{j=1}^{n} w_{ij}(y_i - \bar{y})(y_j - \bar{y})}{\left(\sum_{i=1}^{n} (y_i - \bar{y})^2 \right) \left(\sum \sum_{i \neq j} w_{ij} \right)}$$

Geary's C—for a spatial proximity matrix $\boldsymbol{W}$ spatial correlation in attribute values y_i is estimated as:

$$C = \frac{(n-1) \sum_{i=1}^{n} \sum_{j=1}^{n} w_{ij}(y_i - y_j)^2}{2 \left(\sum_{i=1}^{n} (y_i - \bar{y})^2 \right) \left(\sum \sum_{i \neq j} w_{ij} \right)}$$

Readers who wish to refer back to the estimation of covariograms and variograms for spatially continuous data in Chapter 5, will see that the first of these correlation measures is closely related to the covariogram whilst the second is more related to the variogram.

Strictly, neither of these statistics is constrained to lie in the $(-1, 1)$ range as is the case for conventional non-spatial correlation. This is unlikely to present a practical problem for most real data sets and reasonable proximity matrices. However if required they can be adjusted by their theoretical bound, to force the $(-1, 1)$ range, although the appropriate correction factor can be rather 'messy' mathematically. For example a theoretical bound for I is given by:

$$|I| \leq \frac{n}{\sum \sum_{i \neq j} w_{ij}} \left(\frac{\sum_{i=1}^{n} \left(\sum_{j=1}^{n} w_{ij}(y_i - \bar{y}) \right)^2}{\sum_{i=1}^{n} (y_i - \bar{y})^2} \right)^{\frac{1}{2}}$$

and so I divided by this bound would be restricted to a $(-1, 1)$ range in the same way as the usual non-spatial correlation coefficient.

Of more interest is the generalisation of either I or C to estimate spatial correlation at different spatial lags and so produce a correlogram. This may be performed by simply calculating either of them using the proximity matrix appropriate for that lag, $\boldsymbol{W}^{(k)}$. In the case of Moran's I, we can estimate spatial correlation at lag k by:

$$I^{(k)} = \frac{n \sum_{i=1}^{n} \sum_{j=1}^{n} w_{ij}^{(k)} (y_i - \bar{y})(y_j - \bar{y})}{\left(\sum_{i=1}^{n} (y_i - \bar{y})^2 \right) \left(\sum \sum_{i \neq j} w_{ij}^{(k)} \right)}$$

where $w_{ij}^{(k)}$ are the elements of the $(n \times n)$ spatial proximity matrix at spatial lag k, $\boldsymbol{W}^{(k)}$.

We can thus construct and plot a correlogram where the spatial correlation at a particular spatial lag is plotted against the lag. Note that values at neighbouring lags of a correlogram are highly correlated, since the correlation at larger lags is in part a function of correlations at smaller lags. Care must therefore be taken in attributing particular significance to peaks in a correlogram at certain lags if there are also peaks at smaller lags. This problem can be avoided by the calculation of a *partial correlogram* in which effects due to smaller lags are eliminated when estimating the correlation at any particular later lag. This gives a more direct measure of spatial interaction at this particular lag. We do not give details of these calculations, but simply note that such techniques exist and may be useful in the interpretation of spatial correlation structure.

We should remind the reader here of a point that has been made both in Chapter 3 and in Chapter 5, in relation to the analysis of spatial correlation or spatial dependence. That is, once we start to examine for spatial dependence over anything but very small scales in $\mathcal{R}$, we are making the implicit assumption that the process can considered to be homogeneous or isotropic over such scales. If this is not the case, then any attempt to estimate second order effects faces the problem that they are not then necessarily constant over the scale considered and may also be confounded with first order variation in mean value. One therefore has no repetition available to estimate them. The above comments imply that if we use a spatial correlogram in a situation where there are large scale first order effects, then any spatial dependence it may indicate could well be due to these first order effects rather than to second order dependence.

Let us now explore some of the case studies and see what information we glean from an understanding of their spatial correlation structure. In so doing, we may also obtain some insights, albeit informal, into why such spatial structure arises. This will prove useful in the subsequent modelling of such data.

The Irish blood group data show high positive spatial correlation at all short lags (Figure 7.6), indicating that counties with high percentages of people with blood group A tend to be located near to counties with similar percentages. But, bearing in mind the comments above, this spatial correlation could arise because of a rather simple trend in the data, from higher values in the south-east to lower values in the west and north. Certainly, the map seems to indicate a trend, or first order effect. Support for this is found in the correlogram, where

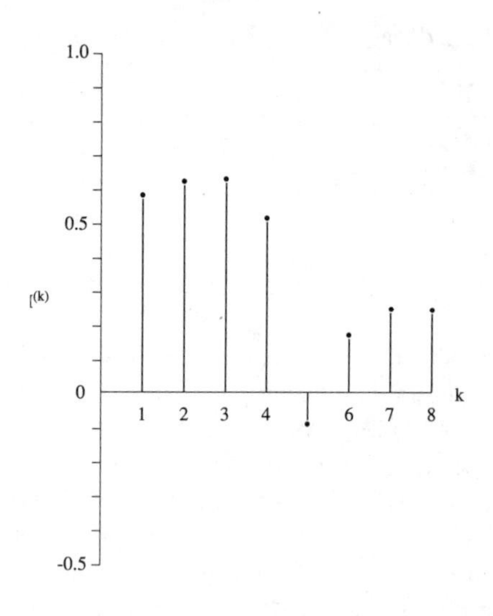

Fig. 7.6 Correlogram for Irish blood group data

correlation does not decline to near zero after one or two lags, but rather 'persists' over several lags. This is indicative of non-stationarity or heterogeneity in the data. Consequently, although the correlogram is conceptually a means of exploring second order effects, in this case it has some (limited) value as a tool for giving evidence of non-stationarity in area data. How might non-stationarity arise in these data? We might want to try to 'explain' this using the modelling techniques which we will introduce later, but, as noted earlier, it is thought that distribution of this blood group reflects historical settlement processes, with Anglo-Norman settlement in the east giving rise to spatial concentration. With local dispersal of population, and historically short-distance migration and marriage distances, we clearly have some putative explanation for such a distribution.

The distribution of the vote for Clinton also shows positive spatial correlation at short lags, though this is somewhat muted. As we saw from the map, the highest data value is in Arkansas, but we do not observe obvious locally elevated concentrations in the vote in adjacent states. Instead, other higher percentages are, to some extent, dispersed, with Maryland, New York, and Illinois having at least 50% of the population voting for Clinton. It would appear that the moderate degree of positive correlation arises from the tendency of states in the west-central USA to 'behave' similarly, voting more conservatively.

With a total of 18 variables in the Chinese data we can expect to find some varying spatial correlation structures, and this is indeed the case. For several variables we detect a high degree of positive spatial correlation at lags 1–3, possibly again indicative of first order trends. Good examples of this are the variables measuring the percentage of irrigated land that is power-irrigated, the fertility rate, and the data on deaths from cancer among males. Other variables, such as that on retail sales per head of population, levels of rural incomes, and the distribution of the urban population, show quite high positive correlation at lag 1 (where only neighbouring provinces are considered) but a 'tailing off' at further lags; see Figure 7.7a. This may be evidence of 'second order' effects rather than broad regional trends.

A further class of variable is that which displays little spatial correlation; for example, per capita output from the coal industry, or hospital beds per head of population; see Figure 7.7b. Here, there is little detectable association between the value in one province and values in neighbouring provinces. Only one variable, students enrolled in specialised secondary schools, shows evidence of negative correlation at lag 1. When mapped, the variable shows a predictable urban bias, but with high values in the two north-west provinces.

Clearly, then, different processes are operating to produce rather different spatial patterns. We should be extremely cautious about making *any* inferences about process simply on the basis of a map pattern. However, we might speculate that some of the highly correlated variables (for instance, those measuring the use of mechanisation and power in agriculture, or grain yields) stem from broad regional trends in agricultural structure and practice. Similarly, the cancer variable shows a swathe of high rates along the southern coastal belt, which may reflect large scale variations in regional diet or exposures to tobacco and alcohol. In other cases, such as per capita gross industrial output, the distribution of rural incomes, or the distribution of urban population, where we see correlation at the first lag, we might suggest that this is a second order effect, reflecting the spatial localisation of economic activity in 'core' areas. The economic landscape is divided up into areas of high and

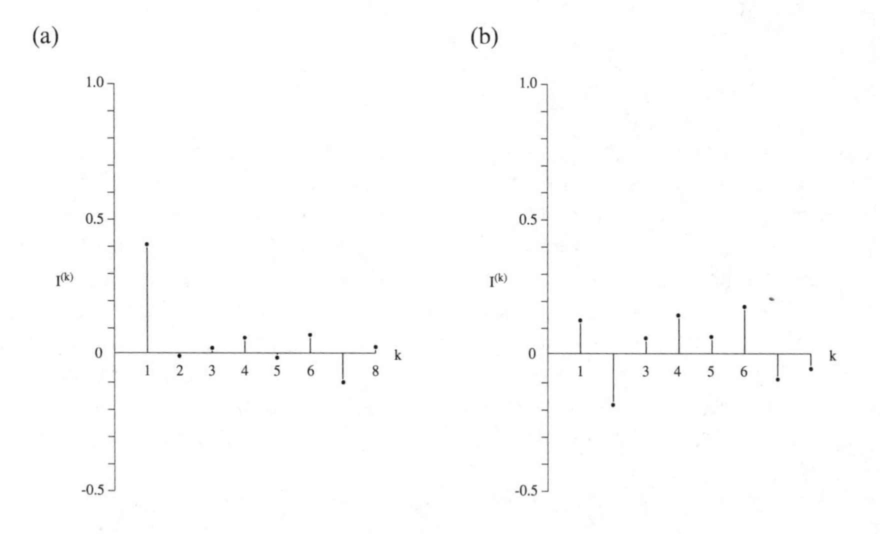

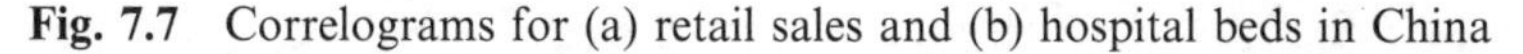
Fig. 7.7 Correlograms for (a) retail sales and (b) hospital beds in China

low activity, reflecting the ways in which goods and services are exchanged over small regions.

In making these speculative remarks we are anticipating the wish to model such areal distributions. We now turn our attention to some more formal ways of explaining spatial variation in a variable of interest.

7.5 Modelling area data

So far in this chapter we have been concerned with exploring area data in a fairly informal way. These methods are often only a preliminary to more formal modelling approaches which seek to establish relationships between the attribute values y_i, the relative spatial arrangement of the $\mathcal{A}_i$, and possibly also the values of other attributes, $\boldsymbol{x}_i^T = (x_{i1}, \ldots, x_{ip})$, recorded for each $\mathcal{A}_i$. In this section we consider the construction of such specific models. In the first instance we shall mostly be interested in models which involve first order variation in the mean value, μ_i, of the attribute. We then move on to consider how to include second order effects into such models.

7.5.1 Non-spatial regression models

An obvious tool to adopt in the modelling of area data is the multiple regression model, the workhorse of much non-spatial data analysis. We considered similar types of model in our discussion of 'trend surface analysis' in Chapter 5. In the particular case which we are now considering, it is convenient to express that model, for all areas together, using matrix notation as:

$$\underset{(n \times 1)}{\boldsymbol{Y}} = \underset{(n \times p)}{\boldsymbol{X}} \; \underset{(p \times 1)}{\boldsymbol{\beta}} + \underset{(n \times 1)}{\epsilon}$$

where $\boldsymbol{Y}$ is the vector of the random variables, Y_i, or more strictly $Y(\mathcal{A}_i)$, in each of the areas $\mathcal{A}_i$; $\boldsymbol{X}$ is the matrix of the values of p explanatory variables, in each area, with row vectors $\boldsymbol{x}_i^T$, $i = 1, \ldots, n$; and ϵ is a vector of zero-mean random variables, ϵ_i, representing fluctuations from the trend or mean value $\mu_i = \boldsymbol{x}_i^T\boldsymbol{\beta}$ in the area $\mathcal{A}_i$. The standard regression assumption is that the errors, ϵ_i, in different areas have constant variance, say σ^2, and that they are independent, so that their covariance is zero. This implies there are no second order effects present in the process $Y(\mathcal{A}_i)$. Under this assumption we may 'fit' the model to our observed data, y_i, by *ordinary least squares*, deriving estimates $\hat{\boldsymbol{\beta}}$ and their associated standard errors, by using the standard regression results:

$$\hat{\boldsymbol{\beta}} = (\boldsymbol{X}^T\boldsymbol{X})^{-1}\boldsymbol{X}^T\boldsymbol{y}$$

$$VAR\left(\hat{\boldsymbol{\beta}}\right) = \sigma^2(\boldsymbol{X}^T\boldsymbol{X})^{-1}$$

where $\boldsymbol{y} = (y_1, \ldots, y_n)^T$, is the vector of observed values in areas $\mathcal{A}_i$. Note that if appropriate some of the explanatory variables $(x_{i1}, \ldots, x_{ip})$ could consist of the coordinates (s_{i1}, s_{i2}) of the centroids of the areas $\mathcal{A}_i$ (and possibly their powers and cross products) in order to introduce the notion of purely spatial effects (a trend surface) into the model as well as dependence on the values of the other covariates.

Details of this standard multiple regression model may be found in most elementary statistical texts, including considerations such as how to evaluate the 'fit' of the model to the observed data and test the significance of either individual β_i values or combinations of them. We have already discussed some basic details concerned with such models in Chapter 5. For convenience we repeat some of those again here. For example, the variance, σ^2, of ϵ_i in the above expression for $VAR(\hat{\boldsymbol{\beta}})$ would normally be unknown and would be estimated from the *residuals* of the fitted model, the differences between the observed values y_i at each sample site and those predicted by the model $\hat{y}_i = \boldsymbol{x}_i^T\hat{\boldsymbol{\beta}}$. The appropriate estimate of σ^2 is given by

$$\hat{\sigma}^2 = \frac{\sum_{i=1}^{n}(y_i - \hat{y}_i)^2}{n - p}$$

Examination of the residuals is also used to assess the 'fit' of the model in various ways. A measure of overall 'goodness of fit' is provided by the *coefficient of determination*, also known as the 'proportion of variation explained', or 'R^2'. This is calculated as:

$$R^2 = 1 - \frac{\sum_{i=1}^{n}(y_i - \hat{y}_i)^2}{\sum_{i=1}^{n}(y_i - \bar{y}_i)^2}$$

'F tests' may be used to compare the fits of alternative models; for example, to ask whether the inclusion of one set of covariates offers a significant improvement over that of another. We refer the reader to one of the many excellent texts on applied regression for the numerous other details of such methods.

As an exploratory tool, a simple ordinary least squares regression model may provide useful insights into overall relationships and spatial trends; but it suffers from exactly the same problem that we encountered when considering ordinary least squares fitting of trend surfaces with spatially continuous data. The standard regression assumptions that ϵ is a vector of independent random errors with constant variance σ^2 are very unlikely to be appropriate for the process Y_i because of the possibility of spatial dependence between the ϵ_i.

In other words, ordinary regression assumes that there is only first order variation and no second order variation in the process modelled. With most real area data this assumption is often violated and residuals from such a regression will be spatially correlated. In addition, it is unlikely that $VAR(\epsilon_i)$ will be constant for all $\mathcal{A}_i$. As a result of one or both of these problems the conventional confidence intervals for regression coefficients and the corresponding assessment of the significance of any of the covariates will be invalid.

As discussed in the case of spatially continuous data, non-constant variance over $\mathcal{A}_i$ may be corrected for by weighted regression (see next section), or by a transformation of y_i to y_i'. A useful class of such transformations is the *Box–Cox* family discussed in Chapter 5. Transformations are particularly useful when the variable of interest is in the form of a count or proportion, a situation which often arises when data are collected over areas.

Where the response variable in a regression is a count of some description, such as numbers of a certain species of plant, or numbers of car accidents, a Poisson distribution for the response may be a reasonable assumption. In this case the variance would be proportional to the mean value. This suggests a square root transformation (Box–Cox with $\lambda = 1/2$), which 'stabilises' the variance in relation to the mean (i.e. the variance of the transformed variable is constant). However, the logarithmic transformation is also often used with count data, as a compromise between wishing to simultaneously stabilise variance and linearise the relationship between the mean value and values of covariates. Strictly, the logarithmic transformation is more appropriate to stabilise variance when this is proportional to the square of the mean, rather than to the mean, as discussed in Chapter 5. However, at the same time we might expect covariates to act multiplicatively or exponentially rather than additively on the response when this is a count. The logarithm is a natural way to linearise multiplicative relationships and it may be argued that in practice many observed counts display an element of 'over dispersion' (their variance is larger than might be expected on the basis of a Poisson distribution).

When the response variable in a regression is a proportion the situation is more difficult. The variance of an observed proportion, with an expected value μ_i, is $\mu_i(1 - \mu_i)/n_i$, where n_i is the denominator on which the proportion is based. At the same time, one might expect covariates, as with counts, to act multiplicatively rather than additively on the response. However, the scale of a proportion is limited to the (0,1) range and therefore a given change in the value of an explanatory variable is unlikely to produce the same change in a proportion that is already towards the extreme end of its scale, as when that proportion is in the middle of its range. Hence we might expect a sigmoid or 'flat S' shaped relationship between response and explanatory variables. This suggests that when y_i is a proportion, appropriate transformations might be the *angular transformation*:

$$y_i' = \sin^{-1} \sqrt{y_i}$$

or the *logistic transformation*:

$$y_i' = \log\left(\frac{y_i}{1 - y_i}\right)$$

Both of these have the required sigmoid shape. The first will also stabilise the variance of observed proportions, but only if they are all based upon denominators of the same magnitude. The second produces a variance–mean relationship which is approximately the reciprocal of that given above for the original proportion. Hence, in practice neither of them is entirely satisfactory and some form of weighted regression (see later) would still be advisable in addition to the transformation to account for non-constant variance. The logistic transformation suffers from an additional problem if the observed proportion is 0 or 1, since the transformed value is then undefined. A modified definition:

$$y_i' = \log\left(\frac{n_i y_i + \frac{1}{2}}{n_i(1 - y_i) + \frac{1}{2}}\right)$$

is useful in this regard. If there is a large number of 0 or 1 values, then this definition can be used throughout, otherwise it can be used for the occasional 0 or 1 values. Given this lack of an entirely suitable transformation, in the informal exploratory analysis of proportions it might be as well to use untransformed proportions directly in regression, or perhaps take a logarithmic transformation on the basis of a multiplicative rather than additive model. It should be appreciated however that no firm conclusions can be drawn from such *ad hoc* analyses.

There are actually far better ways to deal with the modelling of count or rate data than using the kind of transformations discussed above. These involve the use of *generalised linear models*, and we will return to this subject in more detail in the next chapter. It should also be appreciated that in general transformations will not help to alleviate the other problem mentioned earlier that arises when regression is applied in a spatial context: that of possible second order effects or residuals being spatially correlated. Before we consider this further, we give some results of applying ordinary least squares regression to two of our case studies.

Consider first, variation in the vote for Clinton in the 1992 US Presidential election. Given our above comments on the problems of applying ordinary regression to proportions, we hesitate to start with an example which involves a response variable which is a percentage. However, in this case none of the percentages are extreme and our analysis is informal in nature; we are not attempting to derive any definitive model of the voting patterns. We will simply proceed with a logarithmic transformation of the percentages on the basis of the likelihood of multiplicative rather than additive effects for covariates. A preliminary examination of simple two-variable correlations between the logarithm of the vote and the possible covariates included in the data set described earlier gives correlations of 0.348 with the percentage of the population that is black, 0.220 with the percentage unemployed, 0.283 with personal income per capita, −0.361 with the percentage that has completed high school education and 0.244 with business failures. Other correlations are very low. However, these variables are themselves intercorrelated and fitting an ordinary least squares model with the logarithm of the vote as the response variable yields only three variables with estimates of regression coefficients significantly different from zero. These are the percentage of unemployed, per capita personal income and the percentage completing high school education. The corresponding estimates of the regression coefficients (and standard errors) are 0.0413 (0.016), 0.00004 (0.00001), −0.01132 (0.00305), respectively. This suggests that the distribution of unemployment, income and education all had a significant effect on the distribution of the vote for Clinton. Together, the three variables account for 43.5% of the variation in the logarithm of the vote (i.e. $R^2 = 0.435$). Adding further covariates fails to improve significantly on this.

When we examine the residuals for spatial dependence we find there is an absence of correlation; Moran's I statistic yields a value of 0.026, a value which is not significantly different from zero. It would appear, then, that the ordinary least squares model is reasonably adequate, given the provisos already made concerning the modelling of proportions in this way. Arkansas, Clinton's home state, is under-predicted (a high positive residual); while in Idaho and Utah, rather conservative states in the west, residuals are large and negative; the vote for Clinton is rather lower than the covariates predict.

Consider another example, that concerned with trying to explain variations in mortality from myocardial infarction (heart attacks) in English Health Authorities. Here the response variable is a standardised mortality ratio (SMR). We shall have more to say about such measures in the next chapter. For the present we simply note that they are a ratio of the number of deaths in each area to that expected in the area on the basis of national age–sex specific rates. In general, a reasonable assumption for such SMRs is that their variance will be proportional to their mean. This follows from the argument that the observed number of deaths in any area, although strictly binomial, is only a small proportion of the population from which they arise and so can be considered to follow a Poisson approximation to the binomial. Our earlier remarks on modelling counts would therefore suggest a square root or possibly a logarithmic transformation. Here we use the logarithmic transformation of the SMRs. If they are indeed Poisson distributed, then the variance of the logarithms will be inversely proportional to the number of deaths in each area, rather than constant. A weighted regression, as described in the next section, may therefore strictly be applicable with this response variable. However, in this case we have access only to the SMRs and not the actual numbers of deaths in each of the Health Districts and so are unable to construct appropriate weights. We therefore proceed with an ordinary least squares regression.

Our main interest in analysing these particular data at this stage is to contrast certain results from ordinary least squares with those which we shall present later in the chapter, when spatial dependence is allowed for in the regression model. We do not therefore consider relationships between the logarithm of SMRs for myocardial infarction and all possible covariates in this data set. We simply report the results of an ordinary least squares regression between this response variable and a single covariate, the logarithm of the Jarman score. We discussed this 'deprivation' measure earlier in the chapter. We obtain an estimated regression coefficient of 0.623 with a standard error of 0.093, corresponding to a t value (ratio of the estimate to its standard error) of 6.6992. This suggests that the logarithm of the Jarman score is significantly related to variations in the logarithm of SMRs for myocardial infarction, as would perhaps have been expected, given the nature of this disease. However, only approximately 19% of the overall variation is accounted for. Furthermore, the residuals are very highly spatially correlated ($I = 0.4477$ and this persists at higher lags). If we map these residuals we find a clear pattern of positive residuals in the north of the country and negative ones in the south. Clearly, while the Jarman index has gone some way to predicting variation in mortality, it is hardly a very adequate explanation and it has not succeeded in capturing the 'north–south' divide that broadly characterises the English social landscape. Also, the degree of spatial correlation that remains in the residuals implies that our estimate of the relationship between the Jarman index and mortality is highly suspect.

Ex 7.8

Clearly, there are spatial effects in the data and we need to correct for them if we are to obtain a 'truer' picture of the relationship between mortality and the Jarman index. If the spatial effects are purely a first order north–south trend, then we could correct for them simply by including an extra covariate into our regression model to 'remove' the effect. If we do this, using a crude measure such as the latitude of the 'centres' of each of the Health Districts, then the situation certainly improves. The new covariate is highly significant, as we might have expected given our remarks on the obvious spatial trend in the earlier residuals. Of more interest is the change in the estimated regression coefficient for the Jarman index. The new value is 0.392 with a standard error of 0.083, still highly significant, but with a very different value to that obtained before. However, the residuals from this new regression remain highly spatially correlated ($I \approx 0.35$). Further, if we map these residuals we do not find an obvious spatial pattern to suggest how we might easily remove this remaining spatial dependence by modification to the simple north–south trend that we have already included.

So, we have improved the situation by trying to 'explain away' the spatial dependence as a first order trend, but we have not completely solved the problem. We still have a high degree of residual spatial dependence and as a consequence our estimated relationship between mortality and the Jarman index is still suspect. The spatial dependence seems more complicated than purely a first order effect. We need to consider whether we can adjust our model to allow for second order spatial effects, a point which we now take up in greater detail.

7.5.2 Generalised least squares

The understanding that ordinary least squares regression does not provide for the possibility of second order effects leads us, as in Chapter 5, to consideration of *generalised least squares* in order to allow us to relax the assumption of spatial independence, inherent in the simple regression model. With this modification the model becomes:

$$\underset{(n \times 1)}{\boldsymbol{Y}} = \underset{(n \times p)}{\boldsymbol{X}} \underset{(p \times 1)}{\boldsymbol{\beta}} + \underset{(n \times 1)}{\boldsymbol{U}}$$

where $\boldsymbol{U}$ is a zero-mean vector of errors with variance–covariance matrix $\boldsymbol{C}$, i.e.

$$E(\boldsymbol{U}) = 0$$
$$E(\boldsymbol{U}\boldsymbol{U}^T) = \boldsymbol{C}$$

Then the least squares estimates for $\boldsymbol{\beta}$ and the corresponding standard errors become:

$$\hat{\boldsymbol{\beta}} = (\boldsymbol{X}^T\boldsymbol{C}^{-1}\boldsymbol{X})^{-1}\boldsymbol{X}^T\boldsymbol{C}^{-1}\boldsymbol{y}$$
$$VAR(\hat{\boldsymbol{\beta}}) = (\boldsymbol{X}^T\boldsymbol{C}^{-1}\boldsymbol{X})^{-1}$$

Weighted least squares are a special case of this where $\boldsymbol{C}$ is a diagonal matrix and would be appropriate in the simple case where the errors, U_i, are heteroscedastic. Each observation in the regression is weighted in inverse proportion to its variance, so giving less weight to observations with large variance.

As in the case of spatially continuous data, generalised least squares gives us a way of modelling both first and second order variation in the observed process. Of course the problem is again that we do not know $\boldsymbol{C}$ (the covariance structure of the process over the areas $\mathcal{A}_i$) and so we therefore need tools to model aspects of covariance structure in area data. This is the subject of the next section.

7.5.3 Tests for spatial correlation

In an earlier section we used the correlogram to explore spatial dependence in area data. It is therefore natural to consider using this to ascertain whether the errors arising from regression models are spatially dependent and second order effects need to be incorporated. An automatic question which then arises is whether the observed values of the spatial correlation measures discussed previously are significantly larger than might be expected to arise by chance from a spatially independent process. In other words, do their observed values indicate significant spatial correlation in the process? So far, we have simply looked at the magnitude of these values in an informal way. Now we need to consider more formally the question of what constitutes a 'large' spatial correlation.

We consider this here purely in the case of Moran's I. There are two main approaches to testing observed values of I for significant departure from the hypothesis of zero spatial correlation.

Random permutation test—This is based on the following simple idea. Suppose we have n values, y_i, relating to areas $\mathcal{A}_i$. Then $n!$ permutations of this map are possible, each corresponding to a different arrangement of the n data values, y_i, over the areas. One of these corresponds to the observed situation. The value of I can be calculated for any one of these $n!$ permutations and hence an empirical distribution built up for possible values of I under random permutations of the n data values.

If the value of I corresponding to the observed arrangement of the y_i corresponds to an 'extreme' value in the permutation distribution then we would interpret this as suggesting that some rule other than random assignment of values to areas had been followed in generating the observed y_i arrangement—in other words as evidence of significant spatial correlation.

In most cases computation of as many as $n!$ arrangements will be infeasible; if so, a close approximation to the permutation distribution can be obtained by using a 'Monte Carlo' approach and simply sampling randomly from a reasonable number of the $n!$ possible permutations.

Approximate sampling distribution of I—If there is a moderate number of areas then an approximate result for the sampling distribution of I under certain assumptions may be used to develop a test. If it is assumed that the y_i are observations on random variables Y_i whose distribution is normal, then if Y_i and Y_j are spatially independent ($i \neq j$), I has a sampling distribution which is approximately normal with:

$$E(I) = -\frac{1}{(n-1)}$$

$$VAR(I) = \frac{n^2(n-1)S_1 - n(n-1)S_2 - 2S_0^2}{(n+1)(n-1)^2 S_0^2}$$

where:

$$S_0 = \sum_{i \neq j}\sum w_{ij}$$

$$S_1 = \frac{1}{2}\sum_{i \neq j}\sum (w_{ij} + w_{ji})^2$$

$$S_2 = \sum_k \left(\sum_j w_{kj} + \sum_i w_{ik} \right)^2$$

We can therefore test the observed value of I against the percentage points of this approximate sampling distribution. An 'extreme' observed value of I indicates significant spatial correlation.

The hypotheses involved in each of the above two tests are somewhat different. The randomisation test embodies the assumption that no values of y_i other than those observed are realisable—the data are being treated as a population and the only question being investigated is how the data values are arranged spatially. The test is therefore a test of pattern or organisation in the observed values relative to the set of all possible patterns in the *given* data values. The approximate sampling distribution test makes the assumption that the observed y_i are observations on (normal) random variables, i.e. they are one realisation of a process and that other possible realisations can occur. The test is therefore one of spatial dependence or correlation (providing the distribution of the Y_i can be assumed to be normal). Which of these two tests is used therefore depends upon what sort of data one is dealing with. In the case of analysing election results one might be justified in considering this an unrepeatable experiment and proceed with the randomisation test. In the case of disease rates the assumption that the observations arise from random variables is more likely to be reasonable and the sampling distribution test might be appropriate (although it is unlikely the disease rates would arise from a normal distribution!).

Care needs to be taken in applying these formal tests of spatial correlation when I has been calculated from residuals which arise from a regression. The problem arises because if p parameters (regression coefficients, β_i) have been estimated in the regression, then the observed residuals are automatically subject to p linear constraints. That is, the observed residuals will be automatically correlated to some extent and so the above testing procedure for Moran's I will not be valid. If $p \ll n$ then we might be justified in ignoring this. If not, then strictly we should use adjustments to the mean and variance of the approximate sampling distribution of I. We do not give details here, referring the reader to references at the end of this chapter which also cover tests for spatial correlation at spatial lags other than the first.

7.5.4 Spatial regression models

Having discussed hypothesis tests for significant spatial correlation we can now revert to our generalised least squares model, and see how we can use these to develop this model. Recall that the model was:

$$\boldsymbol{Y} = \boldsymbol{X\beta} + \boldsymbol{U}$$

where $\boldsymbol{U}$ is a zero-mean vector of errors with variance–covariance matrix $\boldsymbol{C}$, i.e.

$$E(\boldsymbol{U}) = 0$$
$$E(\boldsymbol{U}\boldsymbol{U}^T) = \boldsymbol{C}$$

In our discussion of the modelling of spatially continuous data in Chapter 5, we were able to estimate a variogram of residuals obtained after preliminary fitting of this model by ordinary least squares. We then proceeded to fit a smooth distance relationship to this variogram, in order to obtain a succinct description of covariance structure, allowing us to estimate the elements of the covariance matrix $\boldsymbol{C}$. This then enabled us to complete our specification of a final model for the observed spatial process as consisting of a first order trend (as reflected through the basic regression relation $\boldsymbol{X\beta}$) plus a stationary, zero-mean second order process (as reflected through the variogram model and the corresponding covariance matrix $\boldsymbol{C}$).

With area data we have some difficulty in trying to follow the same approach. The models we fitted for the covariance structure of spatially continuous data were based on a belief that the second order effects were stationary and could be modelled by a smooth function of distance between point locations. Stationarity of the second order component is a somewhat more questionable assumption for processes operating on a set of areas. Even if there exists an underlying continuous space process which is stationary, indirect observation of this using aggregated values over irregular areas will result in variances and covariances in the areas that will not be the same for all areas. Further, we do not have a simple measure of 'distance' between areas. We have already argued that proximity measures other than Euclidean distance between centroids may in many cases be preferable for area data.

Thus, although we could attempt direct modelling of covariance structure for area data based on the residual correlogram or a related measure, assuming data values to occur at area centroids, it is preferable to look at other indirect ways of specifying a covariance model. This is where the correlogram and tests for significant spatial correlation developed earlier fit in. We can use these to check that whatever models we propose have succeeded in explaining the observed second order variation by demonstrating an acceptable degree of spatial independence in the residuals from the final model.

In other words, instead of trying to specify $\boldsymbol{C}$ directly in the above generalised least squares model, we specify it indirectly by an *interaction scheme*. This is done by including, in the model, relationships between variables and their neighbouring values (usually involving a few extra parameters which need to be estimated) and which indirectly specify particular forms of $\boldsymbol{C}$. Such variate interaction models need not assume stationarity for the second order component, nor are they necessarily restricted to covariance structures which are smooth functions of Euclidean distance.

One simple variate interaction model is:

$$\boldsymbol{Y} = \boldsymbol{X\beta} + \boldsymbol{U}$$
$$\boldsymbol{U} = \rho \boldsymbol{W U} + \epsilon$$

where ϵ is a vector of independent random errors with constant variance σ^2 i.e.

$$E(\epsilon) = 0$$
$$E(\epsilon\epsilon^T) = \sigma^2 \boldsymbol{I}$$

and where $\boldsymbol{W}$ is a proximity matrix as discussed previously. For convenience we will assume that $\boldsymbol{W}$ has been standardised to have row sums of unity.

This model is an example of a *simultaneous autoregressive* model (SAR), in this case with just one interaction parameter ρ—more complex forms of this class of model are possible, which involve several interaction parameters and proximity matrices at different lags.

The model can be written as

$$\begin{aligned} \boldsymbol{Y} &= \boldsymbol{X\beta} + \rho \boldsymbol{WU} + \epsilon \\ &= \boldsymbol{X\beta} + \rho \boldsymbol{W}(\boldsymbol{Y} - \boldsymbol{X\beta}) + \epsilon \\ &= \boldsymbol{X\beta} + \rho \boldsymbol{WY} - \rho \boldsymbol{WX\beta} + \epsilon \end{aligned}$$

Hence $\boldsymbol{Y}$ is expressed as a response to several influences; Y_i in area $\mathcal{A}_i$ depends on the surrounding values Y_j $(j \neq i)$, through the term $\rho \boldsymbol{WY}$; it also depends on the general trend through $\boldsymbol{X\beta}$; and on neighbouring trend values through $\rho \boldsymbol{WX\beta}$. This particular SAR model is often referred to as the *autocorrelated errors model.*

We mentioned that variate interaction models indirectly imply a covariance structure $\boldsymbol{C}$. In this case we can derive this directly, with some matrix manipulation:

$$\begin{aligned} \boldsymbol{C} &= E(\boldsymbol{UU}^T) \\ &= E\left((\boldsymbol{I} - \rho \boldsymbol{W})^{-1} \epsilon \epsilon^T \left((\boldsymbol{I} - \rho \boldsymbol{W})^{-1}\right)^T\right) \\ &= (\boldsymbol{I} - \rho \boldsymbol{W})^{-1} E(\epsilon \epsilon^T) \left((\boldsymbol{I} - \rho \boldsymbol{W})^{-1}\right)^T \\ &= (\boldsymbol{I} - \rho \boldsymbol{W})^{-1} \sigma^2 \boldsymbol{I} \left((\boldsymbol{I} - \rho \boldsymbol{W})^{-1}\right)^T \\ &= \sigma^2 \left((\boldsymbol{I} - \rho \boldsymbol{W})^{-1} \left((\boldsymbol{I} - \rho \boldsymbol{W})^T\right)^{-1}\right) \\ &= \sigma^2 \left((\boldsymbol{I} - \rho \boldsymbol{W})^T (\boldsymbol{I} - \rho \boldsymbol{W})\right)^{-1} \end{aligned}$$

To ensure invertability of $(\boldsymbol{I} - \rho \boldsymbol{W})$ so that we obtain a valid covariance matrix $\boldsymbol{C}$, we need restrictions on the value of ρ. If $\boldsymbol{W}$ is standardised to have row sums of unity then these restrictions effectively amount to $-1 \leq \rho \leq 1$.

Note that $\boldsymbol{U}$ is not necessarily a stationary process under this interaction scheme. Consider the simple example for $\rho = 0.5$ and $n = 3$ areas with proximity matrix:

$$\boldsymbol{W} = \begin{pmatrix} 0.0 & 0.3 & 0.7 \\ 0.2 & 0.0 & 0.8 \\ 0.7 & 0.3 & 0.0 \end{pmatrix}$$

Then:

$$\left((\boldsymbol{I} - \rho \boldsymbol{W})^T (\boldsymbol{I} - \rho \boldsymbol{W})\right)^{-1} = \begin{pmatrix} 1.81343 & 0.96528 & 1.30767 \\ 0.96528 & 1.68751 & 1.15136 \\ 1.30767 & 1.15136 & 1.89931 \end{pmatrix}$$

So the diagonal elements of $\boldsymbol{C}$ (the variances) are not the same in different areas. Also, covariances differ between areas with the same spatial proximity, such as areas 1 and 2 and areas 3 and 2. So the residual process, $\boldsymbol{U}$, indirectly implied by the scheme is not a stationary process.

Having arrived at one possible model specification for our area data, how are the parameters of this model going to be estimated? The autocorrelated errors model could be fitted to observed data if the value of ρ were known; then, $\boldsymbol{C}$ could be estimated easily and generalised least squares could be used on the model $\boldsymbol{Y} = \boldsymbol{X\beta} + \boldsymbol{U}$. However, in general ρ is not known and must be estimated from the data. The simultaneous estimation of $\boldsymbol{\beta}$ and ρ is not straightforward and involves use of a computationally intensive maximum likelihood procedure.

A pragmatic way to avoid this simultaneous estimation would simply be to guess a value of ρ. For example, if we assign the maximum possible value $\rho = 1$, the model becomes:

$$\boldsymbol{Y} = \boldsymbol{X\beta} + \boldsymbol{WY} - \boldsymbol{WX\beta} + \boldsymbol{\epsilon}$$

which may be written as:

$$(\boldsymbol{I} - \boldsymbol{W})\boldsymbol{Y} = (\boldsymbol{I} - \boldsymbol{W})\boldsymbol{X\beta} + \boldsymbol{\epsilon}$$

This amounts to an ordinary least squares regression of $(\boldsymbol{I} - \boldsymbol{W})\boldsymbol{Y}$ on $(\boldsymbol{I} - \boldsymbol{W})\boldsymbol{X}$ and these quantities are just the *spatial differences* of the original variable values—the difference between the values in each area $\mathcal{A}_i$ and the weighted averages (as reflected by $\boldsymbol{W}$) of values in surrounding areas.

Of course, spatial differencing does not really solve our problem. Setting $\rho = 1$ is sensible if we believe that there is significant second order variation, but it is purely arbitrary. There are many other possibly better values for ρ (or indeed interaction schemes other than $\boldsymbol{U} = \rho\boldsymbol{WU} + \boldsymbol{\epsilon}$). Using differencing only results in valid estimates of $\boldsymbol{\beta}$ if ρ is indeed unity, in the same way as ordinary least squares on the original values only produces valid estimates of $\boldsymbol{\beta}$ if $\rho = 0$.

However, spatial differencing is simple to implement and may be a useful preliminary approach to improving the inferences obtained from an ordinary least squares regression on the original values. If $0 < \rho < 1$ then fitting the model $\boldsymbol{Y} = \boldsymbol{X\beta} + \boldsymbol{U}$ by ordinary least squares will tend to produce standard errors for estimated coefficients which tend to inflate the significance of the regression and make certain covariates appear more significant than they actually are. Fitting the spatially differenced form by ordinary least squares will tend to correct for this to some extent and is therefore the 'better of two evils'. However, if $-1 < \rho < 0$ then this situation is reversed and ordinary least squares on the original values would be preferable to spatial differencing.

An extension to the pragmatic use of spatial differencing in order to avoid simultaneous estimation of $\boldsymbol{\beta}$ and ρ avoids the assumption that $\rho = 1$ by regressing $(\boldsymbol{I} - \rho\boldsymbol{W})\boldsymbol{Y}$ on $(\boldsymbol{I} - \rho\boldsymbol{W})\boldsymbol{X}$ by ordinary least squares, for a sequence of different ρ values and choosing that which corresponds to the most acceptable residuals.

These sorts of *ad hoc* approaches to fitting the autocorrelated errors model are unsatisfactory from a formal statistical standpoint. One really requires simultaneous estimation of $\boldsymbol{\beta}$ and ρ which will give standard errors and associated confidence intervals for $\hat{\rho}$.

In order to do this we must resort to a maximum likelihood approach which, in turn, involves making some explicit distributional assumptions for the observed process. We proceed by assuming that y_i are observations on n random variables Y_i which are jointly normally distributed with mean $\boldsymbol{X\beta}$ and covariance matrix $\boldsymbol{C}$ (in fact the assumption of normality has already been implicitly made, in that it underlies the validity of most of the statistical tests associated with standard non-spatial regression). Under these assumptions the log likelihood function, $l()$, for the n observations $\boldsymbol{y}$ is:

$$l() \propto -\log|\boldsymbol{C}| - (\boldsymbol{y} - \boldsymbol{X\beta})^T \boldsymbol{C}^{-1} (\boldsymbol{y} - \boldsymbol{X\beta})$$

where, under the autocorrelated errors model:

$$\boldsymbol{C} = \sigma^2 \left((\boldsymbol{I} - \rho \boldsymbol{W})^T (\boldsymbol{I} - \rho \boldsymbol{W}) \right)^{-1}$$

The log likelihood thus depends on parameters $\boldsymbol{\beta}$, σ^2 and ρ. We then use an iterative approach to maximum likelihood estimation as follows:

1. Obtain an initial estimate $\hat{\boldsymbol{\beta}}$ by use of ordinary least squares on the model $\boldsymbol{Y} = \boldsymbol{X\beta} + \epsilon$.
2. Maximise the log likelihood $l()$ with respect to σ^2 and ρ, assuming $\boldsymbol{\beta}$ take the values of their most recent estimates. In other words maximise $l()$, conditional on $\boldsymbol{\beta} = \hat{\boldsymbol{\beta}}$, so obtaining $\hat{\sigma}^2$ and $\hat{\rho}$.
3. Obtain new estimates $\hat{\boldsymbol{\beta}}$ by use of generalised least squares on the model $\boldsymbol{Y} = \boldsymbol{X\beta} + \boldsymbol{U}$ taking:

$$\boldsymbol{C}^{-1} = \hat{\boldsymbol{C}}^{-1} = \frac{1}{\hat{\sigma}^2} \left((\boldsymbol{I} - \hat{\rho} \boldsymbol{W})^T (\boldsymbol{I} - \hat{\rho} \boldsymbol{W}) \right)$$

4. Iterate on steps 2 and 3 until there is an appropriate convergence of parameter estimates $\hat{\boldsymbol{\beta}}$, $\hat{\sigma}^2$ and $\hat{\rho}$. Approximate standard errors of the parameter estimates may then be derived using the standard maximum likelihood method, involving the second derivative of the log likelihood.

Step 2 is computationally intensive; we do not give details of maximisation methods here, but bear in mind that $\hat{\boldsymbol{C}}$ is an $(n \times n)$ matrix and that the function is being maximised with respect to possibly several parameters $\boldsymbol{\beta}$.

The discussion above has concerned parameter estimation for just one type of SAR model, the autocorrelated errors model. Clearly, once parameters have been estimated the important question is whether this model is an adequate representation of the observed process. This will involve exploration of the covariance structure of the final residuals via Moran's I and the associated correlogram. Unfortunately, the formal tests for spatial correlation which we discussed in relation to Moran's I are strictly not valid in the case of residuals from an autocorrelated errors model and so if we use them with such residuals we should be wary of the results we obtain. In practice, informal residual analyses using significance tests as guides and not dogma will usually suffice. We can examine the appearance of the correlogram and also use other simple exploratory devices—for example, plots of residuals, $\hat{\boldsymbol{u}}$, against neighbouring residuals, $\boldsymbol{W}\hat{\boldsymbol{u}}$. Ultimately, the objective is to achieve a set of residuals which do not exhibit marked spatial dependence and which justify acceptance of the model as a reasonable representation of the observed spatial process. If this is the case then we have ended up with a model which incorporates both first order and second order effects and has 'correct' parameter estimates and standard errors, unlike those provided by the earlier ordinary least squares regression which assumed no second order effects.

If we return to one of our earlier examples of regression models we can illustrate some of the possible advantages to be gained by using SAR models. In the Clinton voting example, the absence of correlation in the residuals from the ordinary least squares model suggests that there is nothing to be gained from fitting any model that builds in spatial dependence.

Turning to the example on standardised mortality ratios for myocardial infarction in England, we noted very significant spatial correlation in the residuals from the ordinary least squares model. We now repeat our earlier analysis, but using a simple one-variable SAR model of the type described above, applied to the logarithm of SMR for myocardial infarction and the logarithm of the Jarman index. This results in an estimated regression coefficient of 0.2233 for the logarithm of the Jarman score, with an associated standard error of 0.0799. This gives a z-value of 2.7929 (ratio of the estimate to its standard error), suggesting that while the Jarman variable still has some explanatory power it is nowhere near as significant as in the ordinary least squares case. The SAR model gives an estimate of ρ as 0.7272 with a corresponding standard error of 0.0579, suggesting that the second order (spatial dependence) effect is highly significant. These results imply that the ordinary least squares model 'inflates' the importance of the explanatory variable; when we take proper account of the spatial correlation among the residuals we achieve a much poorer fit. Here we have a clear illustration of the misleading conclusions we may reach when regression analysis fails to acknowledge the spatial arrangement of the observations. The ordinary least squares model can only represent 'first order' effects, while the SAR model allows for 'second order' effects. However, the pattern of residuals from the SAR model continues to show a 'north–south' divide and it is clear that additional variables, not currently part of our data set, would be needed to better explain variations in deaths from heart attacks.

Of course, the validity of any final proposed model will depend upon first, the choice of an appropriate form of first order component and, second, the choice of an appropriate variate interaction scheme. As we emphasised in

Chapter 5 the decomposition of a process into first and second order components is always to some extent arbitrary. In reality the two effects are confounded—certain types of second order variation can lead to trend-like effects and so the two effects may not be objectively distinguishable for a 'real life' process. But if a variate interaction model of the type described above helps us to understand the behaviour of the process and gives us insights into possible explanations for this behaviour then the modelling will have achieved its objective.

If the simple SAR model that we have discussed so far is not adequate, then we will need to look for alternative model specifications. We have only examined one class of interaction model, the autocorrelated errors model. We started with a discussion of this model because it is a simple generalisation of the ordinary least squares model and reasonably easy to understand. However, it is by no means the only SAR model that one might use, and SAR models are not the only class of spatial regression models which have been proposed. For example, there are theoretical arguments which favour an alternative class of models known as *conditional autoregressive models* (CAR).

Firstly, the basic autocorrelated errors model can be generalised to more complex SAR interaction schemes:

$$\begin{aligned} \boldsymbol{Y} &= \boldsymbol{X\beta} + \boldsymbol{U} \\ \boldsymbol{U} &= \rho_1 \boldsymbol{W}^{(1)} \boldsymbol{U} + \rho_2 \boldsymbol{W}^{(2)} \boldsymbol{U} + \ldots + \boldsymbol{\epsilon} \end{aligned}$$

where $\boldsymbol{W}^{(k)}$ is a proximity matrix relating to spatial lag k. Clearly, the numerical estimation problems associated with such extensions can become formidable.

Alternatively, we may wish to simplify the model. Recall that the autocorrelated errors model may be written as:

$$\boldsymbol{Y} = \boldsymbol{X\beta} + \rho \boldsymbol{WY} - \rho \boldsymbol{WX\beta} + \boldsymbol{\epsilon}$$

This could be simplified by removing the term $\rho \boldsymbol{WX\beta}$ to obtain:

$$\boldsymbol{Y} = \boldsymbol{X\beta} + \rho \boldsymbol{WY} + \boldsymbol{\epsilon}$$

The second order variation is then represented purely as an *autoregressive effect*. This results in maximum likelihood estimates which are simpler to compute than for the full autocorrelated errors model, although the same general iterative approach is adopted.

A further simplification results if the first order term is also dropped, yielding the *pure autoregressive model*:

$$\boldsymbol{Y} = \rho \boldsymbol{WY} + \boldsymbol{\epsilon}$$

This assumes that all the variation in the process is captured through second order variation; there is no mean dependence on $\boldsymbol{X}$. It is not a realistic model for a process with first order mean variation, although the maximum likelihood parameter estimation is relatively straightforward.

An alternative to the *pure autoregressive model* is the *moving average model*:

$$\boldsymbol{Y} = (\boldsymbol{I} + \rho \boldsymbol{W})\epsilon$$

If $0 \leq \rho < 1$ this is a model for a stationary, zero-mean process, which exhibits a theoretical correlogram with contrasting behaviour to that from pure autoregression. In the case of pure autoregression, correlation is fairly persistent with increasing distance, whereas in the moving average case it decays to near zero very quickly.

As stated earlier, there is also a variety of alternative model specifications, based upon *conditional autoregressive* (CAR) models. These have some theoretical advantages over the kind of SAR schemes we have described above. We do not give details of such models here. The important point that we wish to emphasise is, rather, the general approach of this kind of spatial modelling. In all cases analysis proceeds by using the correlogram to explore the data for spatial dependence and then identify a possible interaction model. This is fitted by maximum likelihood and the residuals examined for spatial dependence to establish the validity of the model. Further adjustments may then be made to the structure, and the model refitted. This process is repeated until a final acceptable model is obtained.

7.6 Summary

In this chapter we have considered a basic 'core' set of tools that may be used in visualising, exploring, and modelling area data. In the early sections of the chapter we examined simple methods for visualising such data, discussing commonly used tools such as choropleth maps and the problems associated with their use. We then went on to introduce methods for exploring spatial variation and spatial structure in such data.

Fundamental to much of that material and, indeed, subsequent modelling methods, was the notion of a proximity matrix, $\boldsymbol{W}$, which captures the spatial relationships between a set of areal units. As we saw, there are several alternative definitions of $\boldsymbol{W}$. We were then able to use the proximity matrix in association with exploratory techniques designed to investigate either first order trend or second order effects. In the first instance, methods such as such as spatial moving averages, median polish for lattice data and kernel smoothing were suggested as being valuable. For the exploration of second order effects we introduced the ideas of spatial correlation and the correlogram.

We then turned our attention to the modelling of area data. We discussed the modelling of first order variation and its relationship with possible covariates using the standard ordinary least squares multiple regression model widely employed in non-spatial data analysis. We then introduced a modification to this, referred to as generalised least squares, in order to allow for second order effects. To fit such a model we had to specify more precisely the form of the second order effects in our process. This led us into a discussion of the use of the correlogram to test for the significance of second order effects and then finally into the specification of a range of spatial regression models designed to incorporate such effects. These were based on the generalised least squares model, using various variate interaction schemes, incorporating again the idea of a proximity matrix to characterise the second order effects.

In the next chapter we discuss more specialised methods for the analysis of area data. We consider in more detail techniques for the analysis of data consisting of rates or proportions. Then we examine a selection of methods directed to the analysis of multivariate area data. Finally, we outline some methods for the analysis of one specialised type of area data, that arising in image processing and remote sensing.

7.7 Further reading

We first list the sources of data and recommend interested readers to consult these for further insight into the case studies.

The data on child mortality in Auckland come from:

Marshall, R. (1991) Mapping disease and mortality rates using empirical Bayes estimators, *Applied Statistics*, 40, 283–94.

Data for Chinese provinces are taken from:

Cole, J.P. (1987) Regional inequalities in the People's Republic of China, *Tijdschrift voor Econ. en Soc. Geografie*, 78, 201–13.

The 1981 Census data for Barnet were extracted from the Small Area Statistics, held on-line at Manchester Computer Centre. They are used with permission of the Office of Population Censuses and Surveys and are Crown Copyright.

For an introduction to 1981 Census data in Britain see:

Rhind, D. (1983) *A Census User's Handbook*, Methuen, London.

but if British readers wish to use 1991 Census data they should consult:

Dale, A. and **Marsh, C.** (1993) *The 1991 Census User's Guide*, HMSO Books, London.

We shall explore the use of such data in a multivariate context in the following chapter, but a useful general reference on *geodemographics* is:

Brown, P.J.B. (1991) Exploring geodemographics, in Masser, I. and Blakemore, M. (eds.) *Handling Geographic Information: Methodology and Potential Applications*, Longman, Harlow, Chapter 12.

The data relating to the vote for Clinton were taken from:

Pomper, G.M. and others (1993) *The Election of 1992: Reports and Interpretations*, Chatham House Publishers, New Jersey.

with the accompanying variables extracted from the 1992 *US Statistical Yearbook*.

Data for English District Health Authorities were extracted from the *Health Service Indicators* and the *Public Health Common Data Set*, for 1990. Such data are made available annually by the Department of Health. For more details on the various deprivation indices the reader is referred to:

Morris, R. and **Carstairs, V.** (1991) Which deprivation? A comparison of selected deprivation indexes, *Journal of Public Health Medicine*, 13, 318–26.

The data on blood types in Ireland have been extracted from the following book, where similar analyses to those we discuss here have been reported:

Upton, G.J.G. and **Fingleton, B.** (1985) *Spatial Data Analysis by Example: Volume 1: Point Pattern and Quantitative Data*, John Wiley and Sons, Chichester.

The data on emissions of nitrogen oxides and ammonia have been extracted from:

Smith, F.B. (1991) Regional air pollution, with special emphasis on Europe, *Quarterly Journal of the Meteorological Society*, 117, 657–83.

But for more up-to-date figures see:

Sandnes, H. and **Styve, H.** (1992) *Calculated Budgets for Airborne Acidifying Components in Europe, 1985, 1987, 1988, 1989, 1990 and 1991*, Norwegian Meteorological Institute, Technical Report 97, Oslo.

The LANDSAT TM data were made available by Mitch Langford of Leicester University, but permission to use was granted by Professor Paul Mather of Nottingham University and the National Remote Sensing Centre. The wider region from which this tiny subset is taken is analysed in:

Mather, P.M. (1987) *Computer Processing of Remotely-Sensed Images*, John Wiley and Sons, Chichester.

On visualisation see:

Monmonier, M. (1993) *Mapping it Out: Expository Cartography for the Humanities and Social Sciences*, University of Chicago Press, Chicago, Chapter 6.

Robinson, A.H., Sale, R.D., Morrison, J.L. and **Muehrcke, P.C.** (1984) *Elements of Cartography*, Fourth edition, John Wiley and Sons, Chichester, Chapter 14.

An excellent overview of cartograms, with examples, is given in:

Dorling, D. (1994) Cartograms for visualising human geography, in Hearnshaw, H.J. and Unwin, D.J. (eds) *Visualisation in Geographical Information Systems*, John Wiley, Chichester.

Key papers on the construction and use of two-variable maps are:

Eyton, J.R. (1984) Complementary-color, two-variable maps, *Annals of the Association of American Geographers*, 74, 477–90.

Dunn, R. (1989) A dynamic approach to two-variable color mapping, *The American Statistician*, 43, 245–52.

On proximity measures see:

Upton, G.J.G. and **Fingleton, B.** (1985) *Spatial Data Analysis by Example, Volume 1: Point Pattern and Quantitative Data*, John Wiley and Sons, Chichester, 176–85.

On median polish see:

Cressie, N.A.C. (1991) *Statistics for Spatial Data*, Wiley, New York.

On kernel estimation as used by Bracken and Martin see:

Bracken, I. and **Martin, D.** (1989) The generation of spatial population distributions from census centroid data, *Environment and Planning A*, 21, 537–43.

For a discussion of, and approach to, the cross-areal interpolation problem see:

Flowerdew, R. and **Green, M.** (1991) Data integration: methods for transferring data between zonal systems, in Masser, I. and Blakemore, M. (eds.) *Handling Geographical Information: Methodology and Potential Applications*, Longman, Harlow, 38–54.

The classic reference on spatial correlation and the correlogram is:

Cliff, A.D. and **Ord, J.K.** (1981) *Spatial Processes: Models and Applications*, Pion, London.

but a gentler introduction is:

Goodchild, M. (1987) *Introduction to Spatial Autocorrelation*, Concepts and Techniques in Modern Geography, Number 47, Geo Abstracts, Norwich.

See also:

Griffith, D. (1987) *Spatial Autocorrelation: A Primer*, Association of American Geographers Monograph, Washington, DC.

Odland, J. (1988) *Spatial Autocorrelation*, Sage Scientific Geography Series, Sage Publications, London.

The literature on non-spatial multiple regression and related methods is huge. A classic general text is:

Draper, N.R. and **Smith, H.** (1981) Second edition, *Applied Regression Analysis*, John Wiley, Chichester.

On the spatial modelling of area data comprehensive overviews are given in:

Haining, R. (1990) *Spatial Data Analysis in the Social and Environmental Sciences*, Cambridge University Press, Cambridge.

Griffith, D. (1988) *Advanced Spatial Statistics*, Kluwer Dordrecht.

but see also Chapter 5 in Upton and Fingleton cited earlier.

We have fitted the SAR models in this chapter using the software SpaceStat. See:

Anselin, L. (1993) *SpaceStat: A Program for the Statistical Analysis of Spatial Data*, NCGIA, Department of Geography, University of California at Santa Barbara, Ca. 93106-4060.

For another application of spatial regression models in a health context, see:

Cook, D.G. and **Pocock, S.J.** (1983) Multiple regression in geographical mortality studies, with allowance for spatially correlated errors, *Biometrics*, 39, 361–71.

7.8 Computer exercises

Here are suggested exercises that you can try on ideas discussed in this chapter using INFO-MAP and our example data sets. These exercises have been referenced at appropriate points in the chapter by numbered symbols in the margin.

Exercise 7.1

Select the 'Auckland child mortality' data. Select `Symbol` map from the `Map` menu and create a map of proportional circles. Select 'symbol size' to represent the childhood population variable and 'symbol colour' to show the number of child deaths. Does the simultaneous mapping of two variables reveal anything of interest concerning child mortality? You may find it useful to experiment with symbol size via `Symbol type` in the `Map` menu.

Exercise 7.2

Select the 'Mortality in English Health Authorities' data, and create a choropleth map of one of the mortality variables. Experiment with different class interval schemes. Now alter the colours on the map using the option in the `Map` menu. Make pairwise comparisons of particular maps by using the clipboard from the `Edit` menu to save and retrieve maps. What differences do you observe in the pattern of the different mortality rates by using these simple analyses? Speculate on some possible reasons for such differences.

Exercise 7.3

Examine the 1981 Census data referring to Barnet in north London. In particular, derive choropleth maps of the percentage of households without cars and the percentage of single parents; an appropriate denominator for the latter is the number of households with dependent children. You will need to calculate these variables yourself.

Now create a further variable, which is the sum of the standard normal deviates ('z-scores') corresponding to each of these percentages. Assuming that the percentage of no-car households is variable 11 in the data file and single parents is variable 12 this may be done using `Calculate` from the `Data` menu with the formula:

```
dep=({11}-mean({11}))/stdev({11)
    +({12}-mean({12}))/stdev({12})
```

Map this new variable, first modifying the class intervals with `Scaling` in the `Map` menu and selecting 'standard deviates' as the classification scheme. You have created your own, very simple, 'deprivation' score, no different in principle to that derived by Jarman in the data set on 'Mortality in English Health Authorities' referred to in the text, except yours involves just two variables and gives equal 'weight' to both.

Exercise 7.4

Derive a smoothed estimate of the spatial variation in the vote for President Clinton. You can do this as follows. Select the data file and choose `Calculate` from the `Data` menu. Create the following new variable:

```
i=1,count():spatav[i]=mean(if(adjac([i]),{1},miss()))
```

where variable 1 refers to the first data item in the file (the percentage vote). The calculation may take some time depending upon the speed of your PC. Plot the smoothed data as a choropleth map and compare your result with the unsmoothed data. The original and the smoothed map can be viewed on the screen together using the 'clipboard' which is accessed through the `Edit` menu.

Repeat this exercise for the 'Mortality in English DHAs' data, using any of the three mortality ratios (SMRs). Try to relate the map that you obtain to what you may know about 'north–south' divides in the health of the English population. Compare results with those obtained for another of the SMRs. Are there significant differences in the spatial pattern of mortality for different diseases?

Exercise 7.5

Select the 'European air emissions' data, create a choropleth map of the nitrogen oxide emissions and overlay the European country outlines by using `Open` in the `Overlay` menu. Note where the peak emissions seem to be. The median smoothed values, referred to in the text, are included in the file along with the residuals. Map both of these variables as well. Do they help you to interpret the nitrogen oxide emissions?

For comparison, a spatial moving average using the four nearest neighbours is also included for the nitrogen oxide variable. This was calculated using the formula

```
spatav=({2}[near(1)]+{2}[near(2)]+{2}[near(3)]
       +{2}[near(4)])/4
```

The calculation takes some considerable time because of the large number of observations in the grid, so we have included the variable along with the data. Compare the spatial moving average with the median polish. Note that it does not suffer from the 'banding' effects present in the median polish, mentioned in the text. However note also that the spatial moving average contains more missing values than in the original data, whilst the median polish does not. In the spatial average, if any of the four nearest neighbours are missing values, then the result has been considered to be a missing value.

You could try comparing the pattern of ammonia emissions with those of the nitrogen oxide. Is there any correlation between the patterns? Consider the skewness in the data by examining the frequency distribution from the `Analyse` menu and, if necessary, implement a transformation of the data.

Try calculating a spatial moving average for the ammonia values, but only if you have a fast PC! (One of the exercises in Chapter 8 discusses how to calculate these filters more quickly for a regular grid of data.)

Exercise 7.6

Returning to the 'Auckland child mortality' data, derive a kernel estimate of child population at risk in 1000s. First calculate a new variable containing these values, say:

```
atrisk={2}*9/1000
```

`Choropleth map` this new variable and then use the `Kernel map` option in the `Analyse` menu, specifying that you want to estimate 'Intensity', which in this case will give the 'density' estimate discussed in the text. Compare the results obtained by using different bandwidths, in order to see how specifying a longer bandwidth leads to more smoothing of the data.

Now form the ratio of two kernel estimates, one of the cases, the other of the childhood population you have just calculated (the 'population at risk'). Choose 'Ratio of intensities' in the `Kernel map` option and select 'cases' as the numerator and 'atrisk' as the denominator. Note that you are required to specify a bandwidth for each in turn. Again, experiment with different bandwidths in order to see how your estimate of mortality risk varies across the map.

Compare the results you obtain with a further kernel estimate, this time of the mean value of the variable 'Raw SMR'. This variable contains the ratio ($\times 100$) of the number of deaths in each area to that expected if the average death rate over all areas applied. `Choropleth map` 'Raw SMR' and then use `Kernel map` with the 'Mean value' option. To what extent are you obtaining, from these various results, a consistent picture of spatial variation in child mortality?

Exercise 7.7

`Choropleth map` selected variables from the data on Chinese provinces and estimate correlograms for these variables, using `Correlogram` from the `Analysis` menu. For example, the fertility rate (variable 14), rural incomes (variable 16), and coal output (variable 8). Compare the different structures of spatial dependence.

The estimates of spatial correlation shown in the correlogram in INFO-MAP are based on a very simple proximity measure, nearest neighbours (at lag 1, 2, etc). Try to estimate the spatial correlation at lag 1, based on adjacency of zones rather than the simple INFO-MAP proximity of whether a zone is the nearest neighbour. Do this using the variable 'consumer sales' which is variable 15 in the file. First estimate the average value in the neighbourhood of a zone, in a way similar to that used in a previous exercise, except that now you need to exclude the zone itself from the average (in INFO-MAP a zone is always considered adjacent to itself). So you need a modified formula:

```
i=1,count():spav[i]=mean(if(adjac([i]) and
                   not(index()=i),{15},miss()))
```

Then, calculate the correlation between the value in a zone (variable 15) and the average in its neighbourhood (the new variable 'spav'). Assuming this is variable 23, you would use `Calculate/display worksheet` from the `Analysis` menu with the formula:

```
work[1]=corr({15},{23})
```

Compare this value with that given in the correlogram in the text at lag 1. What differences do you observe, and what is the explanation for this?

You may repeat this exercise using other variables from this data set. Alternatively, you could explore spatial correlation in similar ways for the mortality rates in the 'Mortality in English DHAs' data set. The number of areas in that data set will mean that the calculations may take some time.

Exercise 7.8

Select one of the SMRs from the English DHA data, other than the myocardial infarction data modelled in the text. Create a choropleth map and, using the `Scatter plot` option in the `Analysis` menu, explore relationships with possible covariates, such as the Jarman index, the DoE index, and levels of owner occupation. Use any transformations you think necessary. Try to establish a 'best fitting' regression model and note the parameter estimates and their standard errors for this model. (Refer back to one of the exercises in Chapter 1 or Chapter 5, for details of INFO-MAP formula for regressions.)

Calculate the residuals from your model. Use exploratory plots of these residuals to check the assumptions of the ordinary least squares model used. Do the residuals appear to be random? Is their spread related to the fitted values, or to any of the other variables in the file?

Next, map the residuals and examine their spatial arrangement. Try plotting residuals against the northing of areas (obtained from the `north()` function), or any other direction (obtained from a linear combination of `north()` and `east()` functions). Plot residuals against their nearest neighbours (using the `near()` function). Examine their spatial dependence by obtaining a correlogram.

To what extent has your model failed to provide an adequate level of explanation of mortality variation? Can you suggest further covariates, not in the data set, which ought to be added to the model? Compare your results with those obtained for another of the SMRs.

Exercise 7.9

Open the data set 'Blood groups in Ireland', produce a choropleth map of the percentage with blood group A, and derive a correlogram for this variable. Note the correlation at the first spatial lag, which has a value of 0.597. In order to formally test the significance of this correlation, we can use the approximate mean and variance given in the text. Use `Calculate/display worksheet` in the `Analysis` menu and build up the various components of the calculation as follows. There are 26 counties involved so the mean value, $E(I)$, is estimated simply as:

```
work[1]=-1/25
```

Next, recall that the estimates of spatial correlation in INFO-MAP are based on the simple proximity measure of first nearest neighbour, hence $\sum_j w_{ij} = 1$ for all $i = 1, \ldots, 26$. Therefore, the quantity S_0 is just:

```
work[2]=26
```

The quantity S_1 is a little more tricky since if j is the nearest neighbour to i it does not necessarily follow that i is the nearest neighbour to j. We need to first 'zero' a worksheet item by:

```
work[3]=0
```

and then accumulate the sum using the rather complex formula:

```
i=1,count():work[3]=work[3]+0.5*sum((if(near(1)=i,1,0)
                    +if(near(1)[i]=index(),1,0))**2)
```

See if you can work out exactly how this works. One element of S_2 is trivial to calculate since, as mentioned above, $\sum_j w_{kj} = 1$ for all k, given our proximity measure. The second is obtained by using similar ideas as in the formula for S_1. We first zero a worksheet item:

```
work[4]=0
```

and then accumulate S_2 as:

```
i=1,count():work[4]=work[4]+(1+sum(if(near(1)=i,1,0)))**2
```

We then calculate the expression in the text for $VAR(I)$ as:

```
work[5]=(26**2*25*work[3]-26*25*work[4]
        -2*work[2]**2)/((27*25**2)*work[2]**2)
```

Finally, we can calculate the required standard normal deviate for the test as:

```
work[6]=(0.597-work[1])/(work[5]**0.5)
```

and compare the result with percentage points of the standard normal distribution. The result is a highly significant positive correlation, as might have been expected. Notice that the value you obtain for $VAR(I)$ should be the square of that given in brackets against the first spatial lag in the correlogram produced by INFO-MAP.

Try repeating the same kind of calculations for a selected socio-economic variable in the China data set, one that has a more marginal first order spatial correlation.

Is it possible to modify the above procedure to test spatial correlation when this is based on adjacency of zones rather than first nearest neighbours, such as that which you calculated in a previous exercise?

Exercise 7.10

This exercise will allow you to experiment with a simple spatial autoregressive model. Consider again the Irish blood group data. First notice that the spatial differences defined in the text as: $(\boldsymbol{I} - \rho \boldsymbol{W})\boldsymbol{Y}$, can be constructed on the basis of a 'first nearest neighbour' proximity matrix quite simply in INFO-MAP. Say we choose a value of $\rho = 0.1$, then a variable of such spatial differences for variable 1 in the data file is defined by the formula:

```
{1}-0.1*{1}[near(1)]
```

We can use a similar idea to calculate a spatially differenced value of any other variable in the file, so that the formula:

```
res0.1={1}-0.1*{1}[near(1)]-regr({1}-0.1*{1}[near(1)],
          {2}-0.1*{2}[near(1)])
```

calculates the residuals (and allows diagnostics to be viewed) of a regression of $(\boldsymbol{I} - \rho \boldsymbol{W})\boldsymbol{Y}$ on $(\boldsymbol{I} - \rho \boldsymbol{W})\boldsymbol{X}$, where Y and X are, respectively, the percentage with blood group A and the density of towns, and where $\rho = 0.1$ and $\boldsymbol{W}$ is a proximity matrix based on 'first nearest neighbours'. Similarly:

```
res0.2={1}-0.2*{1}[near(1)]-regr({1}-0.2*{1}[near(1)],
          {2}-0.2*{2}[near(1)])
```

would do the same thing for $\rho = 0.2$. Start with a value of $\rho = 0.0$, which is equivalent to an ordinary non-spatial regression of the two variables. Then increase this in steps of 0.1, simply changing the name of the new variable and the value of ρ in the formula as you go. After each regression examine the `Regression diagnostics` in the `Analysis` menu, map the residuals just derived and estimate a correlogram for them. How does the regression parameter (and the corresponding standard error) for urban density change? How does the spatial dependence in the regression residuals change? Can you explain this, and what implications does it have for trying to get a 'true' picture of the significance of the relationship between the percentage with blood group A and urban density?

You could also try more complex proximity structures. For example

```
resnew0.1={1}-0.1*({1}[near(1)]+{1}[near(2)])/2-
               regr({1}-0.1*({1}[near(1)]+{1}[near(2)])/2,
               {2}-0.1*({2}[near(1)]+{2}[near(2)])/2)
```

would use $\rho = 0.1$ but with a proximity matrix based upon the first and second nearest neighbours.

8

Further methods for area data

Here we consider more specialised topics concerned with the analysis of area data. We begin by examining methods designed particularly for handling counts or rates. Next, we return to some of the multivariate methods first considered in connection with spatially continuous data in Chapter 6 and see how they can also be put to good use in analysing area data. Finally, we devote a section to a special class of area data, that relating to a fine regular grid, such as arises in the processing and interpretation of remotely-sensed and other types of image data.

8.1 Introduction

In the previous chapter we were concerned with trying to detect spatial pattern in values of a single 'response' variable over areas and to model this in terms of covariates measured over the same areas. However, the models employed there assumed that the 'response' variable of interest, or perhaps some simple transformation of it, could be considered to be approximately normally distributed. Whilst this assumption can often lead to useful results, even when its distribution is unlikely to be strictly normal (particularly in preliminary analyses or where the object is simply to uncover major trends and relationships in a set of data), there are cases where more careful consideration of alternative distributions for the 'response' variable is necessary. We begin this chapter by focusing on one kind of area data where this is the case—data in the form of rates or counts recorded over areas. We look at ways of deriving useful maps of the variations in such data and then discuss briefly the modelling of possible explanations for such variations. This involves more specialised methods than those considered in the previous chapter.

We then consider a different kind of problem. Instead of focusing on variations in a single 'response' variable and treating others as covariates, we endeavour to explore variations in more than one variable taken simultaneously. In other words, we return to the kind of multivariate methods introduced in Chapter 6 in relation to spatially continuous data, and we consider how such techniques might usefully be applied to area data. We shall be concerned again with methods which 'collapse' a set of variables to a smaller set of combinations of such variables, so helping us to better understand their mutual behaviour. One example of a multivariate problem which is of particular interest in relation to area data is that of classifying areas according to their values on several variables and we shall look at how multivariate techniques may be used to address this problem.

Finally, we consider methods oriented towards a particular type of area data, that arranged on a fine regular grid or lattice. While few, if any, socio-economic data are of this form, such data do arise frequently in specialised applications, notably those that involve 'image' data. Image data might be derived from satellites (using remote sensing) or, at a very different scale, from medical scanning. Regardless of where the data originate, specialised 'image processing' techniques are often of value when data are arranged on a fine lattice. We introduce our reader to some of these methods.

8.2 Analysis of counts and proportions

Our models for area data in the previous chapter assumed that the data concerned (or some transformation of it) could be taken to be approximately normally distributed. This assumption was involved, to varying degrees, in most of the spatial regression techniques that we considered there. However, it is often the case with data collected over zones or areas, that the 'response' variable of interest is a count or a proportion. In such cases the assumption of normality is strictly not tenable; rather, we would expect models for such data to involve probability distributions such as the Poisson or binomial.

The important point here is not so much the actual distribution of the 'response' variable, but rather the fact that its variance may be related to its mean value. In the previous chapter all of our models assumed that the covariance structure in the data was unrelated to the mean value which we were attempting to model. The variance of both the Poisson and binomial distributions is related to their mean value and it is not therefore possible to separate the estimation of variance from that of the mean when such distributions are involved.

Transformation of the original variable when this is a count or proportion may help to alleviate this problem, and we discussed such transformations in the last chapter. However, we saw that they do not provide an entirely satisfactory solution when counts or proportions are involved. An alternative is to use generalisations of classical non-spatial regression methods, which provide models for the analysis of counts, proportions and categorical data in general. This broad area is referred to as *generalised linear modelling* and includes models for Poisson and binomial 'responses' as well as log-linear models for the analysis of contingency tables (multi-way tables of counts). We do not discuss these methods in any detail here, but we feel we should at least introduce our reader to the idea of some simple generalised linear models that may be of use when data comprise counts or proportions.

Before discussing the generalised linear modelling of counts and proportions in more detail, we wish first to draw attention to some particularly useful techniques concerned with preliminary mapping and exploration of spatial variation when data take the form of rates or proportions.

8.2.1 Probability mapping

When the attribute of interest is a rate or proportion, exploratory mapping of the rates to display geographical variability is an obvious first step in any analysis. However, using the raw observed rates may be misleading, since the variability of such rates will be a function of the values of the 'population' to which they relate, and this may differ widely from area to area. Several alternatives have therefore been developed to attempt to better highlight anomalous areas whilst accounting for the variability of the underlying 'population' in each of the areas.

One simple approach, particularly common in epidemiology or medical geography where the rates relate to mortality or morbidity from some disease, is to map some measure of *relative risk*. Suppose that the observed count in each of the areas is y_i and this arises from a corresponding population n_i. Suppose further that the overall rate of occurrence is in fact constant for all areas and that areas are independent. Then a reasonable assumption is that counts y_i are observations on independent Poisson random variables with expected values μ_i. In that case a sensible estimate of the mean (expected) count in each area, μ_i, is given by:

$$\hat{\mu}_i = n_i \left(\frac{\sum y_i}{\sum n_i} \right)$$

Instead of mapping the raw rates $r_i = y_i/n_i$ for each area we map the *relative risk*. That is, we divide the observed count, y_i, by its estimated expected value, $\hat{\mu}_i$, and multiply by 100. Thus a value of 200 suggests that the observed count, y_i, in area i, is double what we would expect, $\hat{\mu}_i$. This simple scheme can clearly be extended to take account of variations in the structure of the population in each area, as well as its size, when the data are available in appropriate subdivisions. For example, we might calculate overall age–sex specific rates and use these to determine the expected count in each area, given the age–sex structure and size of the population there. This of course leads to the *standardised mortality ratios* favoured in official health statistics, an example of which we encountered in the last chapter with the data on English District Health Authorities.

We may illustrate the idea with reference to the Auckland child mortality data introduced in Chapter 7. Since the deaths were accumulated over a nine year period we take the population 'at risk' in each area to be approximately nine times the recorded population under five years in the 1981 census, which is the midpoint of the period concerned. We begin by calculating the expected number of deaths in each area, the set of $\hat{\mu}_i$. This is done simply by multiplying the population 'at risk' in each area by the child mortality rate for the region as a whole (0.00263), as in the above formula. So for each zone we now have an observed and expected number of deaths and can calculate a crude measure of relative risk (observed divided by expected, multiplied by 100). The resulting map is shown in Figure 8.1.

We can get some absurdly variable results using this approach. For example, one zone has only one recorded death, with 0.14 expected, yielding a relative risk of 703.2! The problem is typical with these kinds of methods when rare events are being considered or populations are highly variable. In general, a 'relative risk' of 200 can be obtained from $y_i = 100$, $\hat{\mu}_i = 50$, or from $y_i = 3$, $\hat{\mu}_i = 1.5$, or from any other y_i and $\hat{\mu}_i$ where the former is twice the latter. We

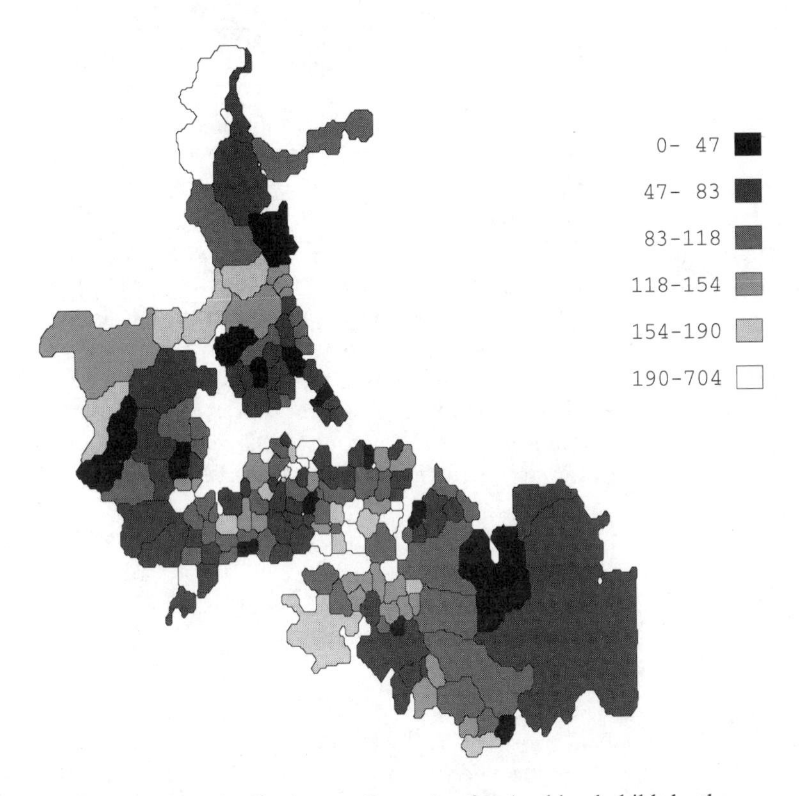

Fig. 8.1 Standardised mortality ratios for Auckland child deaths

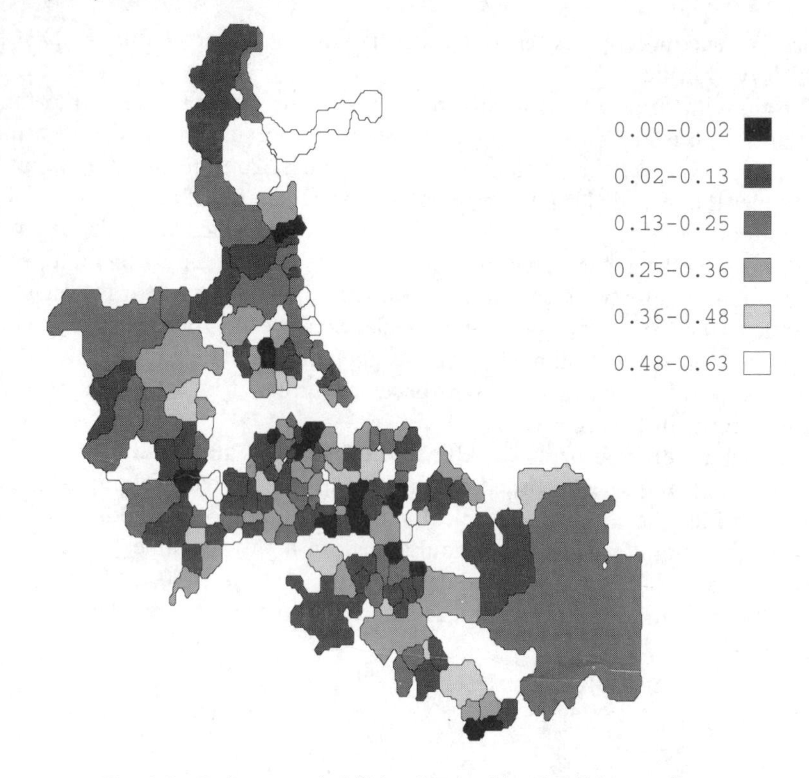

Fig. 8.2 Poisson probabilities for Auckland child mortality

thus see that in areas with small populations it is possible to get wildly varying relative risks; the addition of just a single observed case can increase the estimate dramatically. The problem of course is that the reliability of our estimates of $\hat{\mu}_i$ varies widely and this is not being taken into account.

One suggestion to counter this problem is to produce a *probability map*. Instead of mapping the ratio of observed to expected counts, we map the probability of getting a count which is more 'extreme' than that actually observed, under the assumption that the count in each area i is Poisson with mean value μ_i. We estimate μ_i in the same way as previously, obtaining $\hat{\mu}_i$. Then we map:

$$p_i = \begin{cases} \sum_{x \geq y_i} \frac{\hat{\mu}_i^x e^{-\hat{\mu}_i}}{x!} & y_i \geq \hat{\mu}_i \\ \sum_{x \leq y_i} \frac{\hat{\mu}_i^x e^{-\hat{\mu}_i}}{x!} & y_i < \hat{\mu}_i \end{cases}$$

Small values of p_i (say < 0.05) would then indicate that an area's rate is unusually high or low.

Again, we illustrate the idea with reference to the Auckland child mortality data. Computing the Poisson probabilities as described above yields the probability map shown in Figure 8.2, which replaces the extreme values in the earlier map of relative risk with more sensible assessments of how extreme these areas are on a probability scale. Note that small probabilities relate to 'extreme' areas that are either significantly high or significantly low risk.

Although the idea of the probability map is to standardise rates onto a probability scale for proper comparison, it is only an exploratory device and extreme p_i should not be interpreted too rigorously. We have made the point several times that spatial variation in attribute values can arise either from first order or second order effects. The model underlying the probability map is essentially one which contrasts spatially independent areas with heterogeneous means against spatially independent areas with a homogeneous mean. No reference is made to the possibility of spatial dependence between the areas. Extreme values of the probabilities p_i may be more due to the lack of fit of a spatially independent model than to heterogeneous rates. Indeed, the simple mapping of probability values can be improved by the techniques we discuss in the following section.

Ex 8.1

8.2.2 Empirical Bayes estimation

As implied by our above use of mortality data as an illustration, probability mapping has been put to considerable use in medical geography and epidemiology. We saw how we could map the probability that the observed count in an area is significantly different from the mean or expected value. However, the probability values depended upon a rather simple and possibly invalid model. The alternative to mapping probability values is the estimation, and mapping, of a measure of 'relative risk', but we have also seen that this can produce wildly varying relative risks in some cases.

There are, then, problems in mapping either Poisson probability values or measures of relative risk. How might we improve the estimate of a rate (such as that of child mortality) in an area? One way is to argue that when the estimate of disease rate for an area is considered somewhat unreliable and highly variable because of the small numbers upon which it is based, then we should in some way 'smooth' our local risk assessment by knowledge gleaned from the overall pattern of rates. We would wish the degree of 'smoothing' to reflect how confident we are in the local estimate of risk.

Such a problem suggests the adoption of a *Bayesian* statistical approach. We do not wish to go into detail here about Bayesian estimation techniques in general. Suffice it to say that Bayesian statistics is concerned with statistical estimation where prior knowledge or beliefs about parameters of interest are taken into account when estimating their values, as well as observed data. In general, an unconditional *prior probability distribution* for the values of a parameter of interest is converted to a *posterior distribution* for the values of that parameter using the data that are actually observed. Bayes theorem is used to derive the posterior distribution by combining the likelihood for the data with the prior distribution and this explains the name given to these techniques in general. The posterior distribution for the parameter is then used to derive an estimate for the parameter (usually the mean value of the posterior distribution) and a standard error for this estimate (related to the variance of the posterior distribution).

With such methods it is acceptable to base the prior distribution for the parameter of interest on the results of the analysis of previous data, or even purely on the background knowledge, intuition, or judgement of the analyst. Another possibility is to base the prior distribution upon certain global aspects of the data currently to hand; if so, such techniques are commonly referred to as *empirical Bayes estimation.*

It is the latter that we will use in the present situation. Our prior knowledge will be represented by considerations based on the overall rate across all areas; we will then use Bayesian techniques to modify what we observe in a given area on the basis of this. Mathematically, we can set the problem up as follows. We suppose that the true, unknown rate in each area is θ_i and let $r_i = y_i/n_i$ be the observed rate. In a non-Bayesian framework the best estimate, $\hat{\theta}_i$, of θ_i, is just r_i. However, suppose now that we have a prior probability distribution for each θ_i, with mean value γ_i and variance ϕ_i. Then it can be shown that the best Bayes estimates of θ_i based on combining these prior distributions with the observed rates are given by:

$$\hat{\theta}_i = w_i r_i + (1 - w_i)\gamma_i$$

where

$$w_i = \frac{\phi_i}{(\phi_i + \gamma_i/n_i)}$$

This is known as a *shrinkage estimate.* If w_i are thought of as weighting factors for each area, then the first part of the expression relates to the emphasis placed on the observed rate for zone *i*; the second part relates to the prior beliefs about the rate. The adjustment factor, w_i, is a function of the population at risk, n_i, in each zone and also the variance, ϕ_i, of the prior distribution, which reflects how confident we are of the prior beliefs about the rate. Where the population is relatively large we do not 'shrink' the estimate for zone *i* towards the prior so much; we have more confidence in the precision of the observed rate. However, where population is small, we have less confidence and we are happier to believe that our estimate of the rate is perhaps rather similar to our prior beliefs. Clearly, as w_i approaches 1 we give increasing weight to the observed rate for area *i*, while as it approaches zero the estimate for *i* effectively approaches the prior mean γ_i.

Thus far we have said nothing about where to obtain values for the prior means and variances γ_i and ϕ_i. In empirical Bayes the idea is to estimate these from the data as well. Of course, in order to do so it is necessary to make some simplifying assumptions, since currently there are as many of them as there are areas, and therefore the problem of estimating these parameters is just as difficult as the original problem of estimating the θ_i! One obvious reduction is to assume that the prior means and variances for all areas are the same. That is, $\gamma_i = \gamma$ and $\phi_i = \phi$. Then if we are further prepared to assume some particular mathematical form for the prior distribution of which these are the mean and variance, it becomes possible to derive maximum likelihood estimates for γ and ϕ from the observed rates.

One simple assumption, which is plausible and does have some theoretical justification, is that a suitable prior distribution is a gamma distribution. This distribution has two parameters, v and α, known respectively as the scale and shape parameters; the mean of this distribution is given as v/α and the variance as v/α^2. So in this case we have $\gamma = v/\alpha$ and $\phi = v/\alpha^2$. The problem of estimating γ and ϕ is therefore equivalent to estimating v and α. Given such estimates $\hat{v}$ and $\hat{\alpha}$, then the estimated weighting factor in our earlier expression for the shrinkage estimator is just:

$$\begin{aligned}\hat{w}_i &= \frac{\hat{\phi}}{\left(\hat{\phi} + \hat{\gamma}/n_i\right)} \\ &= \frac{\hat{v}/\hat{\alpha}^2}{(\hat{v}/\hat{\alpha}^2 + \hat{v}/n_i\hat{\alpha})} \\ &= \frac{n_i}{(n_i + \hat{\alpha})}\end{aligned}$$

and therefore the Bayes estimates of the rates are:

$$\begin{aligned}\hat{\theta}_i &= \hat{w}_i r_i + (1 - \hat{w}_i)\hat{\gamma} \\ &= \hat{w}_i r_i + \frac{(1 - \hat{w}_i)\hat{v}}{\hat{\alpha}} \\ &= \frac{y_i + \hat{v}}{n_i + \hat{\alpha}}\end{aligned}$$

When n_i is large w_i is close to 1 and so most weight in the estimate of θ_i is given to r_i. Conversely, when n_i is small more weight is given to $\hat{v}/\hat{\alpha}$, our estimate of the prior mean.

We have still not said how the estimates of α and v are derived. Earlier, we commented that if we had a specific distributional form for the prior distribution then maximum likelihood could be used to obtain estimates of the prior parameters based on the observed rates. In this case we have assumed a gamma distribution for the prior so we are able to proceed in this way. However, this is not trivial. Solution of the resulting equations requires an iterative numerical approximation. We do not give details here, providing references to these at the end of the chapter.

It is worth noting that an approximate alternative estimation procedure exists to such iterative maximum likelihood estimation. This is simple to apply and may be particularly useful in preliminary analyses. It also produces good starting values for the full maximum likelihood iterative procedure. The technique considers direct estimation of γ and ϕ, by the method of moments, as opposed to specifying a gamma prior and proceeding to maximum likelihood estimation for v and α. The idea is simply to estimate γ by the pooled mean of the observed rates, that is:

$$\hat{\gamma} = \frac{\sum y_i}{\sum n_i}$$

and then to estimate ϕ, based upon a weighted sample variance of observed rates about this mean as:

$$\hat{\phi} = \frac{\sum n_i (r_i - \hat{\gamma})^2}{\sum n_i} - \frac{\hat{\gamma}}{\bar{n}}$$

where $\bar{n}$ is the average population across all the areas, and the convention is adopted that $\hat{\phi} = 0$ whenever the above expression is negative.

With these estimates $\hat{\gamma}$ and $\hat{\phi}$, the shrinkage weighting factor is then estimated as:

$$\hat{w}_i = \frac{\hat{\phi}}{\left(\hat{\phi} + \hat{\gamma}/n_i\right)}$$

and therefore the Bayes estimates of the rates are:

$$\hat{\theta} = \hat{\gamma} + \frac{\hat{\phi}(r - \hat{\gamma})}{\left(\hat{\phi} + \hat{\gamma}/n_i\right)}$$

We can apply these various ideas to the Auckland mortality data, used earlier. An original analysis of these data by Roger Marshall is referenced at the end of this chapter, and gives maximum likelihood estimates $\hat{\alpha} = 4.336 \times 10^3$ and $\hat{v} = 11.55$; the corresponding prior mean and variance estimates are therefore $\hat{\gamma} = 2.664 \times 10^{-3}$ and $\hat{\phi} = 0.614 \times 10^{-6}$. The simpler method of moments procedure yields $\hat{\gamma} = 2.633 \times 10^{-3}$ and $\hat{\phi} = 0.728 \times 10^{-6}$. From either of these, the empirical Bayes estimates of the rates in each area $\hat{\theta}_i$ may be calculated from the expressions given earlier. In this case, differences in the results from either method are not great. The map of the empirical Bayes estimates of the mortality rate per 1000 under-five population, based on the maximum likelihood estimates, is shown in Figure 8.3. The spatial pattern remains, very broadly, the same as a map of 'raw' rates, but high rates on the fringes of the region are shrunk towards the overall mean, while high rates in the main urban areas are shrunk to a much lesser degree. The mortality pattern reflects

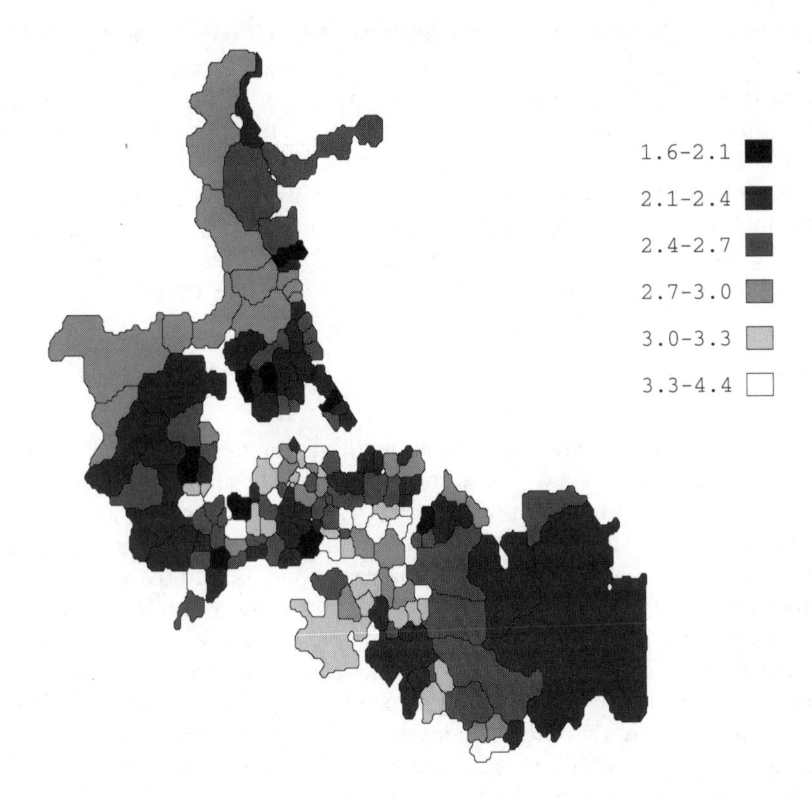

Fig. 8.3 Bayes estimates of Auckland child mortality (rate/1000)

the socio-economic gradients within the region. Mortality is higher in the less affluent suburbs of the city, situated within the most densely populated urban area, the isthmus bounded by the Waitemata and Manukau harbours.

The prior distribution for θ_i in all the empirical Bayes estimates discussed above is 'aspatial'; that is, the prior mean and variance is assumed equal over all areas. An alternative would be to require our adjusted estimate for an area to be shrunk towards a 'neighbourhood' rather than a global mean. This would seem more sensible, since the empirical Bayes estimates presented above are invariant to the spatial configuration of areas and this does not seem intuitively reasonable. We can achieve this by simply modifying the prior distribution for θ_i to allow for a mean and variance which are related to a 'neighbourhood' of i, rather than being constant for all areas; then an observed rate based on small numbers is 'shrunk' towards a neighbourhood, rather than a global, mean. Full maximum likelihood estimation using such neighbourhood priors is computationally prohibitive; however, the method of moments approach may be easily modified. One simply computes $\hat{\gamma}$ and $\hat{\phi}$ over neighbourhoods rather than over the whole region. Such neighbourhoods might be defined, for example, on the basis of areas sharing a common boundary.

We have only referred to empirical Bayes estimation of rates for a specific age group; that is, we have corrected only for differences of size in the population in different areas and not its structure. In the Auckland example this was not an issue since we were using the appropriate under-five population. However, in some applications comparison of disease rates over all age groups may be necessary. This requires additional adjustment for age structure in the population in each area. A similar adjustment for gender may also be necessary. Further details of these techniques, as well as other aspects of empirical Bayes methods are given at the end of this chapter.

8.2.3 Generalised linear models

So far in our discussion of specialised methods for the analysis of count and rate data across areas we have been concerned purely with exploratory methods for obtaining more precise estimates of the rates in each area. We may of course be interested in modelling these counts or rates, seeking relationships between the value of the count or rate in each zone, y_i, and the values of other attribute values, $\boldsymbol{x}_i^T = (x_{i1}, \ldots, x_{ip})$, recorded for those areas. We have already discussed the modelling of area data in Chapter 7. Our starting point was the standard non-spatial linear regression model, and we then extended this to incorporate the possibility of spatial dependence in the data, through the idea of variate interaction models. What, then, is the problem with applying exactly the same approach to a 'response' variable which is a count or a proportion?

We have already touched on this in our discussion of transformations of the response variable in Chapter 7. Let us leave aside for the moment the question of spatial dependence or second order effects in the data and simply consider applying standard non-spatial linear regression to a set of counts or proportions. We need to appreciate more clearly what problems are involved and why it may not be particularly sensible.

In essence, there are really two distinct components of most statistical models. Firstly, there is a proposed mathematical relationship between the mean value of the response variable, μ_i, and the values of the explanatory variables. This usually involves parameters whose values are unknown and need to be estimated. Secondly, there are some assumptions about how one expects observed values of the response variable to be distributed about this mean value; in other words, a specification of the form of the probability distribution for the response, often referred to as the *error distribution*.

In general, for any particular set of data the modeller wishes to specify both an appropriate mathematical relationship for the mean value and a suitable error distribution. The problem with standard regression models is that they restrict these choices severely. In the simplest regression case, the mean value is restricted to be a linear function of the explanatory variables and the error distribution to be normal with a constant variance. It is only in that case that parameter estimation by use of ordinary least squares is justified and all the familiar regression results follow concerning standard errors and so on, as outlined in Chapter 7. Once we change either of these components of the simple linear regression model then the same results will not necessarily be valid.

For example, we know from Chapter 7 that if we drop the assumption of constant variance for the error distribution we must use weighted least squares (a special case of generalised least squares) to estimate parameters. We also know that if we move away from the situation where the mean value of the response is expressed as a linear function of the unknown parameters, then resulting least squares equations will be non-linear and the standard regression solution will not apply. Our suggestion in Chapter 7 was to use a transformation of the response variable to try to linearise the mean relationship and stabilise the variance. However, we should appreciate that we may not be able to do both at the same time. If we model a transformation of the original variable, chosen to 'linearise' the mathematical relationship for the mean, then we will be lucky if the error distribution of the same transformed variable can be considered normal with constant variance, which is the assumption inherent in using the transformed variable in a standard regression.

So, the regression models of Chapter 7 place restrictions on the choices that the modeller can make. In many situations such restrictions may be adequate to accommodate the data concerned, particularly if they are not applied too rigorously. Standard regression results are in fact fairly robust to reasonable departures from normality. For example, we felt it adequate in Chapter 7 to model the logarithms of standardised mortality ratios in the English Health Authorities within the standard regression framework, even though the assumptions which this implied were dubious. We mentioned there that a reasonable assumption for such SMRs might be that their variance will be proportional to their mean. This follows from the argument that the observed number of deaths in any area, although strictly binomial, is only a small proportion of the population from which they arise and so can be considered to follow a Poisson approximation to the binomial. The logarithmic transformation is appropriate to stabilise variance when this is proportional to the square of the mean, rather than the mean. Nevertheless, for the purposes of our analysis, regression of a logarithmic transformation was considered a 'reasonable' way to proceed. It represented a compromise between wishing to stabilise the variance and at the same time linearise the relationship between the mean value and the explanatory variables, since we expected multiplicative rather than additive effects on the original scale. We were forced into this compromise because within the standard regression framework one cannot transform the relationship between the mean of the response and the explanatory variables independently of the error distribution that applies when the transformed variable is used in the regression. Furthermore, we must aim to make the latter normal with constant variance.

One can fall back on a weighted regression to try to correct for remaining heteroscedascity in a transformed response; but such *ad hoc* approaches are not always satisfactory. There are limits as to how far we can 'bend' the standard regression framework to accommodate different types of data. For example, as we noted in Chapter 7 there is really no entirely acceptable way of dealing with either count or rate data by using simple transformations within a standard regression framework. There is of course an alternative which does not impose such restrictions; that is, to use the general method of maximum likelihood to estimate parameters. However, then virtually every model becomes a separate computational problem. In practice, one of the most valuable aspects of standard multiple regression models is that they can all be handled by general purpose and widely available software. Ideally we require a less restrictive class of models, but one with sufficient general structure to enable them all to be handled within the same computational framework. Such a class of models do exist and are referred to as *generalised linear models*.

Such models include not only standard multiple regression models, but also those for Poisson and binomial error distributions as well as log linear models for the analysis of multi-way tables of counts. They are related as being general models for probability distributions belonging to the exponential family. At a practical level they are unified by virtue of the fact that maximum likelihood parameter estimation for such models can be implemented through a technique known as *iterative re-weighted least squares* (IRLS). This has allowed general purpose software packages, such as GLIM, to be developed to handle such models very flexibly, in much the same way as standard statistical packages handle more familiar multiple regression models. Many other major statistical packages now also have routines for handling such generalised linear models.

We cannot go into the broad subject of generalised linear models in much detail here, but we can at least outline the basic idea. A generalised linear model consists of two components, an error distribution for the response variable within the exponential family of distributions, and a monotonic *link function*, $g()$, such that:

$$g(\mu_i) = \boldsymbol{x}_i^T \boldsymbol{\beta}$$

where μ_i is the mean value of the response variable. The link function thus allows a non-linear relationship between the mean value of the response, μ_i, and the linear function of the explanatory variables $\boldsymbol{x}_i^T \boldsymbol{\beta}$, familiar from standard regression. Further, the link function is allowed to be specified separately from the error distribution. Such models are clearly not completely general; the error distribution has to be within the exponential family and the mathematical form of the relationship between the mean value and the explanatory variables has to be expressed in the form given above. Nevertheless, they represent a far broader class of models than the standard multiple regression models, the sub-class of generalised linear models where the error distribution is normal and the link is the identity function.

Our original reason for introducing such models was an interest in modelling counts or proportions. How does this relate to the above framework? A typical generalised linear model for count data consists of an error distribution which is Poisson and a link function which is logarithmic:

$$\log(\mu_i) = \boldsymbol{x}_i^T \boldsymbol{\beta}$$

A commonly used model for proportions is a binomial error distribution with a logistic link function:

$$\log\left(\frac{\mu_i}{1 - \mu_i}\right) = \boldsymbol{x}_i^T \boldsymbol{\beta}$$

Other link functions may be specified in both these models if thought more suitable in any particular application.

As mentioned, one of the advantages of generalised linear models is that they can all be fitted by the use of IRLS. In the case of normal errors and an identity link, this is equivalent to ordinary least squares regression. In other cases it means that parameter estimation reduces to a weighted regression, but one where the weights involved depend upon the parameters, $\boldsymbol{\beta}$, which are being estimated. In this case initial 'guesses', $\hat{\boldsymbol{\beta}}^{(0)}$, for the parameter estimates are used to compute the weights for a first weighted regression. This then results in revised estimates, $\hat{\boldsymbol{\beta}}^{(1)}$, which are then used to compute revised weights for a second weighted regression and so on, until adequate convergence is achieved in the parameter estimates; hence the name 'iterative reweighted least squares'. General and very efficient computational algorithms for such iterative procedures may be developed and these are incorporated into packages such as GLIM (Generalised Linear Interactive Modelling).

Standard errors for parameters produced from generalised linear models may be interpreted in much the same way as in standard regression models. However, the overall fit of the model is assessed through a quantity known as the *deviance*, rather than through the residual sum of squares. The deviance is a log likelihood ratio statistic; if the model is an adequate explanation of variations in the response the deviance should theoretically be distributed as χ^2, having degrees of freedom equal to the difference between the number of observations and the number of parameters in the model; if the model is poor the deviance will be significantly larger than this. The relative fit of two alternative models may be similarly assessed by looking at the difference between their deviances; an insignificant difference in fit corresponds to a difference in deviance which should be distributed as a χ^2 distribution having degrees of freedom equal to the difference between the number of parameters in the two models.

Some of these ideas can be illustrated by fitting a very simple model to the Irish blood group data first encountered in Chapter 7. We consider modelling the proportion of the adult population with blood group A as a simple function of the categorical variable that denotes whether or not a county is within or beyond the 'Pale', that part of the country that was settled and dominated by the Anglo-Normans. Recall that for this data set we also have the sample size for each county, upon which the proportions are based. The simple generalised linear model applicable here has a binomial error distribution with a logistic link function:

$$\log\left(\frac{\mu_i}{1-\mu_i}\right) = \beta_1 + \beta_2 x_i$$

where μ_i is the mean proportion in each county i with blood group A, and x_i has the value 1 if the county is within the 'Pale' and 0 otherwise. To fit such a model in GLIM we need three variables: y_i, the observed proportions; x_i, as described above; and n_i, the sample size in each county. The model is then fitted by declaring the response variable to be the number with blood group A in each county, $n_i y_i$, the error to be binomial with n_i as the vector of denominators, and the link function to be logistic. The resulting parameter estimates (and standard errors) are $\hat{\beta}_1 = -0.9634$ (.01693), and $\hat{\beta}_2 = 0.2255$ (.02027), showing a highly significant difference between counties within and outside the 'Pale'.

We may contrast the results of this 'correct' analysis with those obtained by simply taking a logistic transformation of the observed proportions, y_i, to form a new response variable:

$$y_i' = \log\left(\frac{y_i}{1-y_i}\right)$$

and performing an ordinary least squares regression of y_i' on x_i. This gives the results $\hat{\beta}_1 = -0.9677$ (.02521), and $\hat{\beta}_2 = 0.2037$ (.03711). The 'Pale' effect remains significant but the parameter estimates and standard errors are some way from the maximum likelihood values obtained from the IRLS procedure used in GLIM.

A better alternative to this naive regression, although still an approximation to the 'correct' generalised linear model, is provided by a technique often referred to as *empirical logistic regression*. This employs a weighted regression of the logistic transformation of the original proportions. The idea is as follows. If Y_i is the random variable representing the original proportions with mean μ_i, then under a binomial assumption $VAR(Y_i) = \mu_i(1-\mu_i)/n_i$, where n_i is the denominator upon which each observed proportion is based. It follows that the variance of the logistic transformation of Y_i is then approximately given by:

$$VAR(Y_i') = \frac{1}{n_i\mu_i(1-\mu_i)}$$

and a crude estimate of this is provided by taking:

$$VAR(Y_i') \approx \frac{1}{n_i y_i(1-y_i)}$$

Therefore we carry out a weighted regression of y_i' on x_i using weights based on these approximate variance estimates. Recall from Chapter 7 that this involves the special case of generalised least squares where the variance–covariance matrix C would have diagonal elements

$$c_{ii} = \frac{1}{n_i y_i (1 - y_i)}$$

and zeros elsewhere. Each value of y_i' in the regression is weighted in inverse proportion to its variance, so giving less weight to observations based upon small sample sizes. In the Irish blood group example the results obtained are $\hat{\beta}_1 = -0.9618$ (.02853), and $\hat{\beta}_2 = 0.2247$ (.03416). These are considerably closer to the exact results obtained by using the generalised linear model than those obtained from an unweighted regression, but the standard errors in particular remain some way from their correct values.

Ex 8.3

Generalised linear models provide the analyst with a very powerful and flexible set of tools to handle data such as counts and proportions, where the assumption of a normally distributed response variable is untenable and simple transformations do not suffice to address the problem. Indeed, we shall come across them again in Chapter 9 when considering the modelling of spatial interaction data. However we should point out, in conclusion to our brief introduction here, that spatial forms of such models are not well developed and modifications do not currently exist which account for residual spatial dependence in the data. When it comes to applying these non-spatial tools to spatial data, probably the best approach at present is simply to be aware that problems may arise with their associated standard tests when there is strong residual spatial dependence. They can be alleviated to some extent by attempting to introduce location into the link function to remove any such effects present. For example, one might introduce the spatial coordinates of observations as additional covariates, or classify regions in terms of their broad location and treat this classification as an extra, categorical, explanatory factor in the model. This assumes of course that one can 'explain away' spatial dependence in terms of a first order spatial trend and the reader should appreciate by now that this will not always be possible.

8.3 Multivariate methods

In the previous chapter we considered situations where several attributes were measured in each area, but so far in our analyses we have been interested in a single 'response' variable; the other variables we have treated on a different footing, as possible 'explanations' for variations in the variable of primary interest.

In this section we discuss multivariate techniques, methods relevant where spatial pattern in several 'response' variables taken together has to be considered. We have already given a fairly full description of multivariate methods in Chapter 6 and so refer the reader back to that discussion. The techniques outlined there are also applicable in various ways to area data. For example, such analyses are relevant to attempts to classify areas on the basis of several measurements for marketing purposes; or to studies concerned with patterns of general health in the population. If we recall the Chinese province data, or the Barnet census data, introduced in Chapter 7, we do not have a variable whose spatial variation is to be explained by other variables in the data. Rather, we seek structure in each of these data sets: a classification or *regionalisation* of China or (at a rather different geographical scale!) Barnet.

As we said in Chapter 6, specifically spatial forms of multivariate analysis are not well developed and those that do exist are somewhat specialised and rather advanced. What is often done is to use conventional multivariate techniques for purposes of reducing the multivariate attributes to one or two dimensions (significant components) and then to look at these dimensions from a spatial perspective, seeking spatial patterns, relationships and classifications. We consider first a multivariate analysis of the Chinese socio-economic data before turning to some census data analysis.

Recall that we have a set of 18 variables for Chinese provinces, representing a set of social and economic indicators. Here we apply principal components analysis, as discussed in Chapter 6, to the standardised variables, with the aim of reducing the set of variables to a smaller set of uncorrelated components that summarise most of the variation in the original data. We shall hope to interpret such components to give us insight into what factors are important in understanding the broad-scale socio-economic landscape.

Performing a principal components analysis on the data yields four components which explain most of the variation in the original 18 measures. The loadings of the original variables on these four components are given in Table 8.1. The variable names are abbreviations of the variables which are listed in full in Table 7.1.

We see that virtually all the variables load negatively on the first component, with fertility having quite a high positive loading. This component can be interpreted broadly as a 'modernisation' component (strictly, it is a 'traditional society/economy' component, given the sign of the loadings; the fertility rate loads positively, while variables indicative of modernisation, such as industrial output, load negatively). The second component has high positive loadings for the secondary education variable, and high negative loadings for fertiliser and grain yield. We can suggest that this is a component that, generally speaking, contrasts urban–industrial areas and agricultural areas. Component three has high negative loadings of power irrigated area and coal output, and expresses other urban–rural contrasts; only the deaths from cancer among men loads highly (again, negatively) on the fourth component.

We might now subsequently use the scores of each of the provinces on two or more of these components to derive a multivariate regionalisation of the country. However, we will look at this idea in more detail in relation to an alternative data set. Readers will recall from Chapter 7 the set of variables extracted from the 1981 Census for Barnet in north London. Because of space limitations (we have 619 EDs), we have only included a modest number of variables in our data set. This somewhat weakens the case for a principal

Table 8.1 Principal component loadings for Chinese provinces

	Component			
	1	**2**	**3**	**4**
Industry	−0.804	0.022	−0.309	0.106
Tractor	−0.842	−0.038	−0.208	0.068
Power irrig	−0.650	−0.319	−0.559	0.282
Rural elec	−0.941	0.172	−0.149	0.112
Fertiliser	−0.511	−0.732	0.075	−0.268
Grain yield	−0.301	−0.843	0.219	−0.190
Industry per capita	−0.926	−0.023	0.108	−0.042
Coal	0.038	0.448	−0.644	−0.343
Hospital beds	−0.756	0.407	0.091	−0.232
Doctors	−0.872	0.349	0.191	0.009
Graduates	−0.879	0.143	0.252	0.048
Second schools	−0.699	0.556	0.183	−0.341
Urban	−0.902	0.236	0.018	0.212
Fertility	0.710	0.304	0.446	−0.263
Retail	−0.970	0.119	0.175	−0.004
Rural income	−0.884	−0.071	0.206	0.083
Rural expend	−0.806	−0.407	0.309	−0.039
Cancer	−0.444	−0.276	−0.417	−0.587

Table 8.2 Principal component loadings for Barnet enumeration districts

	Component		
	1	**2**	**3**
Owner occupied	−0.9059	0.1498	0.2175
Local authority	0.7671	−0.5025	−0.1619
No car	0.9314	0.1309	0.1323
Two cars	−0.8436	−0.2628	−0.0069
Single parents	0.7797	−0.2176	0.0202
Lone female pensioners	0.4294	0.4223	0.7364
Migrants	0.2296	0.7226	−0.6084

components analysis and a subsequent classification based on a subset of these components, since we have not chosen the subset of variables with this objective particularly in mind. Nevertheless, using just these variables will sufficiently illustrate the way in which similar, more extensive, multivariate data are widely used in 'geodemographics'. We first express the original counts as percentages of appropriate denominators to obtain seven new percentage variables. Performing a principal components analysis after standardising these seven variables yields three components that together account for 84.9 per cent of the total variation. The loadings matrix is given in Table 8.2. Variables loading positively on the first component are those that often appear in deprivation indices; measures of material disadvantage, such as lack of a car, or social disadvantage, such as single parent families. By way of contrast, households with two or more cars, and rates of owner occupation, load negatively. We might interpret the component, a little crudely, as a 'deprivation' index. The second component measures population turnover; the migration variable, loading positively on this component, records the proportion of households with residents living elsewhere one year earlier. The proportion of households rented from the Local Authority (Council) has a negative loading on this component. Finally, the proportion of pensioner households with women aged over 75 years and living alone, has a high positive loading on the third component.

The scores on each component may be obtained and mapped, as shown in Figure 8.4 in the case of the first component. The shading here is in six equal intervals from dark for the most negative scores (−4.01 being the minimum) to light for the most positive (+6.01 being the maximum).

The alphabetic prefixes for EDs in these data are codes for larger electoral 'wards', the smallest units used in Local Government in Britain. Some wards, such as that prefixed with a 'C' (it is called Burnt Oak), are virtually entirely characterised by high positive scores. Similarly, zone 'U', to the south-west (West Hendon), has generally high positive scores. These we might characterise as rather deprived areas. In contrast, Hale ward ('P'), just north of Burnt Oak, Brunswick Park ('B') and Totteridge ('T') wards have a preponderance of negative scores, suggesting these are zones of relative affluence. Yet most wards are rather heterogeneous, containing a mix of positive and negative scores; often attached to small areas that are very close neighbours. This is significant, since some planning, particularly by Health Authorities, takes place at a ward level, or even higher. Mapping scores on the second and third components reveals a greater level of spatial fragmentation. The results show that wards are far from homogeneous spatial units. The same picture of spatial heterogeneity at particular scales will apply to any country where census data are supplied within the framework of a nested hierarchy of small and larger units.

Having performed a principal components analysis such as this, it is typical to then go on and attempt to derive a classification of areas using the scores on two or three components. Commonly, clustering methods such as the '*k*-means' or other such algorithms are used for this purpose. We have described these methods in Section 6.3.5, and refer the reader there for a discussion. Note that there are arguments both for and against applying clustering methods to a subset of principal components as opposed to the original variables. The fact that the principal components are uncorrelated seems attractive in that one avoids 'double counting' factors which are measured by more than one of the original variables. On the other hand, selecting only the first few principal components may leave out components which, while not important in terms of the overall variance in the data, are useful in terms of classification. Where

Fig. 8.4 First component PCA scores for Barnet EDs

principal components scores are used for clustering and are themselves derived from standardised variables, one should consider very carefully whether it is sensible to standardise again before using them in a cluster analysis. Doing so effectively attaches equal weight to each component, and works against their order of importance in terms of explaining variance as determined by the PCA analysis.

If we apply the 'k-means' clustering technique described in Chapter 6 to the Barnet data using the first three principal components described above (without standardisation), then we do find evidence of homogeneous social areas, corresponding broadly to those identified earlier by examination of scores on the first component; for instance, in the west of the borough, and to the north. However, regardless of the number of clusters we choose, to some extent the picture resembles a patchwork quilt. In general, zones that are spatially adjacent are not necessarily assigned to the same cluster. This illustrates a general point, already made in Chapter 6: there is nothing in these clustering methods that forces areas close in space to cluster together in socio-economic space. This is perfectly right and proper; why should we force adjacent zones to be in the same class? Classifications are by their very nature partitions of data; why should we force yet another constraint onto them? This is a powerful argument against such a 'contiguity constraint'. However, there may be some instances where contiguous regions *are* required or desirable. Perhaps the most persuasive example arises in the political division of space, where an administrative unit or electoral division such as a constituency is usually required to comprise an amalgamation of smaller units that are spatially contiguous. To do this requires that the classification takes account of both the multivariate nature of attribute data *and* relative location. Relative location may be incorporated via an $(n \times n)$ adjacency matrix for all zones, with elements set to 1 if zones are adjacent, 0 otherwise. Zones are then permitted to 'fuse' only if they are similar in terms of the attributes being used in the classification and also adjacent. There are several other ways in which one can incorporate contiguity or other spatial criteria into classification methods. We do not discuss these here, referring the reader to one of the texts listed at the end of this chapter. To repeat, these techniques should not in general be forced on a set of data; they have the disadvantage of separating into different classes zones that are some geographical distance apart, even though they are located close together in attribute space.

More generally, we should be constantly critical of *any* classification of data. In deriving three, four, or five or more classes from one of the clustering techniques there is no guarantee that separate classes are internally homogeneous. Clustering methods will always produce clusters. Small areas that may be quite dissimilar may be forced into the same class. We can always gain some sense of the extent to which this is the case by plotting pairs of variables used in the classification. For instance, if we plot scores of EDs in Barnet on the second component against those on the first component, it is difficult, visually, to recognise obvious 'clusters' (Figure 8.5). It is, however, possible to spot EDs that 'behave' rather differently from others; outliers that may be, nonetheless, forced into a class with other areal units! Such problems have not prevented a vast amount of applied work being carried out in order to attach convenient labels to areas, for the purpose of 'targeting', as we discussed in Chapter 7. These sorts of classification can often prove useful in geodemographics and marketing; in identifying areas that are socially similar, perhaps for the purposes of marketing some product. But we should not let the fact that such classifications are widely used blind us to their shortcomings and the need to exercise critical scrutiny.

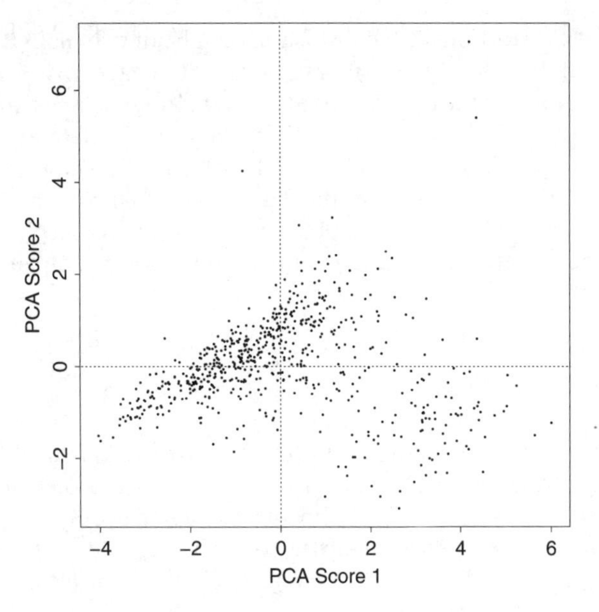

Fig. 8.5 Second v. first component PCA scores for Barnet EDs

8.4 Image analysis

In the final section of this chapter we turn to specialised methods that are useful in handling *image data*. Image data are digital data that come to us on a fine regular grid or lattice. We mentioned such data briefly in Chapter 7 and in Chapter 2, when we contrasted such 'raster-based' data with that in vector form. They represent a special form of area data which arises in a number of different applications. Although we shall focus here on environmental remote sensing, the techniques we discuss could also be applied in medical imaging; for instance, where we wish to explore, understand and interpret images that result from the scanning of body organs. Readers will be pleased to know that our case studies do not include scans of any of the authors' organs!

In order to do justice to this topic we would need to write an entire book; fortunately, others have done so and so we refer the reader to their far more detailed texts, listed in our section on further reading at the end of this chapter. In skimming the surface of a complex topic we shall separate a discussion of tools for *enhancing* images from a brief consideration of the use of multivariate methods, whose use for exploring area data we have already looked at in a different context.

Before doing so we should make it clear that the sorts of methods we discuss assume that some *pre-processing* of the data will have already taken place. We do not discuss this in detail. Suffice it to say here that this is an attempt to remove known distortions. Such distortions can be of two main types. The first is *geometric* and relate to the handling of satellite imagery, which needs to be corrected to allow for the earth's curvature and also to be registered to match a coordinate system. This is of obvious importance if the data are subsequently to be used in a GIS. The second type of distortion is known as *radiometric* distortion and this has to do with the digital data itself, which may be subject to obvious sources of noise, generated for example by atmospheric conditions such as haze. There is a variety of methods, some simple, others quite sophisticated, for treating these effects.

8.4.1 Image enhancement

Let us assume that we already have some pre-processed data. What are we to do with it in order to 'add value' to it? The first set of techniques we shall discuss have to do with the enhancement of image data. By this we mean the modification of pixel values in order to highlight information and to aid interpretation. We shall consider three general classes of technique for doing this.

The first set of techniques is known as *contrast enhancement*. These are essentially visualisation techniques. For example, when analysing LANDSAT TM data (a tiny example of which we discussed as one of our case studies in the previous chapter), the pixel values can, in principle, take on values from 0 (no reflectance) to 255 (maximum reflectance). Yet, typically, the values in a particular band are spread over a much smaller range (perhaps 30 to 100 in band 4, for example). If this image is displayed on a machine capable of handling an 8-bit colour resolution (i.e. $2^8 = 256$ colour values are capable of being displayed) the image (using a grey scale) would not show much contrast; missing from the display would be very dark values (those with reflectance < 30) or medium-bright values (> 100). Consequently, we might 're-allocate' or 'stretch' the range of values in our data to match the 0 to 255 colour range. Thus the value 30 would be assigned a value of 0, that of 100 given the value 255 and intermediate values interpolated onto the $(0, 255)$ interval. There are various ways of performing this interpolation or stretching. The simplest is a linear scaling, such that the middle value of 65 (in our little example) is given the new value 127, and so on. When the 'stretched' image is displayed we will see much greater visual contrast; details that did not appear on the 'raw' image will now be revealed.

As noted, this is a visualisation technique; the original values are, in a sense, only temporarily modified and any subsequent analysis will always be done on the raw data. Moreover, the enhancement of the data does not depend on spatial context. By way of contrast, a second class of enhancement method is that which creates a new set of image data, based on spatial proximity. These are *filtering* methods and are exploratory, rather than visualisation, tools. We have already encountered similar techniques in Chapter 7 and the principles outlined there are the same here. We saw how we could use moving averages to smooth out local variations; the value in one area (now a pixel) was modified by taking the value in that area, together with those in neighbouring areal units, and then (in the simplest case) taking the mean of the data values. This local mean became the new, filtered, value in an areal unit. Such averaging, in the context of image analysis, is known as applying a *low-pass* filter. What this means is that the low frequency components in the data are retained ('passed' through the filter) while the more erratic, higher frequency components are removed or filtered. In image processing such filters are often used in the pre-processing stage, for example to remove the 'striping' effects apparent in raw data because of the way the data are collected or scanned. General patterns of variation are thus revealed, blurring the image so that deviations from a more regional trend are removed. In essence, such low-pass filters are a means of revealing first order variation in the data.

In the context of the regular grids of data obtained in remote sensing, such filters can be thought of as a 'moving window' that visits each pixel in turn, modifying its value. The size of window may be varied to achieve different effects. It might be a 3×3 filter, such that the pixel value is replaced by the average of itself and eight surrounding pixels. More smoothing will occur if the filter window is widened to a 5×5 matrix of weights; the weights need not be identical but could be devised to decline with distance such that more weight is given to nearest neighbour pixels than to those two lags away.

A related form of smoothing uses the median value rather than an unweighted mean of the neighbourhood. This is, of course, less sensitive to extreme values and as a result it does a better job of preserving boundaries or edges that may be smoothed by a classical mean filter. Another form of smoothing is to use the median polish technique discussed in the previous chapter. This too removes the first order effects, leaving a set of residuals for subsequent analysis of the second order variation in the data.

The opposite of low-pass filtering, *high-pass* filtering, sharpens the image, revealing edges and boundaries (marking clear breaks in reflectance values from one pixel to its neighbours) that may aid interpretation. Now, it is the high frequency components or changes in values that one wants to emphasise. The simplest way of obtaining the high frequency component in an image is to apply a low-pass filter and then to subtract this from the original data. Other methods use differencing techniques that are the discrete equivalent of derivatives in calculus. For example, one common image 'sharpening' technique employs the *Laplacian* of a digital image, defined as:

$$\nabla^2 y_{ij} = (y_{i+1,j} + y_{i-1,j} + y_{i,j+1} + y_{i,j-1}) - 4y_{ij}$$

Such second differencing shows where the gradient of the image is changing and thus will pick up areas where pixel values change rapidly (as at a coastline, for example). A common way to use this is to subtract the digital Laplacian from the original image. If we do this we obtain:

$$y_{ij} - \nabla^2 y_{ij} = 5y_{ij} - (y_{i+1,j} + y_{i-1,j} + y_{i,j+1} + y_{i,j-1})$$

Now if the pixel (i,j) is in the middle of a region of no, or slow, change in values then $y_{ij} - \nabla^2 y_{ij} \approx y_{ij}$, that is the pixel value is unaltered. On the other hand, if (i,j) is on the low side of an edge, then some of the neighbours' values are higher than y_{ij} and none are lower, so $y_{ij} - \nabla^2 y_{ij} < y_{ij}$. The opposite situation would apply if (i,j) is on the high side. Thus subtracting the Laplacian from the original image has the effect of sharpening edges. Looking at the above expression this process simply corresponds to passing a moving filter across the map, where each cell value is replaced by a combination of its own value and those of its immediate neighbours according to the scheme:

$$\begin{pmatrix} 0 & -1 & 0 \\ -1 & 5 & -1 \\ 0 & -1 & 0 \end{pmatrix}$$

Strikingly effective enhancement of edge detail may result from such filtering. One problem of course is that 'noise' in the image is also sharpened. Various other filtering schemes may be developed, using similar kinds of ideas.

Finally, we draw attention to, but do not give details of, a third class of image enhancement methods that draw upon the same kinds of Bayesian approaches discussed in an earlier section of this chapter. Recall that in Bayesian statistics we seek to combine both observed data and prior knowledge in arriving at estimates of parameters of interest. In the context of image analysis we have a set of data on the pixel values, and we can postulate some prior probability distribution for any 'signal' we might expect in the image. In most cases this will consist of general statements about the probability of observing particular features in the image and will be tailored towards the particular application in question—so-called 'scene-specific' knowledge. We can also assume some general probability distribution for the measurement errors ('noise') that will be present in the data. This may include a covariance structure for these errors. For example, a typical assumption might be a multivariate normal distribution with some simple local correlation between errors. The prior distribution for 'signal' may then be combined with the observed data via Bayes theorem to provide a posterior distribution for the 'signal' in the image. The posterior distribution may then be used to 'reconstruct' the 'signal' on the basis of maximising the posterior probability of the value of any pixel. This is often referred to as a MAP (maximum *a posterior* probability) estimate. The procedure can be computationally very intensive for images of realistic size, and a number of numerical algorithms have been proposed to approximate the maximisation. Perhaps the most commonly used of these techniques is that of *iterated conditional modes* (ICM). We direct the reader to references at the end of this chapter for further details.

8.4.2 Multivariate analysis of image data

The above techniques are of use in the handling of an image of a single variable. But much image data (notably that used in remote sensing) takes the form of multivariate data. In other words, our data comprise a set of $m \times n$ pixels arranged on a lattice with m rows and n columns. These $m \times n$ spatial units, in turn, form the rows of a data matrix, the p columns of which are the different, multi-spectral, bands at which electromagnetic reflectance is captured by the sensor.

Given multivariate (i.e. multi-band) data there are a number of potentially useful operations we can perform. Some of these use multivariate statistical methods, but others are quite simple. For instance, there are some useful arithmetic manipulations of multi-band data. The most widely used of these is the division of reflectance values in one band by the corresponding pixel values on another band; this is known as *ratioing*. One use of ratioing is in eliminating the effects of varying illumination in areas of changing slope. Another use is to highlight particular land use characteristics. For example, if we form the ratio of two TM images (such as bands 4 and 3) and then filter the resulting image, we obtain a good picture of spatial variation in biomass. This is because healthy green vegetation reflects strongly in band 4, the near-infrared, but absorbs energy in the visible red portion of the spectrum, band 3.

More complex arithmetic operations are regularly used when analysing remotely sensed imagery collected from other satellites. For example, the AVHRR sensor has a spatial resolution of 1.1 kilometres and may be used for monitoring land use at continental scales. If reflectances from bands 1 and 2 of these data are extracted, the ratio of the subtracted to the summed values on these bands yields a widely used measure of vegetation cover, the so-called *Normalised Difference Vegetation Index*.

These sorts of simple arithmetic operations are of considerable value. But given that multi-spectral sensors yield multivariate data, there is also merit in bringing to bear on such data some of the multivariate techniques discussed in Chapter 6 and in earlier sections of this chapter. For example, one important use of techniques such as principal components analysis (PCA) in the handling of TM data is in collapsing the seven correlated bands to a smaller set of uncorrelated components. To illustrate this we consider a PCA analysis of the data we described in Chapter 7. As we said when discussing these data, they represent but a tiny subset of a typical LANDSAT TM scene and it will certainly be the case that the results of analysing an entire scene using PCA would look rather different.

We standardise each of the variables before performing the analysis, so that effectively we are working with the correlation matrix of the original seven variables. This is as follows:

$$\boldsymbol{R} = \begin{pmatrix} 1.00 & \cdots & & & & & \\ 0.950 & 1.00 & \cdots & & & & \\ 0.958 & 0.934 & 1.00 & \cdots & & & \\ 0.571 & 0.704 & 0.473 & 1.00 & \cdots & & \\ 0.844 & 0.891 & 0.809 & 0.815 & 1.00 & \cdots & \\ 0.460 & 0.439 & 0.474 & 0.337 & 0.585 & 1.00 & \cdots \\ 0.900 & 0.902 & 0.901 & 0.620 & 0.941 & 0.660 & 1.00 \end{pmatrix}$$

Note the high level of inter-correlation, though note too that the thermal band (6) departs from this general pattern, the correlations being much lower than for other bands. The eigenvalues of this matrix lead to corresponding eigenvectors with loadings on each of the original seven variables as shown in Table 8.3.

We find high loadings for all bands on the first component, except for band 6. This suggests that the first component simply represents a measure of 'average reflectance'. Most bands, with the exception of band 5, have moderately high loadings on the second component as well, although band 4 loads negatively. However, the highest positive and negative loadings come from bands 3 and 4 respectively, a contrast that is indicative of the vigorous growth of vegetation. It can thus be thought of as a 'greenness' component. Band 6, the thermal band, is the only one to load highly on component 3, suggesting that this component is picking up reflectance mostly in the thermal band. The last four components do not explain much of the total variability in the data. This will generally be the case in a PCA of such data, but it does not necessarily follow that such components are uninterpretable or uninformative. They may reveal particular patterns of land use, for example. Since we are only examining a very small scene it is not worth making too much of the analysis reported here. In an analysis of the wider scene from which our data were extracted, the geographer Paul Mather notes that while the first component is a weighted average of all seven bands, the second is dominated by band 6, while the third is dominated by band 4 (the infrared), contrasting land and water bodies. Together, these three components accounted for 96 per cent of the variance in the seven bands.

How are such components commonly used, given the fact that PCA is essentially a 'data reduction' technique? Of course, each component can be mapped, either on a colour or black–white monitor. However, another technique arises from the fact that a particular colour composite can be made by 'mixing' the first three components. If they are associated with, for example, the colours red, green and blue, then a 'false colour composite' may be obtained and displayed on a monitor. It is called a 'false colour' composite because the three components do not correspond to the three visible bands and

Table 8.3 Principal component loadings for LANDSAT TM data

	Component						
	1	**2**	**3**	**4**	**5**	**6**	**7**
Band 1	0.827	0.383	−0.281	0.223	0.043	−0.105	−0.099
Band 2	0.896	0.228	−0.256	0.199	0.031	−0.102	0.162
Band 3	0.783	0.488	−0.295	0.237	0.050	0.101	0.004
Band 4	0.886	−0.462	−0.003	0.054	−0.010	0.003	−0.002
Band 5	0.989	0.120	0.017	−0.080	0.019	−0.001	0.000
Band 6	0.558	0.365	0.702	0.265	0.037	−0.003	0.002
Band 7	0.907	0.401	−0.004	0.026	−0.130	0.001	−0.001

the colours are not therefore 'natural'. Clearly, this idea of combining three components can be, and frequently is, extended to treatment of triples of the original bands. Such composites are widely used in all sorts of applications, from geological reconnaissance to environmental assessment.

Another use of multivariate methods in the analysis of image data is concerned with associating or labelling particular pixels, or groups of pixels, with particular land characteristics (or, indeed, varying water quality if that happens to be the subject of interest). This is the same type of classification problem discussed in an earlier section of this chapter, but this time in the context of image processing. It is typically approached in one of two ways.

One way, known as *supervised classification*, requires us to build up a kind of template for each feature of interest. For example, we might know that the ground cover at a particular location (known as a 'training site'), is 'linseed'. We then note the reflectances on different bands for various types of 'training site'. Because of the inherent variability in reflectance, not all pixels corresponding to a particular 'training site' will have exactly the same pattern of values on the different bands. Indeed, the amount of variability for pixels relating to the same 'training site' will vary depending on the type of site involved. For example, built-up areas may be spread over a wide range of reflectances, on one or more axes (bands or components), while other classes, such as sand, may be more homogeneous. We can thus think of each pixel as being located in a p-dimensional multivariate space according to its values on the different bands and at the same time falling into a known group as identified by the 'training site' to which it corresponds. This is referred to as a 'training set'. We can now attempt to use the 'training set' to develop rules that allow us to classify pixels in new images of unknown areas as falling into one or other of the different land uses, depending on their values on the different bands. The simplest such rule is to calculate the centroids of each of the different types of pixels in the 'training set' and then label the unknown pixel with the land class of the centroid to which it is closest, using p-dimensional Euclidean distance. However, more sophisticated multivariate approaches are available to develop rules for the classification of new pixels based on the 'training set'. These involve the kind of discrimination techniques which we mentioned in Chapter 6.

With *unsupervised* classification we do not rely on prior ground knowledge or a 'training set'. Instead, we proceed very much along the lines of the multivariate cluster analyses we have used earlier in this chapter. All unknown pixels are located in a multivariate space, either the original p-dimensional data space, or perhaps a subspace of this determined for example by using the data reduction method of PCA. Pixels are then allocated to one of a chosen number of arbitrary classes and then re-allocated (on the basis of their squared distance to class centroids, for example) in an iterative fashion until the re-allocation ceases to change the position of the class centroids. Of course, the procedure assumes that the number of classes is known *a priori*, and relates to the 'k-means' clustering described in Chapter 6. Alternative clustering schemes discussed there could also be used. Once this procedure is complete, interpretation of the classes must rely on checking against field evidence, or by consulting maps or air photographs. More detail on both supervised and unsupervised classification is provided in standard texts on image processing or remote sensing, such as those we list at the end of this chapter.

8.5 Summary

In this chapter we have tried to set out some further, rather specialised, techniques that spatial analysts use for understanding area data. These have fallen into three distinct groups.

Firstly, we have examined some methods for the analysis of area data in the form of counts or rates, methods such as probability mapping, empirical Bayes estimation and generalised linear models. Such methods are now being put to increasing use for better and more reliable understanding of disease risk, for example. Next we have seen how a range of multivariate methods, such as principal components analysis and cluster analysis, can be used in conjunction with area data, to classify places and derive combinations of variables that serve to summarise the structures in multivariate area data. Such methods are widely used to define area-based classifications for the targeting of goods in the private sector or services in the public sector. Lastly, we have given a brief introduction to the analysis of a special class of area data, that arising in image analysis in general and remote sensing in particular.

We now leave the subject of area data and, in the final part of our book, look at another type of spatial data, that concerned with spatial interaction, such as flows of people or goods between locations.

8.6 Further reading

On the general statistical modelling of counts and proportions see:

Aitkin, M., Anderson, D., Francis, B. and **Hinde, J.** (1989) *Statistical Modelling in GLIM*, Clarendon Press, Oxford.

Dobson, A.J. (1988) *An Introduction to Statistical Modelling*, Chapman and Hall, New York.

There is also a good account of modelling spatial categorical data in:

Upton, G.J.G. and **Fingleton, B. (1989)** *Spatial Data Analysis by Example: Volume 2: Categorical and Directional Data*, John Wiley and Sons, Chichester. Chapters 6–7.

For an alternative introduction to modelling categorical data with specific reference to applications in geography see:

Wrigley, N. (1985) *Categorical Data Analysis for Geographers and Environmental Scientists*, Longman, Harlow.

Probability mapping has been quite widely used in geographical research, notably in medical geography. For further discussion of the method see:

Norcliffe, G.B. (1980) *Statistics in Geography: an Inferential Approach*, Hutchinson, London.

For a critique of probability mapping and a brief discussion of the empirical Bayes approach see:

Sheldon, T.A. and **Smith, D.** (1992) Assessing the health effects of waste disposal sites: issues in risk analysis and some Bayesian conclusions, in Clark, M., Smith, D. and Blowers, A. (eds.) *Waste Location: Spatial Aspects of Waste Management, Hazards and Disposal*, Routledge, 158–86.

An introduction to the empirical Bayes method, together with a MINITAB 'macro' for calculating empirical Bayes estimates of relative risk, is given in:

Langford, I. (1994) Using empirical Bayes estimates in the geographical analysis of disease risk, *Area*, 26, 142–90.

For applications and extensions to the empirical Bayes methods see the following collection of essays:

Elliott, P., Cuzick, J. and others (1992) *Geographical Epidemiology*, Oxford University Press, Oxford.

For further details of the application of empirical Bayes to the Auckland mortality data see:

Marshall, R. (1991) Mapping disease and mortality rates using empirical Bayes estimators, *Applied Statistics*, 40, 283–94.

and for a general review of methods for the analysis of spatial patterns of disease see:

Marshall, R.J. (1991) A review of methods for the statistical analysis of spatial patterns of disease, *Journal of the Royal Statistical Society, Series A*, 154, 421–41.

A good reference on geodemographics is:

Brown, P.J. (1991) Exploring geodemographics, in Masser, I., and Blakemore, M. (eds.) *Handling Geographic Information: Methodology and Potential Applications*, Longman, Chapter 12.

For further interpretation of the Chinese data see:

Cole, J.P. (1987) Regional inequalities in the People's Republic of China, *Tijdschrift voor Economische en Sociale Geografie*, 78, 201–13.

The literature on image analysis is substantial. A good introduction is provided by:

Mather, P.M. (1987) *Computer Processing of Remotely-Sensed Images*, John Wiley, Chichester.

Mather's book contains full analyses of the entire LANDSAT TM scene from which we have extracted a very small subset. See, in particular, pages 11–16 and Chapters 5–7. See also:

Lillesand, T. and **Kiefer, P.** (1987) *Remote Sensing and Image Interpretation*, 3rd edition, 1994, Wiley, Chichester.

The GIS IDRISI, referred to briefly in Chapter 2, has excellent functionality for the handling of remotely sensed imagery and the exercises that come with the software include an invaluable set for gaining a deeper understanding of the processing of such data. The exercises are based on LANDSAT TM data. The principal components analysis that we report in this chapter was conducted using IDRISI.

On Bayesian techniques in image enhancement see:

Besag, J. (1986) On the statistical analysis of dirty pictures, *Journal of the Royal Statistical Society, Series B*, 48, 259–302.

Besag, J. (1989) Towards Bayesian image analysis, *Journal of Applied Statistics*, 16, 395–407.

These kinds of approaches and others are also discussed in:

Ripley, B.D. (1988) *Statistical Inference for Spatial Processes*, Cambridge University Press, Cambridge, Chapter 5.

8.7 Computer exercises

Here are suggested exercises that you can try on ideas discussed in this chapter using INFO-MAP and our example data sets. These exercises have been referenced at appropriate points in the chapter by numbered symbols in the margin.

Exercise 8.1

Construct a choropleth map of child mortality in Auckland, using the variable 'Raw SMR'. This variable was calculated from the raw data by using the formula

```
smrs=100*{1}/({2}*sum({1})/sum({2}))
```

You can check the calculation for yourself if you wish. Find, on the map, the zones labelled 141, 142 and 195. You may use the `Profile locations` in the `Analysis` menu to do this; specifying values in `Descending order` is useful here. Highlighting each of these zones in the displayed list and using <SHFT> + <CR> will enable you to inspect the observed and expected numbers of child deaths for each of these zones. Do the values give you any guidance about the problems of mapping rates on the basis of small numbers?

Now create a map of Poisson probabilities (variable 5 in the file), paying particular attention to zones whose counts are significantly higher than expected. A new variable is useful here, such as:

```
sig=if({4}>100 and {5}<=.05,1,0)
```

Map this variable, first specifying appropriate (discrete) class intervals using the `Scaling` option in the `Map` menu. Repeat the exercise to examine zones where the counts are significantly low. Speculate on why certain zones have 'significantly' high or low counts.

Exercise 8.2

Remaining with the Auckland data, in the text we discuss the derivation of empirical Bayes estimates of the death rates by using maximum likelihood estimates of the mean and variance of the prior distribution. We quote these estimates as $\hat{\gamma} = 2.664 \times 10^{-3}$ and $\hat{\phi} = 0.614 \times 10^{-6}$. From these we can calculate the empirical Bayes estimates of the death rate, $\hat{\theta}_i$, for each area from the expressions given in the text. An appropriate corresponding INFO-MAP formula in this case is:

```
ebrates=2.664+({3}-2.664)
          *(0.614/(0.614+(2.664/({2}*9/1000))))
```

where variable 3 is the raw death rate per 1000 under-five population, variable 2 is the under-five population in 1981 and it is to be remembered that the deaths relate to a

9 year period. The result of this calculation is included in the file as the variable 'Bayes rate per 1000'; you can check the calculation for yourself if you wish.

Map these Bayes rates as a choropleth map and compare this with the 'Raw rates per 1000'. Notice how the Bayes estimation has 'shrunk' the raw rates towards the overall global rate in some areas (where numbers of deaths and child population are small) more than in others (where numbers of deaths and 'base' populations are larger).

We also mentioned, in the text, simpler 'method of moments' estimates of the prior mean, γ, and variance, ϕ. These estimates can be calculated directly in INFO-MAP. The 'method of moments' estimate $\hat{\gamma}$ corresponds to a worksheet item:

```
work[1]=sum({1})/sum({2}*9/1000)
```

This should reproduce the value quoted in the text of $\hat{\gamma} = 2.633$. The 'method of moments' estimate $\hat{\phi}$ is given by a worksheet item:

```
work[2]=sum(({2}*9/1000)*({3}-work[1])**2)/sum({2}*9/1000)
        -work[1]/mean({2}*9/1000)
```

which should reproduce the value quoted in the text of $\hat{\phi} = 0.728$.

Based on these you should now be able to calculate empirical Bayes estimates of death rates in each area, by using the same formula as that discussed earlier, but replacing the maximum likelihood estimates with the values of `work[1]` and `work[2]`. Do this and compare your approximate empirical Bayes estimates with those based on the maximum likelihood estimates. Is there any significant difference between them? (You may like to use `Modify` from the `Data` menu and inspect the data values in tabular form to see differences in detail.)

In the text we also mention that the approximate 'method of moments' estimates can be adapted to allow 'neighbourhood' empirical Bayes estimation, by calculating their values within the 'neighbourhood' of each area instead of globally over all areas. If we define a 'neighbourhood' to be zones sharing a common boundary, then such calculations may also be performed in INFO-MAP, by using the `adjac()` function. First, we calculate three new data items as: the total number of deaths in the 'neighbourhood' of each area; the corresponding under-five population (1000s) in that neighbourhood; and the number of 'neighbouring' areas involved:

```
i=1,count():ndeaths[i]=sum(if(adjac([i]),{1},0))

i=1,count():npop[i]=sum(if(adjac([i]),{2}*9/1000,0))

i=1,count():n[i]=sum(if(adjac([i]),1,0))
```

Each of these calculations will take some time to compute, depending on the speed of your PC. Suppose that these three new variables are numbers 8, 9 and 10, respectively, in the file, then 'neighbourhood' estimates of $\hat{\gamma}_i$ are simply calculated as:

```
gamma={8}/{9}
```

We assume this new variable ends up as variable number 11. 'Neighbourhood' estimates of $\hat{\phi}_i$ are then more awkward, but the following formula should do the trick:

```
i=1,count():phi[i]=sum(if(adjac([i]),({2}*9/1000)*
                    ({3}-{11})**2,0))/{9}-{11}/({9}/{10})
```

where variable 3 is the raw death rate in each area. We assume this new variable is variable 12. The convention mentioned in the text is to take $\hat{\phi}_i = 0$ if its 'method of moments' estimator is negative, so we need to ensure this by a further adjustment:

```
{12}=if({12}<0,0,{12})
```

Finally we are in a position to calculate the empirical Bayes estimates, $\hat{\theta}_i$, based upon these 'neighbourhood' estimates of prior mean, $\hat{\gamma}_i$ and variance, $\hat{\phi}_i$. The required formula is:

```
nebrates={11}+({3}-{11})*({12}/({12}+({11}/({2}*9/1000))))
```

We have included these results in the file as the variable 'Neigh. Bayes rates per 1000', so that you can check that you have got all these calculations correct (or we have got ours wrong!).

Map the 'neighbourhood' Bayes estimates and compare them with those which use shrinkage towards a global mean and variance. Notice how there is a greater tendency for the 'neighbourhood' rates to merge areas into homogeneous spatial blocks. What other differences do you notice between the neighbourhood and global Bayes estimates? You can also compare the 'neighbourhood' Bayes estimates with the unadjusted 'neighbourhood' estimates $\hat{\gamma}_i$ which represent a simple spatial smoothing of the raw rates.

Exercise 8.3

We consider here applying the 'empirical logistic regression' method discussed in the text, to the 'US Presidential election (1992)' data set. In Chapter 7 an analysis of this data using an ordinary regression model applied to a logarithmic transformation of the proportion voting for Clinton, suggested three variables with a particularly significant relationship to the vote; these are: 'unemployment', 'income' and 'high school education'. We see here whether those results hold when we consider a more appropriate model for the proportions.

Open this data set and first calculate a logistic transformation of the proportions voting for Clinton in each of the states, using

```
logits=log({1}/(100-{1}))
```

We assume this new variable is variable 11 in the file. Calculate the residuals of an ordinary regression of the logits on the 'unemployment', 'income' and 'high school education' variables as follows:

```
reslogits={11}-regr({11},{5},{6},{7})
```

Use `Regression parameters` from the `Analysis` menu and note the estimated coefficients, and their standard errors. Are all three variables significant in the regression? Does their significance differ in any way from that determined in Chapter 7 using logarithms of the vote and ordinary regression? Map the residuals and explore their spatial dependence using a correlogram. What are your conclusions?

Now attempt to improve the model by accounting for unequal variances in the observed proportions, by using the weighted regression technique referred to in the text as 'empirical logistic regression'. Calculate a suitable weighting variable as

```
wts={2}*({1}/100)*(1-{1}/100)
```

Note that variable number 2 is an estimate of the voting population in each state. Assuming this new variable is number 13 in the file, calculate the residuals of a weighted regression of the logits on the 'unemployment', 'income' and 'high school education' variables as follows:

```
reswtlogits={11}-wreg({11},{13},{5},{6},{7})
```

Use `Regression parameters` from the `Analysis` menu and note the estimated coefficients, and their standard errors. How have these changed from the earlier values, in particular that for 'unemployment'? Map the new residuals and check their spatial dependence using a correlogram.

What are your final conclusions concerning variables that have a significant effect on the distribution of the vote for Clinton?

Exercise 8.4

Map, in turn, the scores on successive principal components for the Chinese provinces and interpret the maps in the light of the component loadings matrix given in Table 8.1. Plot a graph of the scores on component 1 against those on component 2 and attempt to identify the 'locations' of particular provinces in 'socio-economic' space. You can do this by examining the scores using the `Profile locations` option in the `Analysis` menu.

Exercise 8.5

Attempt a multivariate classification of areas for the Chinese provinces using two or more of the principal component scores, and the `kmean()` function in INFO-MAP. This is done as follows:

```
clus1=kmean(4,1,,{19},{20},{21},{22})
```

where the first parameter is the number of classes you wish to obtain, the second specifies whether you do or do not (0 or 1) wish to standardise the data before classifying, and the third represents an initial set of clusters (specify ,, as above, if a default is to be used). Note we suggest here that you do not standardise the data if you are using principal components, since they have themselves already been derived from standardised variables. These parameters are then followed by the list of variables to be used in the classification.

Set the 'discrete' option in `Scaling` from the `Map` menu and then `Choropleth map` the classes you have derived. Experiment with using different numbers classes in the `kmean()` formula and compare your solutions.

Compare your solution with that obtained from single linkage clustering (as discussed in Chapter 6). For example:

```
clus2=slink(4,1,{19},{20},{21},{22})
```

where parameters are as before, save for the fact that there is no argument for an initial classification.

Overall, what different sorts of classifications of China do you obtain? To what extent are homogeneous, contiguous, socio-economic regions created? You may find it of value to consult John Cole's original paper, cited in the section on further reading.

Exercise 8.6

Produce component score maps for the Barnet census data; scores on the three components discussed in the text are given in the data file 'Principal component scores'. Use either the first two, or all three, components to create your own multivariate classification of census data in this London borough by using *k*-means clustering, experimenting with a variety of numbers of clusters and mapping the solutions. Refer to the loadings of each of the principal components on the original variables, given in Table 8.2. How easy is it to recognise distinctive social areas in the borough?

Continuing with the Barnet data, repeat the clustering exercise but this time working from the raw data rather than the principal components. Load the data file 'Census counts'. Use a carefully selected subset of the original variables and convert them to appropriate percentages. Do not attempt to generate too many new variables in this file otherwise you will run out of data space. Overwrite original variables with percentages rather than create a new variable and use `Modify` from the `Data` menu to delete unwanted variables from the data file. Then attempt to cluster areas on the basis of your selected percentages using '*k*-means'. Think about whether you need to standardise them or not. How do your results compare with those obtained using the earlier clustering on the first three principal components of all variables?

Consider ways in which you might test to see how homogeneous the classes are in one of your clusterings and what interpretation they have. For example, plot a variable such as percentage car ownership, or percentage migrants, against the cluster variable concerned. Is there any obvious discrimination between the classes?

Exercise 8.7

Open the 'LANDSAT TM data for the High Peak' and load the data file entitled 'Band 1 filters'. Choropleth map the band 1 data and then map the median polish filter of this data ('low frequency' component). Compare the 'smoothed' image with the original. The 'clipboard' may be useful here. Derive and map the residuals from the median polish ('high-frequency component'). Do you consider that the median polish technique is useful here and if not then why not?

INFO-MAP is not designed for image processing and has no specific functions related to this area. However it is instructive to see how some simple filters may be constructed using INFO-MAP even if the process is rather 'clumsy' and computationally inefficient. The order of the High Peak LANDSAT data is arranged on the (30×30) grid starting at the upper left hand corner and running down the columns starting at the top of each column. Thus, in INFO-MAP `{1}[i-1]` refers to the band 1 value in the grid cell above observation *i*, whilst `{1}[i+1]` is that below it, `{1}[i+30]` is that to the right and `{1}[i-30]` is that to the left and so on; except of course at the edges of the grid. This allows us to construct a simple filter in INFO-MAP corresponding to the (3×3) moving average 'low-pass' filter mentioned in the text. For the band 1 data a suitable formula would be:

```
i=32,869:movav[i]=({1}[i]+{1}[i+1]+{1}[i-1]+{1}[i+30]+
  {1}[i-30]+ {1}[i+29]+{1}[i-29]+{1}[i+31]+{1}[i-31])/9
```

Assuming this is variable 4, we should clean up the remaining 'edge effects' in the top and bottom row of the grid by:

```
{4}=if(round(north())=1 or round(north())=30,miss(),{4})
```

Map this 'low-pass' filter. How does it compare with the median polish? Derive and map the residual 'high frequency component'.

We can construct an image sharpening filter as described in the text, although with the very small area covered by this image one would not necessarily expect striking boundary effects. A suitable formula to calculate the original image minus its Laplacian for the band 1 data would be:

```
i=31,870:delsq[i]=5*{1}[i]-{1}[i+1]-{1}[i-1]-
                              {1}[i+30]-{1}[i-30]
```

Assuming this is variable 5, we then clean up the 'edge effects' in the top and bottom row as before:

```
{5}=if(round(north())=1 or round(north())=30,miss(),{5})
```

Try calculating this variable and then mapping it. How successful is it in sharpening edges or boundaries in the original image? You might be able to improve the result by subtracting some multiple of the Laplacian from the original image; experiment with this idea.

You might try repeating similar filtering for the band 2 variable. If you do, then do not attempt to generate too many new variables in this file otherwise you will run out of data space. Overwrite old variables with new calculations rather than create new variables and use `Modify` from the `Data` menu to delete unwanted variables from the data file.

Exercise 8.8

Open the 'LANDSAT TM data for the High Peak' and load the data file of the same name. Create scores on the first two principal components for the High Peak LANDSAT data. For example, to calculate scores on the first component, using all seven bands, do the following:

```
comp1=prco(1,0,{1},{2},{3},{4},{5},{6},{7})
```

where the first parameter is the number of the component whose scores you wish to obtain, the second specifies whether or not (0 or 1) to standardise the data, and the remaining parameters are the variables to be analysed. The loadings may be viewed from within the `Analyse` menu, specifying `PCA diagnostics`. Repeat the procedure for the second component. Map the results for each component in turn, and plot scores of one against the other. Try to interpret the results in the light of the loadings matrix provided in Table 8.3 in the text.

Experiment with other components, but overwrite old variables with new calculations rather than create new variables, otherwise you will run out of data space in this file.

Part E

The Analysis of Spatial Interaction Data

9

Methods for spatial interaction data

Here we consider the analysis of spatial interaction data—numbers of people, goods or services 'flowing' between 'origins' and 'destinations'. In general, our objective is to model the pattern of flows in terms of the geographical accessibility of 'destinations' relative to 'origins', and in terms of the 'demand' at 'origins' and the 'attractiveness' of 'destinations'. Such analyses are relevant in studies of health services, shopping behaviour, migration, transport planning, and in many similar areas. When flows are restricted to travel along a network, forms of mathematical optimisation and network analysis common in Operations Research (OR) also have application here, and we discuss these briefly. We also touch on location problems, since there is a reciprocal relationship between the optimal location of new public and private facilities and the understanding of observed flows to existing facilities.

9.1 Introduction

In this final chapter we turn attention to the analysis of data which, rather than being located at points or in areas, are related instead to pairs of points, or pairs of areas. Such data are usually referred to as *spatial interaction data* and consist of *flows* of some description from a set of *origins* to a set of *destinations*.

A typical example arises in a health context, where the origins may be areas of residence, which give rise to particular numbers of patients requiring treatment; these patients are then referred (or 'flow') to a set of destinations such as hospitals or clinics. Another example comes from retailing, where we again have a set of origin zones, generating demand for consumable goods; these demands are satisfied by retail centres and shoppers travel from origins to retail destinations. In other cases our interest might be in freight flows, population migration, or journeys-to-work. In all these cases something (people or goods) moves from one set of places to another. But we might also consider examples where less tangible things (for example 'information' or ideas) 'flow' across space.

The primary objective in the analysis of interaction data is to understand or model the pattern of flows. If we can model such data we might then be in a position to use such a model for planning purposes. For example, if we can adequately capture the way in which patients 'interact with' hospitals we can perhaps experiment with modifications to the health care delivery system; what is likely to happen to the pattern of flow if we close a particular hospital or if we add a new service to a hospital? If we can achieve a good description of the current configuration of flows from consumer origins to retail destinations then we can speculate, with some confidence perhaps, on what might happen if a new superstore were to be opened in one of the zones, or if a new residential estate (generating further retail demand) were to be built. Both examples emphasise that our interest in spatial interaction models is motivated by the need both to understand and explain flows, and also to aid planning. As our examples make clear, the modelling and planning can take place both in a private and a public sector context.

We need to emphasise that it is not the primary concern here to understand flows at the individual level. Rather, we want to model aggregate interaction patterns. We acknowledge that there will be a myriad of reasons why individuals might choose to shop in one set of locations rather than another; similarly, there will be reasons (and health service constraints) why particular individuals visit, or are referred to, particular health centres. Our interest here lies in examining the spatial behaviour of collections of individuals, rather than individuals themselves.

How, then, do we explain such aggregate spatial interaction? Typically, we can recognise three important sets of factors:

- The spatial separation of origins from destinations
- Various characteristics which determine the volume of flow from each origin
- Various characteristics of the destinations which relate to their 'attractiveness'.

Let us briefly consider each of these sets of factors. The first relate to the way in which spatial separation or distance constrains, or impedes, movement. In general we will refer to this as the 'distance' between an origin i and a destination j and denote it as d_{ij}. At relatively large scales of geographical inquiry (such as the regional scale) this might simply be the straight-line (Euclidean) distance separating an origin from a destination. In other cases (perhaps within an urban area) it might be the travel time, cost of travel, perceived journey time, or any other sensible measure of the separation of origins and destinations. It is sometimes the case that interaction modelling is carried out in situations where origins and destinations are *nodes* on a transport network and flows are restricted to the *arcs* of that network. A sensible measure of distance from an origin node to a destination node would then be the shortest distance between them, whether measured as arc length or travel time. In order to compute such shortest paths we may need to draw upon some of the forms of network analysis commonly used in Operations Research (OR). We shall have something more to say about these methods in the final section of this chapter. Overall, what we require is an appropriate measure of 'distance' which we can then use to study the extent to which spatial separation impedes movement. We shall usually expect to observe an inverse relationship between interaction and distance, known as a 'distance decay' or 'distance deterrence' relationship.

The second set of factors concerns the ability of the origin zones to generate or produce a flow. This might be measured as a single variable. For example, in a retailing context we might have data on the known, or estimated, expenditure of residents on retail goods, while in a health context we might use some measure of population size, suitably modified to allow for the fact that (depending on the particular health care application) some types of people will demand the service more than others. The important point to note is that before we begin to model the flows we may need to model 'trip generation'. We shall expect to observe a positive relationship between the volume of flow generated from an origin and some suitable measure of the size of the origin.

A similar point may be made of the third component in the spatial interaction problem, the representation of the 'attractiveness' of destinations. We could measure this by variables such as the volume of floor space in the retailing context, or number of hospital beds in the health care context. Again, we shall expect to observe a positive relationship between volume of flow to the destination and the size of that destination. But there may be other factors related to the attractiveness of destinations. For example, in modelling the migration of people among Canadian provinces, or French Départements, we could measure the 'attractiveness' of destinations in terms of factors such as job opportunities (or unemployment rates), crime rates, climatic characteristics, and so on. Precisely what factors should be incorporated into such a model again depends on the problem; clearly, the unemployment rate is unlikely to be a determinant of retirement migration, where climate and the social landscape (involving such things as spatially varying crime rates) are likely to assume more significance.

We may not always have a matrix of flows between all possible origins and all possible destinations. In some cases we shall be interested in modelling flows from a subset of origins to a subset of destinations. For example, we might want to model the flows to one particular hospital (as in one of our case studies) or the trips made by holidaymakers to one or two particular resorts. Another example would be where we wanted to investigate the catchment area of a college or university; only one destination is the focus of interest, together with the propensity of different areas to generate students attending that institution. Both of these are examples of 'destination specific' problems, requiring destination specific models. Corresponding 'origin specific' models are available for related problems, where there may only be a subset of origins that are of interest. We return later to such models.

The basic structure of the spatial interaction problem, then, is to express the volume of flow in terms of origin, destination, and spatial separation factors. Mathematically, the situation we are considering in this chapter is one of a series of observations y_{ij} $(i = 1, \ldots, m; \; j = 1, \ldots, n)$, on random variables Y_{ij}, each of which corresponds to a movement of people (or cars, goods, telephone calls, and so on) between spatial locations i and j, where these locations consist of m origins and n destinations. Using the notation of our previous chapters, such origins and destinations may be point locations $\mathbf{s}_i$ or $\mathbf{s}_j$, or alternatively they might be zones or areas $\mathcal{A}_i$ or $\mathcal{A}_j$. Of course, they may also be a mixture of these; for example, if we are interested in the use of recreation facilities such as swimming pools, the origins are likely to be residential areas, while the destinations will be the pools, which are fixed point locations.

In general, we will be interested in models of the form:

$$Y_{ij} = \mu_{ij} + \epsilon_{ij}$$

where, as usual, $E(Y_{ij}) = \mu_{ij}$ and ϵ_{ij} is an error about this mean. We will wish to develop suitable models for μ_{ij} involving parameters which reflect characteristics of the origins i (relating to the propensity to flow from each origin); characteristics of the destinations j (relating to their attractiveness); and the deterrent effect of the 'distance' d_{ij} between origin i and destination j.

The structure of the remainder of this chapter follows the usual lines established in earlier parts of the book. We begin by discussing a range of case studies or examples of data that lend themselves to the kinds of methods we will describe. Next, we consider the visualisation and exploration of such spatial interaction data. We then discuss more formal statistical models. Finally, we consider further and related methods which are of use in the study of spatial interaction.

9.2 Case studies

Throughout this chapter, we shall use a selection of data sets involving spatial interaction to illustrate various forms of analyses. We have provided copies of these data sets on disk. They include:

- Business trips made by air within Sweden
- Migration data for Dutch provinces
- Patients treated from different districts at a large London hospital
- The relative 'attractiveness' of different shopping centres as branch sites for a large UK financial institution.

We begin by describing these data sets in more detail; as usual, references to the sources of the data are given at the end of this chapter.

The first case study concerns journeys by air made by Swedish business travellers in 1966. Data are available on flows among 19 Swedish regions, in each of which an airport is located (Figure 9.1). A full interaction matrix is available, showing the volume of traffic from one region to another. These data represent one way of beginning to understand the pattern of functional linkages that bind different parts of the country together. We do not have details of what such business trips entailed; whether, for example, they represent trips between members of the same, or different, organisations. But, clearly, no modern organisation can function without a flow of information, internally or externally. Whether or not the flow of information is constrained by geographical distance is, of course, a moot point. With the rapid growth in new forms of telecommunication, and the 'convergence' of places in travel time (due to technological improvements in travel) we can expect to see geographical distance shrink in importance as a factor explaining business contacts. However, such contacts in Sweden thirty years ago would indeed have been affected by distance, as well as the spatial arrangement of manufacturing and tertiary activity in the country (origin and destination factors). We should like to be able to understand, and model, the configuration of flows among these Swedish airports. In this particular data set there is a possibility that we might see the opposite relationship between flow and

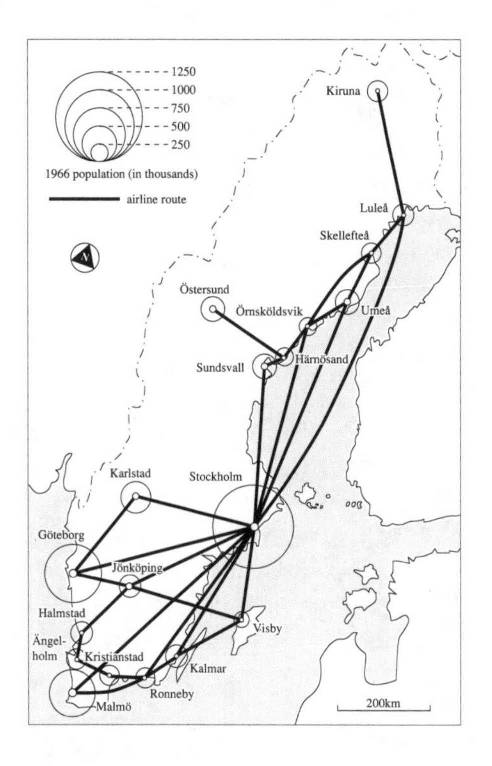

Fig. 9.1 Swedish airports in 1966

distance than that we would normally expect in studies of spatial interaction, since the data relate to air travel only and longer distances may be associated with larger volumes of business air travel. Alternative forms of travel may be used for shorter distances.

The next set of data we include relate to migration among the 11 provinces of the Netherlands in 1974 (Figure 9.2). The data set also includes information on the population size of these provinces in 1974. Each item in the data set comprises the volume of flow *from* a particular province *to* the set of all provinces. This includes internal migration (within the province itself). The data thus comprise a full spatial interaction matrix.

In seeking to understand population change we obviously need information on birth and death rates, and on migration. In many developed countries it is migration that assumes particular significance when trying to understand how population change varies spatially. Since 1945 the influence of natural increase (births minus deaths) on population change has declined throughout the Netherlands. As a result, the relative contribution of internal net migration has grown. In general, what has happened in the Netherlands has been a movement of population from the highly urbanised west of the country to other areas; this 'counter-urbanisation' is typical of many western European countries.

Fig. 9.2 Dutch provinces

Economic factors have declined in importance, while social factors (for example, the provision of good housing, and the quality of the environment) have assumed more significance. Thus we see some important flows from provinces such as Zuid-Holland (in which the major port complex of Rotterdam is located) to Noord-Brabant further south, along with movement west from Utrecht to Gelderland. Of course, these provinces are quite large, and as so often in spatial analysis we need to be aware of scale effects; our data allow us to say nothing about quite local, short-distance moves within provinces.

These data comprise total flows and are not disaggregated by age. In other contexts we might want to focus on particular classes of mover. For example, one important category of migrants comprises retired people. Distance may well be one significant factor, but factors relating to the social and climate characteristics of the possible destinations will assume major significance. As we have said already, while we are interested here in aggregate flows there will be a host of individual factors, such as proximity to (or maybe distance from!) younger relatives that are part explanations of the individual's reasons for a move. We simply want to make the point that the modelling of migration is a substantial topic in its own right and that there are interesting sub-groups of movers to be studied, within the general population.

Our third set of data relates to patients treated at a large London hospital. For each of a set of 14 specialities we have information on the flows of patients from 30 Health Authorities to the hospital, along with a crude measure of total population in each Authority. The data do not comprise a complete spatial interaction matrix of course; rather, they relate to a problem involving a single destination. Our interest centres on examining the way the effect of distance varies among the set of specialities. Does the drawing power, or catchment, of the hospital vary with type of speciality? Are patients with certain types of medical problem being referred to the hospital from much further away than for other medical specialities? Of course, we cannot hope to shed too much light on this problem, mainly because we lack information on other destinations. Under the current National Health Service in Britain, most hospitals (known as 'providers') now buy health care from Health Authorities and General Practitioners (the 'purchasers'). In other words, hospitals must compete within an internal NHS market to provide high quality care at minimum cost. Because of this competition we cannot hope to successfully explain movement to the hospital under consideration without knowing something about competing destinations. We shall return later to the question of how such competition might be incorporated into the modelling of spatial interaction.

The application of spatial interaction analysis to health care is an important one, since it raises issues concerned with the provision of an equitable health care delivery system. What services should be provided, and where? Some medical specialities are clearly very expensive. For example, it is unrealistic to expect every district hospital to offer treatment for cancer; this is more a regional than a district level responsibility. But where should such a service be located? Questions of equity then arise, since some patients will have to travel substantial distances, incurring costs of overcoming distance that are not only financial but also psychological. Such issues are of major concern in the developing world, where resources are limited and the need for a system of health care delivery which is both efficient and equitable is of paramount importance. We shall return later to such locational issues.

Our final case study involves data that result from the kind of models described in this chapter, rather than the interaction matrix of flows from which such models may be developed. It concerns the relative 'attractiveness' of different major shopping centres across the country as branch sites for a large UK financial institution. There are 298 shopping centres in our example, in each of which the organisation currently has a branch. There may be more than one branch in a particular centre. A competitiveness measure is included for the shopping centres which is the percentage of the number of branches of the company, in that centre, to the total number of financial organisations in that centre offering related services. The shopping centres are not the entire set of 600 such centres within which the organisation is represented, but simply the 298 most 'attractive' as determined by a measure specific to the organisation's business and developed from the kind of spatial interaction models we will be describing later in this chapter. We include the value of this 'attractiveness' measure for each of the centres. In addition we include a generally available geodemographic score for each of the centres that represents the potential drawing power of the shopping centre for general retail business rather than banking in particular.

As background we also include the volume of accounts held with the institution in question and that estimated to be held with all comparable institutions (including the institution in question) in postcode areas covering Britain. There are 118 such areas; the original study was carried out at the next lower hierarchical level of postal subdivision, and involved some 8500 zones. The figures we include were aggregated from that level.

We shall give further details later of how these data were derived and used. Suffice it to say here that the organisation was interested in 'rationalising' their branch distribution and using such results to decide to develop a new site or to close a rather unattractive one. The example is of interest in a number of general respects. First, it illustrates the fact that spatial interaction models arise in the commercial world. They are not simply 'toys' that geographers or applied statisticians like to play with. They are actually rather useful in a whole variety of retailing contexts. It is thus no surprise to find all major retailers, both in Britain and abroad, with specialist departments that undertake research into the flows of customers to retail centres, mainly with a view to predicting the likely effects of opening or closing branches. Regardless of whether these retailers are the major supermarkets, the leading motor manufacturers, or large financial organisations such as the one we are considering, they all make use of the sorts of models we shall review later.

A second reason why this particular example is of interest is that it illustrates that in some applications many of the flows we observe in practice will be zero!

Traditionally, analysis of spatial interaction data was based on rather small numbers of origins and destinations. In the larger study from which the results we include were derived, 'flows' of accounts were modelled between 8500 areas of residence and over 600 branches, throughout Britain. We would obviously expect vast numbers of the interactions between origins and destinations to be zero, simply because we would not expect current accounts to 'flow' over distances of, say, more than about 50 kilometres. Account holders in Aberdeen, in north-east Scotland, are unlikely to use branches in London! Such flows we might deem as 'infeasible' and the fact that they are observed not to occur does not contribute any useful information to our understanding or modelling of those that do. On the other hand, there will be some observed zero flows that are of interest to the organisation; for instance, where there is business arising in one area that does not seem to be finding its way to reasonably close centres. These 'feasible' zero flows do contribute to the understanding of the pattern of existing flows; an observation of zero is still an observation. They need to be identified and included in the data upon which models are based.

These are but a few case studies. They are sufficient, however, to allow us to reiterate that spatial interaction problems arise in a variety of contexts, both social and economic, and within both the commercial and public facility arenas.

9.3 Visualising and exploring spatial interaction data

Given an $(m \times n)$ matrix of observed flows, y_{ij}, how might we represent this graphically? Clearly, we run the risk of producing rather cluttered maps if we endeavour to display all $m \times n$ flows! Where interest centres on origin specific problems (or, conversely, destination specific situations) it is quite straightforward to map the flows using line symbols, where the width of the line is proportional to the volume of flow, using a sensible unit width. This is similar in spirit to the mapping of point or area data using proportional symbols that we encountered in earlier chapters. Arrowheads may be added to show the direction of flow. The American cartographer Waldo Tobler has created computer software for displaying such data (Figure 9.3). Note that the example shows *net* flows (here relating to migration among the coterminous states in the US) rather than the entire interaction matrix. Display is clearly more problematic when dealing with an entire interaction matrix, though Tobler's software also includes ways of doing this.

Before we examine more formal models for spatial interaction we should also look briefly at some exploratory techniques for representing such data. Some of these involve drawing simple scatter diagrams, others involve some exploratory statistical techniques linked to visual representations.

One simple way of exploring the data is to plot the volume of interaction from (or to) a particular place against the size of the origin (or destination),

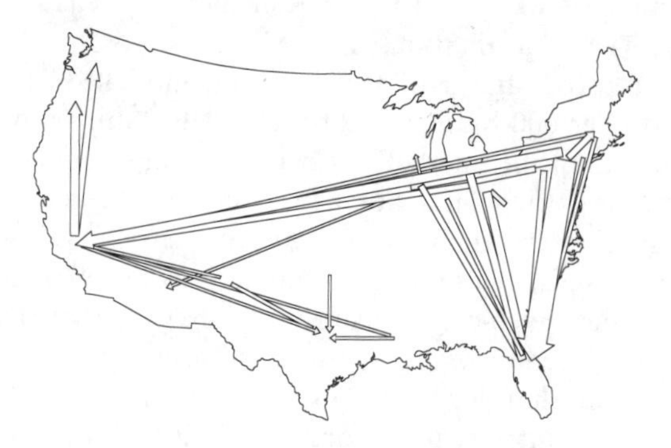

Fig. 9.3 Computer mapping of flow data

and against the chosen measure of spatial separation. As we have already seen, we can expect to observe a positive relationship between the level of interaction and 'size' of origin or destination. Similarly, we can anticipate a negative relationship between interaction and distance. A simple approach is to plot the proportions of flow, $y_{ij}/\sum_j y_{ij}$, out of a particular origin i to different destinations j, against d_{ij}. We could do the same with the proportions, $y_{ij}/\sum_i y_{ij}$, into a destination j from different origins i. Alternatively, we can correct for size effects by dividing y_{ij} by independent measures of origin or destination 'size'. Unless we do this, plots of interaction against distance, without simultaneously accounting for 'size' effects, are unlikely to reveal the true relationship because of the effect on interaction of large centres some distance from the origin or destination.

For example, with migration in the Dutch provinces, an obvious candidate for such a 'size' effect is the population of a destination. So we might try plotting y_{ij}/P_j against d_{ij} for selected origins i, where P_j is the population of destination j. If we do this for the number of migrants leaving Drenthe, in the north-east of Holland, for other destinations (one of which is Drenthe itself) we see that there is a strong distance decay effect as shown in Figure 9.4a. The relationship is far from linear. Taking logarithms of both variables generates a plot that is nearly linear, shown in Figure 9.4b, suggesting a power relationship with distance. An alternative would be to take the logarithm not of distance, but only of the other variable, exploring for an exponential relationship. Note that here we have taken the intra-zone distance (Drenthe to Drenthe) to be half the distance to the nearest neighbouring zone, to avoid taking the logarithm of zero. From this graph we can pick out zones whose interaction appears to deviate from the general trend. In this case, Groningen, to the north of Drenthe, seems to be rather more attractive as a destination than its distance might predict. This may genuinely be the case; on the other hand, bearing in mind that the zones are rather large, and our distance measure is crude, it may

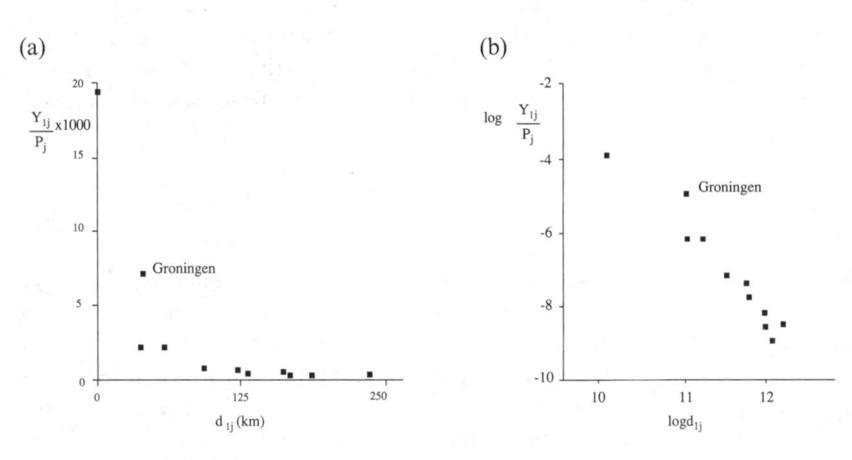

Fig. 9.4 Distance decay effect in Dutch migration

simply be an artifact of the scale and nature of our zoning system. This seems a particularly appropriate place at which to point out that results from spatial analysis in an interaction context are every bit as affected by the nature of the zoning system as they are when analysing the spatial arrangement of values in the areas themselves.

What other procedures are available to us to explore the structure of interaction matrices? The methods we describe in the remainder of this section assume that the set of origins is the same as the set of destinations; thus we are dealing with a square interaction matrix. This is not always the case, as seen in our earlier case study involving 'flows' of bank accounts. Note that when it *is* the case, the interaction matrix is not necessarily symmetric. Indeed it is exactly the question of whether y_{ij} is the same as y_{ji} that some of these methods are designed to highlight and explore.

Some exploratory tools seek to uncover evidence of hierarchical structure in interaction data. For example, if we are dealing with a flow matrix in which the origins and destinations are urban areas, clearly some of these will be 'located' at higher positions in the urban hierarchy than others. Some centres will therefore come to 'dominate' others; where interaction is strongly constrained by geographical distance this dominance may take the form of a single centre acting as a 'magnet' for the surrounding area. More generally, however, a centre will dominate others, not in a simple geographical way but in terms of functional linkages. We can use the flow data in an interaction matrix to reveal the patterns of dominance. A simple way to do this graphically is to represent each place as a point on a topological map and to link i to j with a directed arc if the largest flow from i is to j and if j is larger than i (where size can be measured by the column totals in the interaction matrix). This procedure can be made more sophisticated, but it yields some evidence of hierarchical structure in the form of a *directed graph*.

Waldo Tobler, mentioned earlier in connection with computer mapping of flow data, has also devised some highly imaginative ways of exploring interaction data. From the asymmetric matrix of interactions we may compute:

$$c_{ij} = \frac{(y_{ij} - y_{ji})}{(y_{ij} + y_{ji})}$$

where c_{ij} may be interpreted as a 'current' that 'aids' flow from i to j if $y_{ij} > y_{ji}$. This current may be displayed as a vector drawn from i to j of length $c_{ij}/2$ (note that $c_{ij} = -c_{ji}$). Repeating this for all flows to and from i yields a cluster of vectors at i, the resultant total vector (summing both directions and magnitude) being interpreted as a 'wind' at i. Repeating this for all locations gives a vector field which Tobler refers to as 'winds of influence'; the vector field may be interpolated to a regular grid if there are sufficient origins/destinations. Tobler develops the meteorological analogy further by deriving, from the vector map of winds, a 'pressure field' that is conceived of as giving rise to the winds. Interestingly, Tobler has applied his method to the Swedish data on air travel, deriving a map that shows a net flow towards Kiruna in the north of the country, and Göteborg and Malmö to the south (Figure 9.5).

An interaction matrix may be thought of as a kind of 'similarity' matrix, so we can also consider using some of the multivariate methods discussed in Chapter 6 designed to explore structure in similarity or dissimilarity matrices (one is just minus the other). The technique of multidimensional scaling described there is sometimes appropriate to analysing interaction patterns. Recall that this technique is concerned with deriving a configuration of a set of locations in a space of minimum dimensionality, in such a way that the Euclidean distance between them in this space approximates as closely as possible a given dissimilarity matrix. In this case we could base our dissimilarity matrix on minus the interaction matrix of flows. As a result, we might seek a representation of the areas such that zones between which a lot of interaction occurs are located close together in 'interaction space', while zones that interact little are distant. This idea has sometimes proved useful in the exploration of interaction data.

9.4 Modelling spatial interaction data

We move on now to consider more formal models for spatial interaction data. As mentioned in the introduction we are interested in modelling observed flows according to a statistical spatial interaction model of the general form:

$$Y_{ij} = \mu_{ij} + \epsilon_{ij}$$

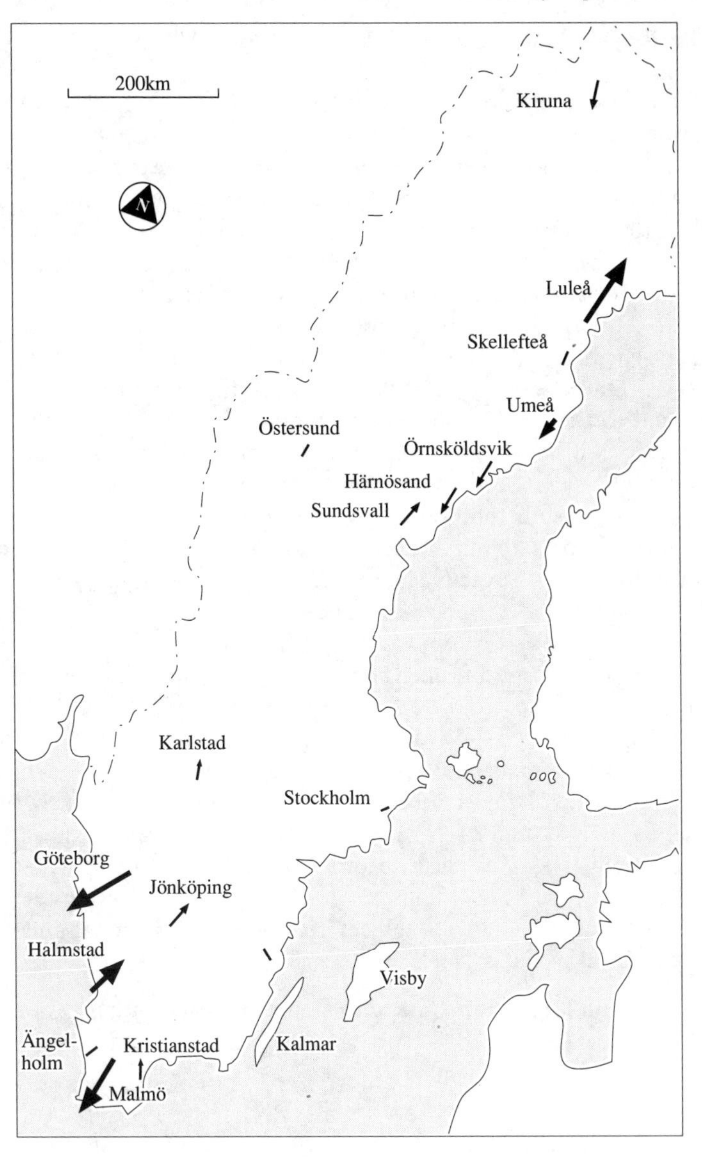

Fig. 9.5 'Winds of influence' in Swedish air travel

where $E(Y_{ij}) = \mu_{ij}$ and ϵ_{ij} is a residual error term. Note that this residual term relates to a pair of locations; we model the individual flows between i and j, not simply the total flow from i or to j.

What specific mathematical form for μ_{ij} can be adopted that will reflect, in a suitable parametric way, the various origin, destination and distance deterrence effects that we referred to earlier?

9.4.1 Basic spatial interaction or gravity models

One of the most common classes of models for μ_{ij}, which simultaneously incorporate the effect of origin and destination characteristics as well as distance, are known as *gravity models*. The term arises because of their analogy with Newton's law of gravity, where force of attraction is proportional to the product of the masses of the two bodies involved and inversely proportional to the square of the distance between them. In our case 'flow' is analogous to force and the origin and destination effects would be analogous to the masses of the two bodies. Such models were originally used simply because they seemed a sensible and convenient mathematical way to represent spatial interaction. Unfortunately, a simple model that predicts interaction as proportional to the product of the origin and destination 'size' and inversely proportional to distance does not always turn out to be adequate. This is because it may generate estimates of interaction that, when summed over rows or columns, give results that are inconsistent with the known number of flows leaving an origin or arriving at a destination and we may wish our model to 'reproduce' these totals exactly. Nevertheless, it can be demonstrated that the simple Newtonian model is one of a wider class of models with a mathematical form that can be theoretically justified on the basis of *entropy maximisation*. This provides a more solid foundation for the widespread use of such models.

The argument behind the *entropy maximisation* approach is as follows. Since it is the mean or expected flow μ_{ij} that we are attempting to model, we consider how we might theoretically expect a set of such mean flows to arise. We start with a situation involving a total number, $N = \sum \sum \mu_{ij}$, of such flows and think of this as a system of N 'average mobile individuals' who are free to arrange themselves as they wish in order to constitute any particular set of average flows, μ_{ij}, between origins $i = 1, \ldots, m$, and destinations $j = 1, \ldots, n$, but we insist that they do so subject to three overall constraints that we consider are fixed in our system:

1. A certain total number, a_i, have to flow from each origin i.e.:

$$\sum_j \mu_{ij} = a_i$$

2. A certain total number, b_j, have to flow to each destination i.e.:

$$\sum_i \mu_{ij} = b_j$$

3. The total 'cost' of travel in the whole system, c, is fixed i.e:

$$\sum_i \sum_j d_{ij}\mu_{ij} = c$$

Recall that we said earlier that we were not interested in the individual mover. We would say that we are ignorant of the details of such a move. Here, however, we are saying that we will have certain information available to us about aggregate flows on a system-wide basis; we know how many trip-makers leave each origin, how many arrive at destinations, and we have added a further piece of information, about the total system-wide 'cost' of travel. We seek a solution as to how we expect mean flows to arrange themselves within the constraints imposed by this global information. We do this by considering the question:

'out of all possible arrangements of the N individuals into sets of flows μ_{ij} that would satisfy these constraints, which is the most likely, or the most probable?'

Using basic ideas of permutations and combinations we can write down the number of different ways that the total number of 'average individuals', $N = \sum\sum \mu_{ij}$, can be assigned to particular flows in such a way that we obtain μ_{ij} of them in the flow from i to j. We get the expression:

$$\frac{\left(\sum_i \sum_j \mu_{ij}\right)!}{\prod_{i,j} \mu_{ij}!}$$

This quantity is often referred to as the *entropy function*. Finding the most likely μ_{ij} out of all possible sets of flows that would satisfy the three constraints outlined earlier, is essentially equivalent to choosing values for the μ_{ij} in such a way as to maximise this entropy function subject to those constraints.

We do not give mathematical details of the maximisation here. Some of the references we give at the end of the chapter give detailed explanations of what is involved. Briefly, the logarithm of the entropy function is maximised, using Lagrange multipliers to take account of the three constraints. The result obtained is that any choice of μ_{ij} that will maximise the entropy function must satisfy the general equation:

$$-\log(\mu_{ij}) - \lambda_i^{(o)} - \lambda_j^{(d)} - \lambda d_{ij} = 0$$

where $\lambda_i^{(o)}$ are the Lagrange multipliers used to take account of the origin constraints; $\lambda_j^{(d)}$ those to take account of the destination constraints; and λ that to take account of the total cost constraint.

Entropy maximisation therefore leads to a model for μ_{ij} of the general form:

$$\mu_{ij} = \alpha_i \beta_j e^{\gamma d_{ij}}$$

where $\alpha_i = e^{-\lambda_i^{(o)}}$, $\beta_j = e^{-\lambda_j^{(d)}}$ and $\gamma = -\lambda$ simply re-express the Lagrange multipliers in a more convenient form.

This is referred to as the general *gravity* or spatial interaction model. In this model, α_i are interpreted as a set of parameters which characterise the propensity of each origin to generate flows, β_j a set of parameters which characterise the attractiveness of each destination and γ a distance deterrence effect.

As stated previously the general form of the gravity model was used extensively to model observed flows purely on intuitive grounds before it was theoretically justified as the general solution to an entropy maximisation problem as described above. Note that if d_{ij} is taken as the logarithm of distance then the model is of the form:

$$\mu_{ij} = \alpha_i \beta_j d_{ij}^{\gamma}$$

and the model in this form resembles the Newtonian law of gravitation in the special case $\gamma = -2$.

9.4.2 Estimating the parameters of gravity models

Having derived a particular functional form for μ_{ij} which has some theoretical justification, we can now return to the problem of modelling a set of observed flows y_{ij}. Recall that we think of these as observations on random variables Y_{ij} with mean value μ_{ij} and the general model proposed earlier was:

$$Y_{ij} = \mu_{ij} + \epsilon_{ij}$$

Incorporating our general gravity model for μ_{ij}, this now becomes:

$$Y_{ij} = \alpha_i \beta_j e^{\gamma d_{ij}} + \epsilon_{ij}$$

From a statistical point of view, fitting this kind of gravity model to observed data is a question of estimating the unknown parameters α_i, β_j and γ. As a first approach it is tempting to take logarithms of this model and write it in a linear form as:

$$\log Y_{ij} = \alpha_i' + \beta_j' + \gamma d_{ij} + \epsilon_{ij}'$$

and then proceed to estimate parameters by using an ordinary least squares regression of the observations y_{ij} on d_{ij} and a set of 'dummy' (0,1) variables to represent the origin parameters, α_i, and the destination parameters, β_j. However, such an approach suffers from two drawbacks.

Firstly, there is no guarantee that flows predicted from the model will necessarily have the property that, when summed by rows or columns, they agree with either the total number observed to flow from each of the origins, or with the total number observed to flow to each of the destinations. Similarly the predicted 'total cost of travel' in the system will not necessarily correspond to that observed. Agreement of these totals would seem to be a desirable property of any 'sensible' estimates for the parameters α_i, β_j and γ, since they are the 'system-wide' fixed constraints which we used to justify the functional form for μ_{ij} used in the model. If we wished to enforce this, then the log-linear regression model given above would need to be modified to incorporate explicitly the required constraints. This would turn it into a non-linear model, and the estimation of parameters then becomes very much more difficult.

Secondly, estimating parameters by the ordinary log-linear regression model given above would only be justified statistically if we believed that flows Y_{ij} were independent and log-normally distributed about their mean value with a constant variance. Such an assumption is patently not valid since flows are discrete counts whose variance is very likely to be proportional to their mean value. Least squares assumptions ignore the true integer nature of the flows and approximate a discrete-valued process by an almost certainly misrepresentative continuous distribution. As a result, ordinary least squares regression estimates and their standard errors can be seriously distorted.

For both of the above reasons, gravity models cannot be fitted by an unconstrained log-linear regression. Maximum likelihood estimation of parameters under more realistic distributional assumptions is generally considered a more suitable approach. The most common assumption is that Y_{ij} follow independent Poisson distributions with expected values $\mu_{ij} = \alpha_i \beta_j e^{\gamma d_{ij}}$. Such assumptions are also open to question, since flows are not strictly independent and, furthermore, a Poisson distribution may not adequately reflect the degree of variation present in many real data sets, since in many applications individuals may tend to 'flow' in groups rather than as individuals. Nevertheless these assumptions are generally regarded as leading to 'reasonable' parameter estimates, at least in the first instance. The alternative would be to use some distributional assumption more able to reflect 'over dispersion'; computational problems involved in parameter estimation can then become considerable. We give some references to work on these sorts of problem at the end of this chapter.

Under the assumption that the Y_{ij} are independent Poisson random variables with means $\mu_{ij} = \alpha_i \beta_j e^{\gamma d_{ij}}$, the basic gravity model can be treated simply as a particular case of a generalised linear model with a logarithmic link and a Poisson error. We discussed such models in some detail in Chapter 8. Recall that maximum likelihood estimates of parameters are derived by means of iterative re-weighted least squares (IRLS), which is implemented as a standard procedure in many statistical packages such as GLIM.

But what of the requirement, mentioned earlier, that a desirable property of parameter estimates should be that predicted flows from the model, when summed over origins and destinations, should agree with observed totals leaving origins and arriving at destinations and that the total predicted 'cost of travel' should agree with that observed? A convenient property of the Poisson assumption is that resulting maximum likelihood parameter estimates $\hat{\alpha}_i$, $\hat{\beta}_j$ and $\hat{\gamma}$ automatically guarantee that fitted flows $\hat{y}_{ij} = \hat{\alpha}_i \hat{\beta}_j e^{\hat{\gamma} d_{ij}}$ satisfy relationships that are entirely consistent with the desirable origin, destination and total cost of travel constraints. This is not too difficult to see since the log likelihood for the general gravity model under the assumption that Y_{ij} are independent Poisson random variables with means $\mu_{ij} = \alpha_i \beta_j e^{\gamma d_{ij}}$, is effectively:

$$\sum_{i=1}^{m} \sum_{j=1}^{n} \left(y_{ij} (\log \alpha_i + \log \beta_j + \gamma d_{ij}) - \alpha_i \beta_j e^{\gamma d_{ij}} \right)$$

To maximise this we need to differentiate with respect to each of the parameters in turn and set the resulting derivatives to zero. The maximum likelihood parameter estimates $\hat{\alpha}_i$, $\hat{\beta}_j$ and $\hat{\gamma}$ must therefore satisfy any equations that result. Setting the derivatives, with respect to each α_i, to zero leads to the estimates having to satisfy:

$$\sum_{j=1}^{n}\left(y_{ij} - \hat{\alpha}_i\hat{\beta}_j e^{\hat{\gamma}d_{ij}}\right) = 0$$

In other words, if we let $\hat{y}_{ij} = \hat{\alpha}_i\hat{\beta}_j e^{\hat{\gamma}d_{ij}}$ be the predicted flow from i to j then:

$$\sum_{j=1}^{n}\hat{y}_{ij} = \sum_{j=1}^{n}y_{ij}$$

The same process using the derivatives with respect to each β_j gives:

$$\sum_{i=1}^{m}\hat{y}_{ij} = \sum_{i=1}^{m}y_{ij}$$

and the derivative with respect to γ gives:

$$\sum_{i=1}^{m}\sum_{j=1}^{n}\hat{y}_{ij}d_{ij} = \sum_{i=1}^{m}\sum_{j=1}^{n}y_{ij}d_{ij}$$

These are exactly the constraints imposed on theoretical mean flows discussed in relation to entropy maximisation earlier where:

$$\sum_{j}y_{ij} = a_i$$

$$\sum_{i}y_{ij} = b_j$$

$$\sum_{i}\sum_{j}y_{ij}d_{ij} = c$$

Thus, maximum likelihood parameter estimates of the general gravity model under the assumption of independent Poisson flows automatically ensure predicted flows that reproduce the observed total flow from each origin and to each destination as well as the observed 'total cost of travel'.

This is a convenient property of the Poisson assumption, since it avoids the need to modify standard maximum likelihood parameter estimation to incorporate explicit constraints on predicted flows. It also suggests a simple iterative algorithm for fitting Poisson gravity models, which may be used as an alternative to direct maximisation of the likelihood and will arrive at the same parameter estimates. This may be particularly useful when the number of origins or destinations is large, since in that case treating a Poisson gravity model as a particular case of a standard general linear model and using a computer package like GLIM to fit it, may present problems. Statistical packages such as GLIM may have difficulty handling very large data matrices and the general IRLS algorithm which they use to find maximum likelihood estimates is not particularly computationally efficient for this particular problem when large numbers of parameters are involved.

This alternative algorithm proceeds as shown in Table 9.1. The algorithm is repeated until convergence to give a final set of parameter estimates $\hat{\alpha}_i$, $\hat{\beta}_j$ and $\hat{\gamma}$, which are entirely equivalent to those that would be obtained from general maximum likelihood parameter estimation. However, no standard errors for the estimates are naturally produced from this algorithm. They may be obtained from the usual large sample result involving the inverse of the second derivative of the likelihood function evaluated at final parameter estimates. This involves further computational steps, details of which we do not give here.

Table 9.1 Iterative estimation of parameters for doubly constrained model

Step 0 Set $s = 0$, where s denotes 'step', and take initial estimates $\hat{\alpha}_i^{(0)} = 1(i = 1, \ldots, m)$ and $\hat{\beta}_j^{(0)} = 1(j = 1, \ldots, n)$

Step 1 Maximise the Poisson log likelihood conditional on the current values of $\hat{\alpha}_i^{(s)}$, and $\hat{\beta}_j^{(s)}$ with respect to the single parameter γ to obtain a current estimate $\hat{\gamma}^{(s)}$. Since there is a single parameter involved this is relatively easy and may be done routinely using a simple numerical optimisation method such as Newton–Raphson.

Step 2 Since it is known that the final parameter estimates must produce predicted flows $\hat{\alpha}_i\hat{\beta}_j e^{\hat{\gamma}d_{ij}}$ which satisfy the origin and destination constraints, iteratively proportionately adjust the current $\hat{\alpha}_i^{(s)}$ and $\hat{\beta}_j^{(s)}$ so that such constraints are satisfied for the current predicted flows. That is, adjust $\hat{\alpha}_i^{(s)}$ according to:

$$\hat{\alpha}_i^{(s)} = \frac{\sum_j y_{ij}}{\sum_j \hat{\beta}_j^{(s)} e^{\hat{\gamma}^{(s)} d_{ij}}}$$

and then $\hat{\beta}_j^{(s)}$ according to:

$$\hat{\beta}_j^{(s)} = \frac{\sum_i y_{ij}}{\sum_i \hat{\alpha}_i^{(s)} e^{\hat{\gamma}^{(s)} d_{ij}}}$$

and keep repeating these calculations until the values of $\hat{\alpha}_i^{(s)}$ and $\hat{\beta}_j^{(s)}$ stabilise to final values which are then designated $\hat{\alpha}_i^{(s+1)}$ and $\hat{\beta}_j^{(s+1)}$. This is known as the *balancing step.*

Step 3 Set $s = s + 1$ and return to *Step* 1.

Before proceeding further it may be useful to illustrate some of these ideas with reference to one of the case studies described earlier. The results of fitting a gravity model of the type described above to the data on migration between Dutch provinces are given in Table 9.2. Here we have used flows in thousands of individuals and a crude measure of separation based on the distance in kilometres between the 'centres' of provinces. For example, the predicted flow (000s) from Drenthe to Groningen, which has a separation of 42.9 km, is given by:

$$\hat{y}_{14} = 10.45 \times 1.065 \times e^{-0.0241 \times 42.9} = 3.96$$

The observed flow is 3.89, so the model over predicts slightly in this case. However, this model will predict total flows to Groningen from all provinces which correspond to observed total flows; similarly with total flows out of Groningen to all provinces. This will also apply for any other of the eleven provinces.

Note that values $\hat{\alpha}_i$ or $\hat{\beta}_j$ from such models are arbitrary in the sense that one could multiply each $\hat{\beta}_j$ by any constant and divide each $\hat{\alpha}_i$ by the same factor and still obtain the same predicted flows. Nevertheless, relative comparisons of $\hat{\alpha}_i$ between origins or $\hat{\beta}_j$ between destinations may be made. Their interpretation is, respectively, the relative ability of origins to generate flows, or destinations to attract flows, after the deterrence effect of distance has been allowed for.

The general form of the gravity model that we have discussed so far and used in the above example is often referred to as a *doubly constrained model*. We have allowed α_i and β_j to be separate parameters for each origin and each destination and have not attempted to 'explain' their values, simply to estimate them in such a way that predicted flows reproduce exactly the total observed

Table 9.2 Parameters of doubly constrained model for Dutch migration

	$\hat{\alpha}_i$	$\hat{\beta}_j$
Drenthe	10.45	0.658
Friesland	16.72	0.845
Gelderland	31.11	1.448
Groningen	20.27	1.065
Limburg	30.00	1.115
Noord-Brabant	33.46	1.482
Noord-Holland	51.54	1.440
Overijssel	20.01	0.944
Utrecht	20.70	0.770
Zeeland	13.53	0.558
Zuid-Holland	59.92	1.615

$\hat{\gamma} = -0.0241$ (0.0009)

flow from each origin and to each destination; hence the name 'doubly constrained'.

In a sense such a model is purely descriptive rather than offering an explanation for the observed flows, and in the next section we will go on to discuss modifications to the model which allow us to incorporate more explanation into α_i and β_j. However, such *doubly constrained* models do offer one element of 'explanation' relating to observed flows. The distance parameter γ quantifies the relative ability of distance (or, more generally, cost of travel) to deter spatial interaction, and this is useful information. As mentioned above, the model also allows relative comparison of the 'generating potential' of origins, or the 'pull' of destinations, after due allowance has been made for distance effects.

We can also use this model to predict what flows will result if certain changes are made in the system. For example, if we wished to see what would happen if twice as many people migrated from Drenthe as at present, but only half as many from Groningen, then we would simply double the α_i estimate corresponding to Drenthe, halve it for Groningen and then use the model to predict resulting flows. Similarly, if the distance measures used in the model relate to travel time or travel costs, then we could use the model to predict what would happen if the travel time between certain origins and destinations were to change. Note that when we use the model in these ways we are assuming that the same origins and destinations are always involved and we know in advance how many will travel in total from each origin and arrive at each destination. We have no way of introducing hypothetical origins or destinations into the system, because we would not know how to assign their α_i or β_j values. Neither can we ask what would happen if the population grows faster in one of the Dutch provinces that in others, because no relationship between population and the numbers migrating is built into the model.

For these reasons the usefulness of doubly constrained models is somewhat limited and we shall take up alternative models in the next section. However, their predictions are useful in some areas. For example, doubly constrained models are often used in transport planning, in modelling the journey to work. Here, we might argue that we know, or can predict by other means, the numbers leaving origin zones and the numbers employed at destinations. What we seek are good predictions of flows from i to j, given these numbers. This would allow us to assess whether the transport network was capable of dealing with existing flows. It might permit us to identify bottlenecks on the road network, for example. The model could also be used to simulate the possible journey-to-work flows if new roads were built, if one-way systems were devised, or if some routes were banned to motor vehicles. This is done by modifying particular elements in the distance, or cost, matrix. For example, such models fitted to our Swedish airline flow data might help to predict where there might be demand for new airline routes and what pricing structure should be employed.

We see that these models, while inevitably simplified, can begin to be used in real planning contexts, even to the extent of informing transport policy.

Further, while we do not consider such extensions, we note that the model can be *disaggregated* to take into account different modes of transport. The introduction of such *modal split* into such models allows us to separate out car users from cyclists, bus travellers, and so on.

9.4.3 Variations of the gravity model

We referred earlier to the general gravity model discussed in the last section as a *doubly constrained model.* To recap, we allowed it to have a separate parameter for each origin and each destination, enabling the fitted model to reproduce exactly the total observed flow from each origin and to each destination.

We repeat that such a *doubly constrained model* is purely descriptive rather than offering much explanation for the observed flows. The distance parameter does quantify the relative ability of distance (or, more generally, cost of travel) to deter spatial interaction, but the origin and destination parameters offer no real explanation of what causes individuals to flow, or why one destination is apparently more attractive than another, when its relative distance from origins is allowed for. They are only of use in making relative comparisons between origins or destinations, not in explaining differences.

It is often the case that one has further attributes measured at either origins or destinations or both, and one wishes to use this additional information to model either the number of individuals that will flow from any origin, or the relative attractiveness of destinations. We saw this in relation to some of the case studies described earlier.

One way to approach this question would be as a two stage process. A doubly constrained model is fitted to the observed flows and, as a separate exercise, the origin or destination parameter estimates, $\hat{\alpha}_i$ and $\hat{\beta}_j$, so obtained, are in turn modelled in terms of the additional covariate information present at either origins or destinations. In preliminary analysis this may well be a useful way to establish what covariates are potentially useful and what form of relationships should be used to incorporate them into the flow model.

Indeed, an examination of how well an interaction model fits individual flows between particular geographical areas should also be emphasised. Studies of selected model residuals (differences between predicted and observed flows) can be performed to examine local effects not accounted for in the model and to investigate possible explanations for the attractiveness estimates $\hat{\beta}_j$ and the source potential estimates $\hat{\alpha}_i$ in more detail. It should be appreciated that the basic spatial interaction model has a very simple structure and undoubtedly masks many complexities in the observed flows. Several characteristics of localities can act as actual or perceived barriers to travel or otherwise influence flows in ways not accounted for by the basic model.

These kinds of approaches are relevant to the case study referred to earlier, concerning the 'attractiveness' of different shopping centres as sites for branches of a large British financial institution. As mentioned, our data are only some of the results of the original modelling exercise. The original study was motivated by wishing to derive, for each major shopping centre throughout the British Isles, a measure specifically oriented to their 'attractiveness' for siting branches of the organisation offering certain kinds of banking services. The flow data comprised the volume of accounts held in each existing branch by people residing in different areas of the country, broken down into some 8500 postal zones. This large and very sparse interaction matrix was used to fit a doubly constrained gravity model to the observed 'flows' of accounts, resulting in relative 'attractiveness' estimates, $\hat{\beta}_j$, for the 600 shopping centres containing branches. The idea was then to model these 'attractiveness' estimates in terms of exogenous measures, such as size of centre, level of competition and so on. If reasonable models could be found to explain differences in 'attractiveness', then it would be possible to extrapolate meaningful estimates of the 'attractiveness', for the organisation's existing customers, of the remaining 1600 shopping centres where the organisation was currently not represented. This would enable them to rationalise their distribution of branches, perhaps closing some down in order to open others in new centres. Further, if the exogenous variables found to explain centre attractiveness were such that they could be adjusted for competitors, it would also be possible to predict the attractiveness of each of the 2200 centres to customers of competitors. Given estimates of competitor customer demand and the location of competitor outlets, it might then be possible to choose optimal sites for expansion. We do not give any details here of the results of such an exercise.

Ex 9.4

Studies such as this are clearly aided by being able to look in detail at various aspects of the model results, in particular to relate them to local geography and infrastructure. Until recently this has been somewhat difficult in practice, particularly in modelling exercises involving large interaction matrices. However, even relatively simple mapping packages provide an excellent vehicle to explore and evaluate interaction models in more detail at a local level. More powerful GIS systems have an even greater potential in this area.

These sorts of analyses of predicted values and origin and destination parameter estimates arising from a doubly constrained model, may thus result in suggesting covariates which might be included into the flow model to explain, rather than simply describe, origin and destination effects. The general form of the doubly constrained model can be naturally adapted to incorporate such covariates. An entire 'family' of spatial interaction models can be conceived, each involving different ways of modelling α_i and β_j and resulting in different constraints on predicted flows. Having looked in detail at the doubly constrained member of this family, let us now briefly consider others.

In the *origin constrained model* the destination parameters, β_j, of the doubly constrained model are replaced by some (usually simple) function of p observed covariates $\mathbf{x}_j^{(d)} = (x_{j1}^{(d)}, \ldots, x_{jp}^{(d)})$ at each of the destinations j. The idea is that the 'attractiveness' of destinations may be characterised in terms of a few measurable attributes of those locations. It is usual to include the covariates in a log-linear way so that the model becomes:

$$Y_{ij} = \alpha_i e^{v(\boldsymbol{x}_j^{(d)};\boldsymbol{\theta})+\gamma d_{ij}} + \epsilon_{ij}$$

where $v(\boldsymbol{x}_j^{(d)};\boldsymbol{\theta})$ is some function (usually linear) of the vector of destination characteristics $\boldsymbol{x}_j^{(d)}$, involving a vector of associated parameters $\boldsymbol{\theta}$. In such a model the total predicted flows to each destination are no longer constrained to equal total observed flows to that destination. However the observed total flow from any origin remains exactly reproduced by the model, hence the terminology *origin constrained* (sometimes known as *production constrained*).

One important area of application of such a model arises in retailing. To take a simple example, we might represent the attractiveness of a destination (which, in a modelling context, would be a zone containing retail outlets) in terms of the volume of, say, floor space. In this case the vector of destination characteristics and parameters would be replaced by only one such characteristic. In the origin constrained model, predicted flows to a particular destination are not constrained to always equal the observed total flow at that destination, so the model allows us to obtain predictions of total flows, or retail expenditure, at the shopping zones, given alternative configurations of floor space in these zones. The origin constrained model therefore allows us to predict total flow at destinations as a function of floor space, given a fixed pattern of demand at the origins. This may be particularly useful in predicting shopping centre usage, and in aiding the evaluation of existing, or new, sites. For example, in a competitive environment we may conceive of using such a model to allocate floor space at retail outlets belonging to our organisation or providing floor space at new outlets in such as way as to maximise revenue to us as opposed to our competitors. We shall return to these kinds of optimisation problems in the final section of this chapter.

Under the Poisson assumption for Y_{ij} the origin constrained model can again be fitted by maximum likelihood to obtain estimates $\hat{\alpha}_i$, $\hat{\boldsymbol{\theta}}$ and $\hat{\gamma}$. GLIM could be used as suggested before, or we could use a variation on the alternative iterative algorithm suggested earlier. In the latter case the algorithm is similar to that used for the doubly constrained model except that the 'balancing step' now only involves stabilisation of the $\hat{\alpha}_i$ parameter estimates, whilst the likelihood maximisation step involves a more complex likelihood function and maximisation is with respect to θ as well as γ. We do not go into details of such parameter estimation algorithms here.

The obvious alternative to an *origin constrained* model is one which is *destination constrained* (also called *attraction constrained*). This is exactly analogous to the origin constrained case except it is the α_i that are now modelled in terms of covariates, $\boldsymbol{x}_i^{(o)}$, measured at each origin. The model is thus:

$$Y_{ij} = \beta_j e^{u(\boldsymbol{x}_i^{(o)};\boldsymbol{\phi})+\gamma d_{ij}} + \epsilon_{ij}$$

where $u(\boldsymbol{x}_i^{(o)};\boldsymbol{\phi})$ is some function of the vector of origin characteristics $\boldsymbol{x}_i^{(o)}$, involving a vector of parameters $\boldsymbol{\phi}$. Total predicted flows to each destination are reproduced by the model, but not total flows from each origin. Fitting is performed in an analogous way to the origin constrained case. This model has been used in a residential location context, where there is a known distribution of employment by zone (thus the destination side is fixed) and the task is to predict the likely distribution of residences. In this case, the covariate factors at the origin might be measures of the quantity and quality of the housing stock. The model can then be used to predict the likely demand for new housing at the origin end if the pattern of job opportunities at the destination end is altered.

Finally, we have the *unconstrained model* where both α_i and β_j are modelled as functions (usually simple) of observed characteristics at both origins and destinations. The idea is that both 'demand' and 'attractiveness' can be 'explained' in terms of a few measurable attributes of origin and destination locations. The model then becomes:

$$Y_{ij} = \mathrm{e}^{u(\boldsymbol{x}_i^{(o)};\boldsymbol{\phi})+v(\boldsymbol{x}_j^{(d)};\boldsymbol{\theta})+\gamma d_{ij}} + \epsilon_{ij}$$

and is not constrained to reproduce the total observed flow either from any origin or to any destination. Again, fitting is by maximum likelihood using a Poisson assumption. Since no constraints are involved no alternative 'balancing' algorithm can be developed.

A particular form of the unconstrained model which has often been used in studies of population migration is where $\boldsymbol{x}_i^{(o)}$ is taken to be simply a single variable, the logarithm of the population, P_i, at origin i, and similarly $\boldsymbol{x}_j^{(d)}$ is taken to be the logarithm of the population, P_j, at destination j. An overall scaling parameter λ is then added to reflect the general tendency for migration in the system. Hence the simple migration model that results is:

$$Y_{ij} = \lambda P_i^{\phi} P_j^{\theta} \mathrm{e}^{\gamma d_{ij}} + \epsilon_{ij}$$

Note that in all the above variants of what we have referred to as a 'family' of doubly, singly or unconstrained spatial interaction models we may introduce more complex distance deterrence functions than $\mathrm{e}^{\gamma d_{ij}}$. We may wish to include non-linear effects by, for example, including a term in $\log d_{ij}$ or effects which are different for different groups of origins and/or destinations. We will return to this point in more detail shortly, and simply note here that any of the above formulations can be rewritten by replacing the term γd_{ij} with some function $w(\boldsymbol{d}_{ij}, \gamma)$ of a vector of distance measures $\boldsymbol{d}_{ij}$ and a vector of parameters γ. No changes are required to the basic fitting algorithms, although these will clearly become more computationally intensive.

Let us now look at applying a selection of these different forms of model to our case studies. First we consider modifying our earlier doubly constrained model for the Dutch migration data, by using an origin constrained model where we model the 'attractiveness' of provinces for migration as a function simply of their total population. That is, we take $\boldsymbol{x}_j^{(d)}$ to be a single variable, the logarithm of the total population (000s), P_j, at destination j; so our model becomes:

$$Y_{ij} = \alpha_i P_j^{\theta} e^{\gamma d_{ij}} + \epsilon_{ij}$$

The results are given in Table 9.3. Although the overall magnitude of the $\hat{\alpha}_i$ has changed, their relative values are broadly as before. The distance parameter also remains the same. The estimated parameter for measuring 'attractiveness' in terms of population is highly significant, and interestingly the overall fit of this model to the observed flows is not significantly worse than that of the previous doubly constrained model (as measured by the change in 'deviance' discussed in Chapter 8).

Table 9.3 Parameters of origin constrained model for Dutch migration

	$\hat{\alpha}_i$
Drenthe	0.404
Friesland	0.648
Gelderland	1.170
Groningen	1.105
Limburg	1.130
Noord-Brabant	1.219
Noord-Holland	1.763
Overijssel	0.725
Utrecht	0.703
Zeeland	0.470
Zuid-Holland	2.033

$\hat{\theta} = 0.4846 \ (0.0749)$
$\hat{\gamma} = -0.0241 \ (0.0009)$

The predicted flow (000s) from Drenthe to Groningen, which has a population of 532.7 (000s), is now:

$$\hat{y}_{14} = 0.404 \times 532.7^{0.485} \times e^{-0.0241 \times 42.9} = 3.02$$

an under prediction of the observed flow of 3.89. Under this model the total predicted flows from a province to all provinces will correspond to those observed, but the total predicted flows into a province from all provinces will not necessarily reproduce the observed numbers.

Finally, we may go one stage further and consider an unconstrained model for these data, where the origin effects are also modelled simply as a function of the total population of a province. That is, we take $\boldsymbol{x}_i^{(o)}$ to be a single variable, the logarithm of the total population (000s) and include an additional scaling parameter, λ, to represent the overall tendency of the population to migrate.

We are then using precisely the simple migration model described earlier. This gives the results below:

$$\hat{\lambda} = +0.0085$$
$$\hat{\phi} = +0.6965$$
$$\hat{\theta} = +0.4654$$
$$\hat{\gamma} = -0.0242$$

All four parameters are highly significant. Our Drenthe to Groningen prediction, given that the population of Drenthe is 393.7 (000s), is now:

$$\hat{y}_{14} = 0.0085 \times 393.7^{0.697} \times 532.7^{0.465} \times \mathrm{e}^{-0.0242 \times 42.9} = 3.59$$

a slightly better fit than before. Under this model neither total predicted flows out of, nor into, provinces will necessarily agree with those observed.

The overall fit of this model is not significantly worse than the singly constrained model given earlier (again as assessed through the change in deviance), or indeed from that of the original doubly constrained model. Since the original doubly constrained model is in a sense the 'best we can do' at explaining the flows (it has a separate parameter for each origin and each destination) and this is not significantly better than the unconstrained model that we have now arrived at, we can conclude that the simple unconstrained model is probably as reasonable an explanation of the migration flows as we are likely to get with these data by using gravity models. Of course, with better separation measures and more data on the characteristics of the provinces, we might well improve the level of explanation.

We may also apply an unconstrained model of a similar type to the data on London patient flows. Here there is just a single destination so that our previous model becomes

$$Y_i = \lambda P_i^{\phi} \mathrm{e}^{\gamma d_i} + \epsilon_i$$

where Y_i represents the flow to the hospital from district i, which has population P_i (000s) and is at a distance d_i from it. Here we will simply use straight line distance in kilometres as the distance separation, with the distance for Hampstead (the district where the hospital is situated) being taken as half the distance of the next nearest district. Obviously, such crude distance measures will not reflect the realistic 'cost of travel' or 'travel time' in the London area, but there will be a broad relationship.

We could fit the above model to the observed flows for any particular speciality (type of treatment). However, this is equivalent to simply rewriting it as:

$$Y_{ik} = \lambda_k P_i^{\phi_k} \mathrm{e}^{\gamma_k d_i} + \epsilon_{ik}$$

where Y_{ik} is now the flow for speciality k and we have provided different model parameters for each speciality. These parameters then characterise the 'catchment population' of this hospital's workload for different specialities k, in terms of a district's population (through $\lambda_k P_i^{\phi_k}$) and the distance of this population from the hospital (through γ_k). Once fitted we can then examine for significant differences between the parameters for the different specialities and possibly simplify the model, perhaps by forming groups of specialities with similar behaviour.

We do not report the parameter estimates in detail here; however, we can summarise the general results. Firstly, there is very little evidence that the parameter ϕ_k helps in this particular case. Its value is not significantly different from unity for any speciality. Hence the λ_k estimates obtained can be interpreted as simply 'distance corrected' workload rates per thousand population. The differences in these values are much as might have been expected, the highest values being in General Medicine, General Surgery and Gynaecology with lower values for more specialised treatments, such as Urology. Generally, the value of the distance parameter γ_k is approximately -0.45 for most specialities, which, given that distances are being measured in kilometres, corresponds to a maximum catchment radius of around 9 km, before the exponential decay function becomes insignificantly small. However there are some specialities with significantly wider catchments of up to 25 km, notably specialist areas such as Neurology, Neurosurgery, Oral Surgery and Radiotherapy; the last two having the highest workload rates among the four. On the other hand, the distance parameter for Geriatrics showed a very much smaller catchment area of no more than 3–4 km.

Notice that in all the examples we have used here we have estimated a single, 'system-wide' distance deterrence parameter for any given type of flow. We have already alluded to the possibility of considering either origin constrained or destination constrained models which involve a distance deterrence parameter specific to each origin or each destination (or to groups of them). In other words, we may wish to generalise γ to γ_i, or γ_j. Details of such models are given in the section on further reading at the end of the chapter; we simply comment here on why they might be of interest. Consider the Dutch migration data. We have already shown how to estimate an overall distance deterrence parameter. Yet with an origin specific model we could estimate such a parameter for each province. This might reveal interesting spatial variation in distance decay effects. However, a word of caution is needed here. When such origin specific models are fitted in practice it is invariably found that the distance deterrence effect is greater in peripheral regions and lower in more central areas. For example, we might obtain a $\hat{\gamma}$ of -0.01 for those leaving Utrecht, in the centre of Holland, and a value of -0.03 for those from Limburg in the south, were we to fit such a model. The interpretation of this might be that people leaving Limburg were more constrained, or deterred, by distance, than those from Utrecht. This seems implausible. What we are observing here is a misspecification of the spatial interaction problem. It has led writers such as Stewart Fotheringham to propose another class of model, those known as *competing destination* models.

In such models, we imagine that destinations 'compete' with each other for interactions, and we build into our spatial interaction model some variable to measure this. The interpretation of such competition depends upon the context. In retailing we might find that an isolated centre captures local trade because of the lack of competition; in migration studies we might imagine that people first select a region (for example, the south of Holland) and then, within that broad region, a particular destination. If, within the broad region, there are relatively few alternatives, we can expect such destinations to draw more migrants than those in regions where there are many possible destinations. As a consequence, we can build into the modelling framework a variable to capture this destination competition. One suggestion is a measure of relative accessibility of a destination, such as:

$$q_j = \sum_k \frac{w_k}{d_{kj}}$$

where q_j is the accessibility of j to all other destinations and w_k is the attractiveness of the kth competing destination.

Although we have made reference already to the measurement of spatial separation used in spatial interaction models it is worth commenting a little more fully on this before we leave the subject. Just as we can envisage a series of factors contributing to the attractiveness of a destination, or the propensity of an origin to transmit flows, so too we can envisage a variety of factors that go to make up some suitable measure of spatial separation. In the transport planning literature such a measure is known as *generalised cost*. This could embrace travel time, t_{ij}, factors that are proportional to actual distance, d_{ij}, and the costs, w_{ij}, of parking, waiting, and so on that all contribute to the cost of overcoming distance. These elements can be combined to give:

$$c_{ij} = \theta_1 t_{ij} + \theta_2 d_{ij} + \theta_3 w_{ij}$$

where c_{ij} is generalised cost. We would then simply use c_{ij} instead of d_{ij} in all our previous models. The parameter θ_2 translates distance into money (for example, the costs of travel by bus, per kilometre), while the other parameters, θ_1, θ_3, represent the valuation of travel and waiting time. We shall return later to the issue of using distances and travel times along networks as measures of spatial separation. But what the generalised cost does make clear is that c_{ij} is not necessarily the same as c_{ji}. When simple Euclidean distances are used in spatial interaction modelling (as they often are) then spatial separation is, by definition, symmetric.

Another problem in using any measure of spatial separation that arises in most practical applications concerns the definition of d_{ii} or c_{ii}: how are we to measure this where we are dealing with zones and there are intra-zonal flows, y_{ii}? In the models we have fitted to the Dutch migration data for illustrative purposes, we have simply taken the intra-zonal distance to be zero (note that this does not imply the predicted flow is zero). However, in general this may not be acceptable, since it does not reflect the fact that there may be more deterrence to intra-zonal flow in one area than in another. There are no hard and fast rules here, though some theoretical results are available for zones that are circular or rectangular. If the zone is approximately circular, one suggestion is to take two thirds of its 'radius' as an intra-zonal distance.

Before we leave this section on the modelling of spatial interaction data, we should also comment on one general feature of the methods that departs markedly from the approach which we have presented earlier for the analysis of other types of spatial data. In previous chapters, whether dealing with point patterns, spatially continuous or area data, we were careful to maintain a balance in our models between both first order and possible second order effects. Indeed, on occasions we have gone to great lengths in some models to ensure that we have adequately allowed for second order effects. That is, we have modelled a spatial covariance structure in residuals about the mean, as well as modelling the mean or first order properties of the process. In this chapter we have focused entirely on modelling first order effects in spatial interaction data. Our family of gravity models are models for μ_{ij}, the mean flow from i to j. We have not commented at all on the possibility of spatial covariance in fluctuations about this mean. In fact, the maximum likelihood methods for estimating the parameters of gravity models make the explicit assumption that fluctuations about the mean are independent; that is, no second order effects are present.

Given our comments on various kinds of spatial models throughout this book, the reader should be somewhat reluctant to accept the proposition that spatial interaction data will have no spatial covariance structure. At the very least, we should explore and analyse the estimated residuals, $\hat{\epsilon}_{ij}$, for possible spatial dependence. In other words, for each predicted flow there is an associated error term, the difference between the observed and the modelled flow. The assumption of no second order effects means that the residual ϵ_{ij} is, for example, uncorrelated with ϵ_{ik}. But it may be the case that if we are over predicting the flow from i to j we are also over predicting that from i to k. As yet, however, there has been little work done on this problem of spatially correlated errors in spatial interaction modelling. Some of the references at the end of this chapter discuss adjusting the Poisson assumption to allow for such effects as well as 'over dispersion' of flows (larger variance than expected under a Poisson assumption). Such modifications increase the difficulties of parameter estimation, and techniques such as *pseudo likelihood* are then required to address such problems.

9.5 Related methods in the analysis of spatial interaction

So far, we have considered a variety of spatial interaction models and have illustrated their uses. In this section we want to consider some further issues related to the subject of spatial interaction. These take us beyond the description and modelling of observed flows. We have alluded already to the possible use of singly constrained gravity models in determining the optimal location for new retail development. In this section we shall take that discussion further and consider in more detail the general problem of optimally locating facilities and investigating their catchment areas. Such problems are known as *location-allocation* problems.

Location-allocation modelling is very often concerned with locating facilities on a network rather than in continuous space. We have already referred to networks on a number of occasions. For example, we noted early in the chapter that spatial interaction modelling is often carried out in situations where origins and destinations are nodes on a transport network and flows are restricted to the arcs of that network. Networks provide a rich source of analytical problems in their own right, many of which are directly related to issues which arise in spatial interaction or location-allocation modelling. Some are concerned with finding a pattern of flows through a network that is in some sense optimal among all sets of flows which would satisfy some overall requirement. For example, we might wish to minimise the total cost of transporting goods between locations, given some requirement to provide defined levels of supply at particular nodes in the network. As we shall see, there turns out to be some relationship between such 'classical' *transportation problems* and the entropy maximisation formulation we have already considered in developing the general mathematical form of the gravity model. Other classes of *network problems* to which we refer are concerned with finding shortest paths, optimal routes, or maximal possible flows through a network.

Although location-allocation, transportation and other network problems are still very much of a spatial analytical nature, we shall see that they take us more into the realms of mathematical optimisation techniques and away from the area of statistical description and modelling of observed spatial data which has been the subject to which we have so far confined ourselves throughout this book. Many of the problems we will discuss, both within a location-allocation and a network context, can be formulated as requiring the optimisation of some objective function, subject to a set of constraints. In general such problems are known as *mathematical programming* problems. Where the objective function and constraints are all linear functions of the variables with respect to which optimisation is required, the problems are known as *linear programming* problems. In cases where such variables are restricted to have only integer (whole number) values, as they will be in several of the situations we will discuss, the problems are referred to as *integer programming* problems.

Mathematical programming in general, and linear and integer programming in particular, are extensively researched areas in their own right. Much of the subject of *Operations Research* (OR), is devoted to the study of such problems and very efficient algorithms have been developed for their solution. However, they rank amongst the most computationally intensive mathematical problems that exist, and exact solutions to many of the problems that we formulate in this section are not achievable with current methodology for anything other than relatively small networks. As a result, such problems are often solved using what are called *heuristics*, sets of rules that give a 'good', but not necessarily optimal, solution to the problem.

Given the substantial nature of this area, combined with the fact that it is somewhat tangential to our main theme throughout the book, we do not deal with location-allocation modelling and network optimisation methods in any detail; indeed, we would require another book to do such methods justice. We simply review and formulate some of the most basic models and point out the links between some of these and the modelling of observed spatial interaction data, considered earlier. We provide references at the end of the chapter for the reader who wishes to follow up the techniques in more detail.

9.5.1 Location-allocation problems

We suggested earlier that an origin constrained spatial interaction model could be used to make predictions, $\sum_i \hat{y}_{ij}$, of the volume of flow arriving at destination j given a pattern of demand at the origins and an attractiveness function for each of the destinations, which might typically include covariates such as retail floor space, parking facilities and so on. In a retail context such predictions allow us to derive estimates of likely sales at centre j under various different scenarios. Such 'shopping' models can then be used in a forecasting context. For example, if we add to the volume of retail floor space at j, a new set of predicted flows is then generated, along with new predictions of the distribution of retail sales and the revenue that suppliers at j can expect to achieve. Alternatively, we can model the changes to the flow of retail expenditure if the demands at the origins are modified, or if the transport network (i.e. distance or cost of travel) changes. The spatial interaction model represents the 'system' and allows us to predict how small changes to one component of the system (such as one origin demand, one change at the destination end, one modification to the transport network) will feed through to possibly have indirect effects at all locations in the system.

This understanding that we might use certain kinds of spatial interaction model to predict 'what might happen if', immediately raises the possibility of using them to help in optimising the placement of retail outlets. Suppose we wish to locate a new retail outlet in an urban area. An origin constrained model can be fitted to flows of customers using both our existing facilities and those of our major competitors. Then, given future patterns of demand at origins perhaps based on estimated expenditure on retail goods by residents in zone i, we could predict likely expenditure on retail goods generated at *any* potential location in the urban area, simply by inserting the set of d_{ij} that represent the cost of residents travelling from homes in zone i to the hypothetical destination j. The predicted total flows of expenditure at all such hypothetical locations can then be mapped as a spatially continuous revenue surface (though one that currently ignores the costs of locating at j); the peak on this surface is then the optimal location, in the sense of maximising revenue.

Many 'classical' location-allocation models, such as those based on central place theory, tend to assume that all individuals who 'flow' (e.g. shoppers) will always visit the closest centre or facility that provides the service they seek. In other words, given a particular placement of facilities, the assignment or 'allocation' of individuals to facilities is then deterministic, or entirely predictable. As a result, the *catchment area* of a facility is unambiguous and well defined. The kind of spatial interaction models we have been discussing throughout this chapter offer a much more natural, probabilistic, assignment of demand to destinations; a better way of characterising the catchment area of a facility. They recognise that while some shoppers in a zone will visit one centre, others will go elsewhere. Distance, along with other factors, *constrains*, but does not necessarily *determine*, the pattern of interaction. For these kinds of reasons a current area of research concerns how to bring together 'classical' location-allocation modelling and spatial interaction modelling, a point we will return to later.

First, however, we need to review briefly the more 'classical' models for optimal facility location. In introducing the subject we considered one possible criterion for determining optimal location of a facility such as a retail centre, that of maximising total possible flows or total possible revenue associated with the facility. There are several other criteria for location that we could consider, and location-allocation modelling in general is concerned with the optimisation of various such criteria, subject to assumptions concerning the factors that determine the 'catchment population' of any potential facility location and constraints concerning demand for the service provided at the various 'origins' involved. As mentioned above, many of the techniques focus on constrained optimisation methods for particular criteria, under deterministic assumptions about catchment populations, usually that individuals will choose the 'nearest' facility. We have already made the point that the 'lessons' of spatial interaction models suggest that the behaviour of flows is potentially more complex and probabilistic. Nevertheless, one might argue that this is essentially a question of how one defines catchment population and does not ultimately alter or detract from the overall location optimisation problem to which much of the work in this area has been devoted.

To gain some insight into location-allocation models it is useful to bear in mind the basic structure of the problem they address. In all cases there is *demand* for some good or service and this demand is allocated, or flows, to particular centres according to a pre-defined set of rules which determine 'who will flow where'. We then wish to locate centres or facilities in such a way that some criterion is optimised, subject to the given demand. Very often there is only a certain discrete set of potential sites at which to locate the facilities and the assumption is made that demand can also be taken to occur at point locations. In other words, the location problem is conceived as being one of placing facilities at certain nodes on a network. This is certainly the case in most of the traditional models, although work has also been done on locating facilities in continuous space.

Reference back to two of our case studies is perhaps instructive here in order to illustrate the problem addressed. In the bank account example, which is very similar to the general retail situation, we might want to locate a new branch that captures as much account business as it can. In the case of patient flow, where we know, or can estimate, the demand for health care, we might wish to determine an optimal location for a new hospital. The reader may view this latter application as somewhat unrealistic in the context of the developed world, where rationalisation (closure!) of hospital sites seems to be the order of the day, but nevertheless, it is of obvious importance in the developing world, where scarce resources must be targeted as carefully as possible when devising optimal location for health care facilities.

In seeking an optimal location we need an *objective function*. In the retail case this might simply be to maximise profits. But in general we can try to optimise (minimise or maximise) other objective functions. This is particularly the case in the context of locating public facilities, where we may wish to locate public libraries, health clinics, fire stations, or recreation centres to best serve the population. One widely used objective function in these contexts is to assume that individuals will always travel to their nearest facility, and then to attempt to choose the location of facilities by minimising the total distance travelled (travel cost) in the system (equivalently, we could think of this as a problem of maximising system-wide accessibility). In this case we may formulate the problem of optimally locating p facilities in a system with n possible nodes as:

$$\min\left\{\sum_{i=1}^{n}\sum_{j=1}^{n} a_i I_{ij} d_{ij}\right\}$$

subject to the constraints:

$$\sum_{j=1}^{n} I_{ij} = 1 \quad i = 1, \ldots, n$$

$$\sum_{j=1}^{n} I_{jj} = p$$

$$I_{ij} \leq I_{jj} \quad i \neq j$$

Here the given demand at i is represented by a_i and as usual, d_{ij} is some measure of the cost of reaching j from i. I_{ij} is a set of indicator variables in respect of which the minimisation is performed, where these take the value 1 if all demand from i is allocated to facility j and 0 otherwise. The second constraint ensures that exactly p facilities are located and the third that no demand is met from j if there is no facility located there. The first set of constraints ensures that a_i, the demand from i, is met in full, since it ensures that demand from each i is allocated to exactly one j. The precise nature of the given demand will depend upon the particular problem. If we are locating fire stations it might relate to the number of domestic and non-domestic properties in a zone; if we are locating public libraries it might simply be related to total population.

When the above model is used to determine the location of a single centre (the case where $p = 1$) it is essentially identical to the so-called Weber problem in industrial location. More generally, however, we shall wish to use it to locate p centres, the so-called *p-median* problem. It seeks the location of p supply centres to minimise the total cost of travel whilst satisfying demand. Solutions to this problem have been developed both in a *continuous* setting (where we assume that centres are free to locate anywhere on a plane) and a *discrete* context, where both centres of demand and the set of feasible locations are points on a transport network (the case formulated above). In general, exact solutions to the latter problem are computationally infeasible for problems of a realistic size. However, good heuristics are available to find acceptable approximate solutions.

As an example of the use of these methods, consider the problem of locating $p = 6$ health centres in the district of Salcette Taluka, in the west of Goa, India. The current health care delivery system already comprises six centres; are these optimally located, according to the objective function defined above? The solution to the p-median problem suggests that two health centres, including that in Margao, the major urban centre in Goa, are optimally located, but that four are not. As can be seen from the mapped outcome (Figure 9.6) the p-median solution suggests that the east of the study area is rather poorly served. Note that the p-median solution also provides the allocation of demand (population) to the optimally located health centres. However, the solution shown here used simple straight-line distance as a measure of spatial separation and took no account of the capacity of centres to deal with patients.

The p-median problem has been widely used in a variety of planning contexts, not simply in health care planning. There are a number of variants on the basic problem. For example, it may be adjusted to take account of varying installation costs of facilities at different nodes, or capacity constraints on the facilities that are located. It can also be extended to include constraints on the maximum distance that individuals will travel to facilities. This last extension raises an important point concerning the p-median formulation. Whilst it generates an 'efficient' solution (in the sense of minimising aggregate travel costs) it does not necessarily generate an 'equitable' solution. In particular, if we are dealing with a situation where demand is *inelastic*, that is, where clients require the service regardless of the cost of getting it—health care or fire cover, for example—we find that the p-median solution tends to ignore the needs of those located furthest away from the chosen optimal location(s).

In order to reduce the variability in the accessibility of origin zones and individuals living therein, alternative objective functions have been suggested. One possibility is to choose a set of centres which minimise the maximum distance separating any user from the nearest facility. This is called a *minimax* solution. Another solution is via what are called *covering objectives*. With these, a standard of service is defined which must be met for all origin zones in terms of a distance (or time) limit from facilities. So, a set of locations is sought such that the entire demand is 'covered', or met, within a critical distance or time. A special and commonly used example of this, the *maximal covering* problem,

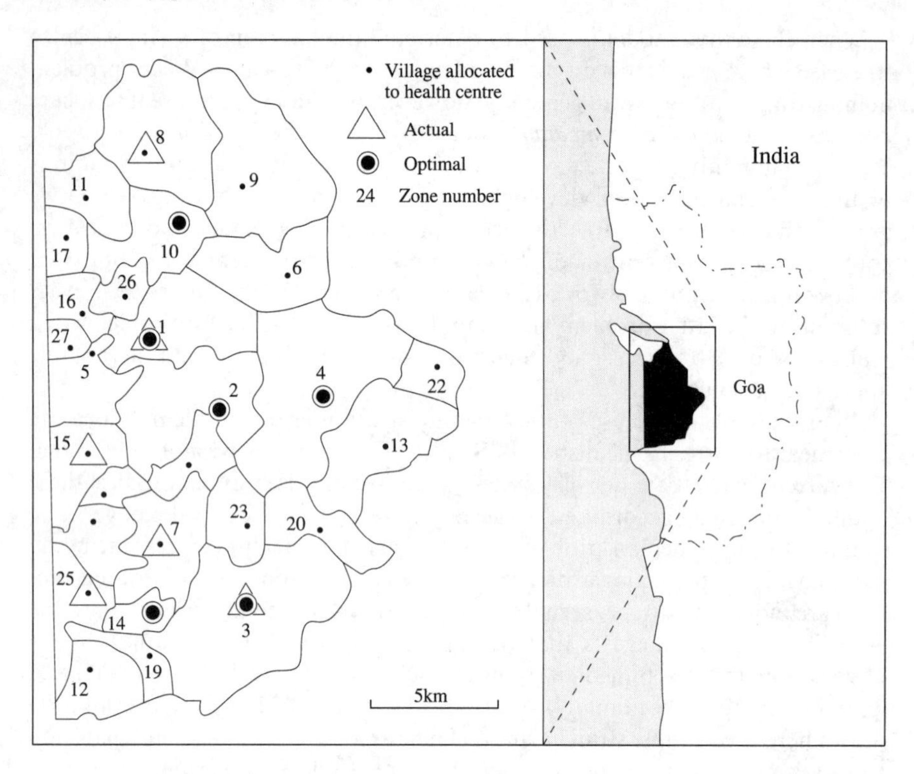

Fig. 9.6 Optimal location of health centres in West Goa. Reprinted from Hodgson, M.J. (1988) *Social Science and Medicine* **26** 153–61, with kind permission from Elsevier Science Ltd, The Boulevard, Langford Lane, Kidlington OX5 1GB, UK.

seeks to find the locations of p facilities so that the number of people within some specified critical distance or time threshold, h, is maximised. Mathematically we may formulate this problem as:

$$\min\left\{\sum_{i=1}^{n} I_i(h)a_i\right\}$$

subject to the constraints:

$$\sum_{j:d_{ij}\leq h} \left(I_j + I_i(h)\right) \geq 1 \quad i = 1, \ldots, n$$

$$\sum_{j=1}^{n} I_j = p$$

As before, the given demand at i is represented by a_i and d_{ij} is some measure of the cost of reaching j from i. However, we now have two sets of indicator variables. $I_i(h)$ is 1 if node i is not within h of any facility located (i.e. it is not 'covered'). I_j is the set of indicator variables in respect of which optimisation is carried out, taking the value 1 if a facility is located at j and 0 otherwise. The objective function is expressed in terms of minimising the amount of demand not covered. The first set of constraints determines whether demand points are covered or not whilst the second ensures that exactly p facilities are located.

The parameter h takes on an interesting role here. Where h is relatively large, the solution to the problem becomes similar to that of the minimax solution; as h shrinks, however, the solution becomes less equitable, since the optimal location is drawn towards the zone with the greatest demand; the demands of other zones are ignored. One of the attractive features of the maximal covering problem is that it can be used to explore just how much cover a certain value of p can provide. Repeated solution can therefore give insight into just how many facilities are required to provide some defined level of 'cover'. This latter problem can also be formulated directly as a modification of the basic problem if required—the so-called *location set covering* problem. There are several other variants of the basic maximal covering problem. As with the p-median problem, exact solutions to the maximal covering problem and its variants are computationally infeasible for networks of even moderate size. Heuristics are available to find acceptable approximate solutions.

How do these sorts of 'classical' mathematical optimisation problems relate to the spatial interaction models we have described earlier in this chapter? We have already touched on this earlier. The optimising criterion in the classical OR models assumes that people visit the closest centre; in other words, that the allocation of demand for any particular facility locations is deterministic. But we have already seen that spatial interaction models generate a probabilistic allocation of demand to centres of supply. Because of this, current research in location-allocation modelling attempts to link these models with spatial interaction models that capture the more complex behaviour of trip-makers. In particular, location-allocation models have been devised that both locate p centres *and* are consistent with probabilistic travel behaviour. One such model sets up an objective function which essentially involves two terms. The first represents average travel costs and is similar to the objective function in a p-median problem. The second involves a distance deterrence parameter and may be interpreted as a measure of the costs or 'disutility' of having a set of possible centres dispersed over space. Together, both these terms provide a measure of overall consumer costs. We do not give details of the explicit formulation here.

In what other ways can we extend location-allocation models, other than by trying to make them more consistent with spatial interaction models? First, we should recognise that, in some cases, we shall want to place constraints on the ability of centres to cope with demand; in other words, we might want to attach *capacity constraints* to centres. These might represent numbers of available bed spaces in hospitals, for example. The 'classical' location-allocation models, such as the p-median problem discussed earlier, generate an 'uncapacitated' solution. However, more realistically we might also wish to incorporate a capacity constraint. This is generally straightforward both for the p-median and for other models we have discussed, although clearly such additions make solutions computationally harder to obtain.

Another possible extension is to develop *hierarchical* location-allocation models. With these, we want to locate not just a single set of facilities, but facilities that occupy different levels of a hierarchy. To return to the problem of locating health centres in part of Goa, for example, one set of health clinics might offer basic nursing and primary care while at a higher level of the hierarchy we might have one or more health centres providing more staff and a fuller range of health care services. It is usual to assume that all low level functions are also catered for in the higher level clinics. Now, in the p-median problem, as we have already seen, people are assumed to make their way to the nearest centre. But this may be rather unrealistic. Might not patients prefer to travel to a larger hospital, one with more staff, a wider range of services and a facility located in a larger urban area, allowing for the combining of a visit to the health facility with other purposes? Location-allocation models have been formulated to deal with this kind of situation. They consider optimal siting of fixed numbers of facilities at each level of a hierarchy of different types. We do not give details of such models here, but simply note that most of the models that we have previously mentioned can be extended to a hierarchical context.

Before leaving location-allocation models we should draw attention to the fact that often in facility location we may wish to optimise not just a single objective function but, rather, a set of such functions. This gives rise to what is known as *multi-criteria* problem solving. Here, a decision maker wishes to determine a configuration of facilities that tries to optimise not only, say, aggregate distance travelled (the p-median problem) but simultaneously also various equity criteria. The extent to which such multi-criteria situations can be addressed by mathematical optimisation techniques is somewhat limited. Certainly it is perfectly possible to set up multi-criteria objective functions, but the real problem is in reliably quantifying and making explicit a decision maker's attitude to competing objectives. To what extent are they prepared to trade off one criterion for another? Are such trade-offs constant over the whole range of possibilities or are there special cases where they would not apply? Quantitative methods do exist to address such questions, but they carry with them a set of behavioural assumptions which some might find hard to accept.

This brings us to a final and related general point. We should emphasise that although location-allocation modelling by its nature focuses on the optimisation of pre-defined criteria, the value of many of these models is rarely the optimal solution they provide, but rather their ability to explore and compare solutions. Facility location in both the public and commercial sectors is a complex decision involving many interacting factors. Optimisation models can provide useful input to that decision making process, but alone they will never provide an ultimate answer. Where they are most useful is in guiding decision makers towards 'good' solutions, allowing the implications of potential solutions to be objectively evaluated and in exploring how sensitive various solutions may be to assumptions or estimates incorporated into the analysis.

9.5.2 Network problems

In the previous section we have reviewed methods for finding an optimal set of locations, where 'optimal' related to various specified criteria. In this section we shall consider the locations of facilities in a network to be given, and consider instead a different class of problem concerned with finding various kinds of 'optimal' paths through the network connecting these facility locations. For example, we might seek one path through the network that minimises the cost of travel between two locations. Alternatively, we might consider the problem of optimally choosing an entire configuration of flows in order to move people or goods between a set of supply nodes in the network and a set of demand nodes in the most 'efficient' way—the so-called 'transportation problem', mentioned earlier. We have already referred to this general class of problem as 'network problems'.

When we considered spatial interaction models earlier, our objective was to obtain sensible and succinct mathematical descriptions of observed spatial interaction. In this section, as in the last, we are not involved with description of a pattern of flows, but rather with optimally assigning a pattern of flows. However, we shall see that there are relationships between network problems and spatial interaction models. One relationship that is immediately apparent is that the analysis of various kinds of paths in a network, such as finding those associated with shortest distance or minimum cost of travel, will have a direct bearing on the sort of spatial separation or distance measures that might be used in spatial interaction models pertaining to that network. However, we shall also see that there are some less obvious relationships, for example between transportation problems and spatial interaction models.

Before considering issues concerned with paths and flows through a network, we first make a few remarks about *network design*. Here, we are given a set of locations to be connected with some transport or communications network. How are we to do this? Again, we need the concept of an objective function. One possibility would be to seek a *maximally connected network*, with each place connected directly to every other; clearly, this would minimise users' transport costs. Conversely, we could design a network that minimised construction costs, a network that connected all places but with the minimum possible length (or cost) of route; the solution here creates a *minimum spanning tree*, which we first encountered in a different context in Section 6.3.5. We refer the reader to the description there of how it is obtained. Of course, in reality, transport networks are invariably designed as a compromise between these various objectives.

Suppose now that the network is given and that each link or arc on this network has a value (distance, cost, or time) associated with it (sometimes this is referred to as the *impedance* on an arc, since it impedes the flow). A whole variety of different optimisation problems may now be posed. For example, at the simplest level, we could ask which is the path or route between a given pair of places, or nodes, on the network that minimises total travel cost (or distance, or time)? Such problems are referred to as *shortest path* problems and one commonly used algorithm for solving such problems is known as *Dijkstra's algorithm* which maps out a minimum cost route (or shortest path if measured as total arc length) from the origin to the destination. We do not give details of the algorithm here; they may be found in any elementary OR text. Another problem involves finding the *maximum flow* that is possible from an origin to a destination, given capacity constraints attached to each arc or link. This problem can be combined with the shortest path problem to ask: what is the least cost transportation arrangement from i to j, given a number of items to transport and a set of capacity constraints attached to each arc? That is, given N items (people or goods, for example) that need to go from i to j, what is the path through the network such that total transport costs (distances) are minimised without exceeding the capacity constraints on any link? This least-cost, maximum-flow problem may be solved using the so-called *out-of-kilter* algorithm which partitions the total flow N that must go from i to j among different possible paths in the appropriate optimal way. Methods such as these are of enormous value in many areas of application; an obvious one being emergency evacuation, where people must be evacuated from a hazard incident zone to one or more reception centres. This must be done in an efficient way, minimising time costs for example, but it will also need to allow for the fact that arcs will be capacitated; that is, limited in terms of the numbers of vehicles with which they can deal.

A further class of network problem involves the derivation of *tours* in the network with various desirable properties. A tour, as its name suggests, is a route from an origin i that passes through all other nodes before returning to i; for less obvious reasons, it is also called a *Hamiltonian circuit*. If we want to find the shortest possible such circuit in a given network, we are involved in what is classically referred to as the *travelling salesman problem*. This can be formulated as an integer programming problem similar to some of those we encountered in the previous section on location-allocation modelling. However, formulation of a problem is not the same as solving it! Exact solutions to this combinatorial optimisation problem are computationally infeasible for all but the smallest of networks, even on the fastest computers. As a result, such problems are solved, like many of our previous location-allocation problems, by using *heuristics*, sets of rules that give a 'good', but not necessarily optimal, solution to the problem. For example, one possible heuristic for the travelling salesman problem is at each stage to visit the nearest unvisited place. This is not a particularly good heuristic, since it results in longer trips in the later stage of a tour; better heuristics can be devised as a modification of this basic idea.

As we have seen, many of the optimisation problems we have considered, both within a location-allocation and a network context, reduce to the constrained optimisation of some objective function. We commented earlier that such problems are known, in general, as mathematical programming problems and come in various varieties such as linear programming and integer programming. The latter refers to the case where variables in respect of which optimisation is performed can only take integer (whole number) values, which has in fact been the case in several of the situations we have discussed. One final class of mathematical programming problems we wish to mention is the so-called *transportation problem*. The structure of this turns out to be not so very different to the spatial interaction problem we began with in this chapter, although the objective is rather different. Recall that, in developing the general spatial interaction model earlier, we imposed a total cost constraint; we wanted to derive the most likely pattern of flows whilst preserving the observed total cost of travel in the system as well as the observed total flow from each origin and to each destination. Suppose now we set up the following constrained optimisation problem:

$$\min\left\{\sum_{i=1}^{m}\sum_{j=1}^{n} y_{ij}d_{ij}\right\}$$

subject to:

$$\sum_{j=1}^{n} y_{ij} = a_i \quad i = 1, \ldots, m$$

$$\sum_{i=1}^{m} y_{ij} = b_j \quad j = 1, \ldots, n$$

The two constraints are identical to the origin and destination constraints that appeared in deriving the general spatial interaction model. However, instead of wishing to derive a most probable pattern of spatial interaction assuming a fixed total cost of travel and given total numbers that must flow from each origin and to each destination, we now wish to find an optimal pattern of flow (one with minimum total cost) subject to the origin and destination constraints. We know the locations of the origins and destinations (so we are not dealing with a location problem); we want an optimal allocation of flows, optimality here meaning minimisation of total transport cost.

This transportation problem arises in all sorts of situations, involving both private and public sector contexts. For example, we might take a set of regions producing known quantities of gas, together with a set of demand zones with known requirements. Or, we might have a set of residential zones, supplying known numbers of students to schools 'demanding' pupils. A solution to the transportation problem will, in both cases, yield a configuration of flows from origin to destination that minimises total transport cost. In the school districting example, it will result in a set of catchment areas for those schools. In some situations, interest in solutions to the transportation problem centres on its use in a policy context to redesign flow patterns; in other cases, however, it may be used simply as a benchmark against which to measure how far actual flows depart from that of system-wide efficiency. Again, we do not give details here concerning the solution to such problems; suffice it to say that most transportation problems are now solved using general purpose packages developed for mathematical programming and, more especially, linear programming.

Before leaving this brief consideration of network problems in general and of the transportation problem in particular, we comment again on the similarities between the structure of the latter and the one we came across in deriving the general spatial interaction model. In one sense we have come full circle in the chapter—the system-wide cost of travel which constitutes the objective function in the transportation problem was one of the characteristics of an observed pattern of flows that we chose to preserve in deriving a general mathematical form for a spatial interaction model. We also chose to preserve total flows to destinations and from origins, the two sets of constraints which also appear in the transportation problem. What, then, is the relationship between these two different problems, the one concerned with a sensible model to describe an observed pattern of flows subject to preserving observed origin and destination total flows, the other with deriving an optimal set of flows subject to the same constraints? The answer lies in a reciprocal relationship between the distance deterrence parameter in the spatial interaction model and the total observed cost of travel in a given system of flows. If we observe a set of flows whose total cost of travel far exceeds the least possible cost as indicated by the solution to the transportation problem with the same origin and destination constraints, then it must be because cost of travel is not at a premium for individuals flowing in the observed system. In other words, we will get an estimated distance deterrence effect for that system which is quite low. On the other hand, if the observed flows have a total cost of travel near to the theoretical minimum represented by the transportation solution, then the distance deterrence effect must be high in the system of observed flows. In this sense the optimal transportation solution for given origin and destination total flows represents an upper bound for the size of the distance deterrence effect in an observed system with those same origin and destination totals. This makes intuitive sense; as distance becomes more and more of a deterrent towards interaction, the observed pattern of flows should become increasingly similar to that which minimises total travel costs.

9.5.3 Software environments for spatial interaction modelling

Given that the focus of our book is on *interactive* spatial analysis we should say something about the kinds of software environments that are currently available for solving spatial interaction, location-allocation, and network problems. As with the situation described in Chapter 2 in respect of more general spatial data analysis, until quite recently software for solving spatial interaction, location-allocation and network problems tended to rely on the running of non-interactive computer programs. That is, data files were prepared (for example, of origin and destination factors and distance matrices), and programs run to generate print-outs that contained little, if any, graphic display. Of course, such software was valuable; it allowed both academics and planners to generate solutions to quite complex problems, and to experiment with different scenarios by submitting another program to the operating system.

However, in the last few years software environments have changed dramatically, and, partly due to the advent of GIS, a number of packages and modules have appeared to greatly enhance the usability of the methods we have described earlier in this chapter. We comment briefly on some examples, recognising that these are merely a few illustrations. Other software is being developed, or will surely develop rapidly, utterly transforming our ability to perform sophisticated analyses. We focus on software developments primarily directed towards spatial applications and leave aside the considerable and continuing advances in general mathematical optimisation software, which also has a significant bearing on our ability to tackle spatial interaction problems.

The GIS software package IDRISI, to which we referred in Chapter 2, offers built-in modules for performing multi-criteria decision making and for solving least-cost routing problems (for example, by attaching weights to different classes of land use so that the route taken by a new transmission line avoids highly weighted areas of prime land). However, the versatility of IDRISI is that it permits users to link in other software. For instance, some researchers have attached simple spatial interaction and location-allocation models to IDRISI. One module determines the accessibility of facilities to a set of demand locations, given a fixed form of distance deterrence and a pre-specified γ parameter. Other modules solve the p-median problem and the maximal covering problem. The solutions to these problems may then be displayed using the graphical capabilities of IDRISI.

Another comprehensive package for spatial interaction modelling and network analysis is called TransCAD. This has facilities for fitting all the spatial interaction models we have outlined, together with various modules for network analysis and location problems. Given its use of detailed digital road network databases it offers considerable functionality for the sorts of modelling we have emphasised in this chapter.

At the time of writing, ARC/INFO, one of the most widely used proprietary GIS, does not offer the ability to solve location problems. It does, however, give functions for solving shortest path and allocation problems, in discrete space. The network modules allow, among other things, functions for attaching impedances to arcs (such as estimates of travel time, and flow capacities), barriers at junctions to prevent movement, and constraints at junctions to reflect restrictions on certain turns at those junctions. Shortest path algorithms may then be invoked to find least-cost routes through the network. Allocation problems are solved by first fixing the locations of facilities at nodes on the network and also attaching to those facilities a capacity on the volume of activity each can handle if this is required. Arcs are then allocated to the nearest facility along the least-cost route, until the maximum impedance limit is reached. For example, if we wished to find all arcs on a road network that were within 30 minutes (the maximum impedance) of a major hospital, the system would find all arcs that could be reached from a hospital within that time. Alternatively, given a set of schools each with a fixed capacity we could allocate 'arc demand' (the number of students living along particular streets) to those schools until the school capacity is met or the maximum impedance (distance or travel time) is met, whichever occurs first.

One of the main advantages of performing this kind of analysis within a GIS environment is that a good system will allow experimentation: the posing of 'what if' questions. What if we convert a set of routes into one-way systems; or if we remove a particular school or hospital at a node on the network; or if we increase the capacity of a particular facility? The interactive nature of modern software means that answers to such questions can be obtained quite rapidly. Given a suitable graphics environment, the analyst can select particular arcs, nodes, or centres for modification and then 'see' the results of such alterations to the database. We emphasise that this is only possible with good, interactive software, as well, of course, as the availability of good digital data, most obviously on the transport network. As we pointed out in Chapter 2, such data are becoming widely available, particularly in the developed world.

The ability to blend good digital data with imaginative software lies at the heart of GIS. But the link to spatial interaction, location-allocation, and network modelling, provides a much more powerful analytical environment. Indeed, these sorts of links give rise to what has become known as a *Spatial Decision Support System* (SDSS). Here, data can be readily interrogated and updated, but the system allows for forecasting, impact analysis and optimisation. Coupled to software that allows for genuine interaction between users and their data, such SDSS may indeed prove to be powerful tools.

9.6 Summary

In the early sections of this chapter we were concerned again with the main theme of our book, that of the statistical description and modelling of observed spatial data, considering in this case data on spatial interaction. We followed the structure established in earlier chapters, talking briefly about methods for visualising and exploring such data before moving on to more formal statistical modelling techniques. We introduced the general spatial interaction or gravity model and discussed methods for fitting such models to observed flows. We considered variants of the basic doubly constrained model and the various potential applications of such models.

In the final sections of the chapter we broadened this discussion to consider various optimisation problems related to the analysis of spatial interaction; problems such as the optimal siting of facilities, optimal paths and tours in networks and the 'classical' transportation problem. In doing so, we moved away from the general theme of the book, which was concerned with spatial *data* analysis and more into the area of spatial analysis in general. Statistical methods moved into the background and the techniques of Operations Research became important; data modelling was replaced with constrained optimisation. This probably means it is time to stop writing!

9.7 Further reading

The Swedish airline data come from:

Törnqvist, G. (1970) *Contact Systems and Regional Development*, Lund Studies in Geography, Series B, 38, Gleerup, Lund, Sweden.

The data on migration between Dutch provinces are taken from the following source. As mentioned in the text, the data we include are total flows; but the publication gives data disaggregated by age, so interested readers might want to create further data sets relating to retirement migration, or movement of those aged under 30, for example.

Drewe, P. (1980) *Migration and Settlement: 5. Netherlands*, RR-80-13, International Institute for Applied Systems Analysis, Laxenburg, Austria.

The data on hospital flows are taken from the so-called Körner Minimum Data Set collected for all hospitals in England and Wales, combined with Hospital Activity Analysis returns, for the Regional Health Authorities embracing the London region. The data are of historical interest only, relating to the mid-1980s.

The data on current accounts are described, and models fitted, in:

Bailey, T.C. and **Munford, A.G.** (1991) A case study employing GIS and spatial interaction models in location planning, in *Proceedings, Second European Conference on Geographical Information Systems*, EGIS Foundation, Utrecht.

On the visualisation and exploration of spatial interaction data, see:

Robinson, A.H., Sale, R.F., Morrison, J.L. and **Muehrcke, P.C.** (1984) *Elements of Cartography*, John Wiley, Chichester, 308–15.

Tobler, W.R. (1976) Spatial interaction patterns, *Journal of Environmental Systems*, 6, 271–301.

Holmes, J.H. (1978) Dyadic interaction matrices, *Progress in Human Geography*, 2, 467–93.

Good introductory material on spatial interaction models is to be found in the following:

Foot, D. (1981) *Operational Urban Models: An Introduction*, Methuen, London.

Haynes, K.E. and **Fotheringham, A.S.** (1984) *Gravity and Spatial Interaction Models*, Sage, London.

Senior, M.L. (1979) From gravity modelling to entropy maximising: a pedagogic guide, *Progress in Human Geography*, 3, 175–210.

Thomas, R.W. and **Huggett, R.J.** (1980) *Modelling in Geography: A Mathematical Approach*, Harper and Row, London, Chapter 5.

Other, more advanced, references on spatial interaction modelling include:

Erlander, S. and **Stewart, N.F.** (1990) *The Gravity Model in Transportation Analysis –Theory and Extensions*, VSP, Utrecht.

Wilson, A.G., (1974) *Urban and Regional Models in Geography and Planning*, Wiley, London.

Fotheringham, A.S. and **O'Kelly, M.E.** (1989) *Spatial Interaction Models: Formulations and Applications*, Kluwer Academic Publishers, Dordrecht.

Upton, G.J.G. and **Fingleton, B.** (1989) *Spatial Data Analysis by Example: Volume 2: Categorical and Directional Data*, John Wiley and Sons, Chichester. Chapter 8.

Batty, M. (1976) *Urban Modelling: Algorithms, Calibrations and Predictions*, Cambridge University Press, Cambridge.

A good reference to general linear modelling, including use of the statistical package GLIM is:

Aitkin, M., Anderson, D., Francis, B. and **Hinde, J.** (1989) *Statistical Modelling in GLIM*, Clarendon, Oxford.

General references on parameter estimation in gravity models include:

Baxter, M.J. (1983) Estimation and inference in spatial interaction models, *Progress in Human Geography*, 7, 40–59.

Baxter, M.J. (1986) Geographical and planning models for data on spatial flows, *The Statistician*, 35, 191–8.

Flowerdew, R. and **Aitkin, M.** (1982) A method of fitting the gravity model based on the Poisson distribution, *Journal of Regional Science*, 22, 191–202.

Openshaw, S. (1979) Alternative methods of estimating spatial interaction models and their performance in short-term forecasting, in Bartels, C.P.A. and Ketellapper, R.H. (eds) *Exploratory and Explanatory Statistical Analysis of Spatial Data*, Leiden, Nijhoff, 201–25.

Excellent introductions to location-allocation models are given in:

Hodgart, R.L. (1978) Optimising access to public services: a review of problems, models and methods of locating central facilities, *Progress in Human Geography*, 2, 17–48.

Ghosh, A. and **Rushton, G.** (eds) (1987) *Spatial Analysis and Location-Allocation Models*, van Nostrand Reinhold Company, New York.

The application of the p-median problem to health care delivery in Goa, together with an outline of the hierarchical model, comes from:

Hodgson, M.J. (1988) An hierarchical location-allocation model for primary health care delivery in a developing area, *Social Science and Medicine*, 26, 153–61.

For a detailed mathematical development of location-allocation models see:

Handter, G.Y. and **Mirchandani, P.B.** (1979) *Location on Networks: Theory and Algorithms*, MIT Press, Cambridge, Massachusetts.

Good general references that cover all or some of the techniques for mathematical optimisation, network problems, and specifically the transportation problem, are:

Smith, D.K. (1982), *Network Optimisation Practice: a Computational Guide*, Ellis Horwood, Chichester.

Williams, H.P. (1990) *Model Building in Mathematical Programming*, Third edition, John Wiley, Chichester.

Gass, S.I. (1975) *Linear Programming: Methods and Applications*, Fourth edition, McGraw-Hill, New York.

Killen, J.E. (1983) *Mathematical Programming Methods for Geographers and Planners*, Croom Helm, London.

Thomas, R.W. and **Huggett, R.J.** (1980) *Modelling in Geography: A Mathematical Approach*, Harper and Row, London, Chapter 6.

In terms of software we have already given some references to GIS products in Chapter 2. An early set of computer routines for solving a variety of location-allocation problems is provided in:

Goodchild, M.F. and **Noronha, V.T.** (1983) *Location-Allocation For Small Computers*, Department of Geography, University of Iowa, Monograph No. 8, Iowa City, Iowa.

The addition of modules to IDRISI for solving location-allocation problems is described in:

Bosque, J. and **Moreno, A.** (1990) Facility location analysis and planning: A GIS approach, *Proceedings, First European Conference on Geographical Information Systems*, EGIS Foundation, Utrecht, 87–94.

And, for a statement of the importance and nature of Spatial Decision Support Systems, see:

Densham, P.J. (1991) Spatial decision support systems, in Maguire, D.J., Goodchild, M.F. and Rhind, D.W. (eds) *Geographical Information Systems: Principles and Applications*, Longman, Harlow, 403–12.

9.8 Computer exercises

Here are suggested exercises that you can try on ideas discussed in this chapter using INFO-MAP and our example data sets. These exercises have been referenced at appropriate points in the chapter by numbered symbols in the margin.

Exercise 9.1

Open the data for 'London patient flows'. First, simply `Choropleth map` flows of patients to the hospital for various different specialities (note the hospital is situated in Hampstead district). Can you identify significantly different patterns for some specialities? Which specialities have the widest catchment areas and which of these constitute significant workloads?

Next, for a speciality of your choice, explore the relationship between the distance of districts to Hampstead (where the hospital is situated) and the ratio of the volume of flow to the population of the district (000s). You will need to calculate new variables to do this. Suppose, for example, that you are interested in Paediatrics, which is variable 2 in the file. Then

```
ratio={2}*1000/{15}
```

will give the ratio of the flow from a district to the population in that district (variable 15). Then the distance of districts from Hampstead (location 13 in the file) is obtained by:

```
dist=dist([13])
```

You may then plot the new variables 'ratio' and 'distance' against each other by using `Scatter plot` in the `Analysis` menu. Go on to plot the logarithm of 'ratio' against 'dist'. If many districts have zeros for patient flow then you may wish to avoid missing values caused by taking the logarithm of zero by adding some small constant to the ratio when taking the logarithms. You could also try logarithm of the ratio against logarithm of 'dist'. In this case you will need to modify the intra-zonal distance for Hampstead. Can you work out how to calculate this to be half that of the distance to its nearest neighbouring district?

Compare the nature of these kinds of relationship for several different specialities. Speculate on why some districts seem to send rather more, and others perhaps fewer, patients to the hospital than you would expect simply on the basis of distance.

Exercise 9.2

Open the file 'Swedish airline flows'. There are actually two locations in this file for each airport. Those indicated as circles have data values relating to business air journeys out of each airport whereas those indicated as triangles relate to business air journeys into each airport. So, for example, if you map the first data item (which relates to Ängelholm) as a choropleth map (where the symbols are treated as 'areas' for mapping purposes), you see simultaneously both the flows to and from Ängelholm for each of the airports in the data set. Explore visually the pattern in flows for different airports, making reference to the discussion in the text concerning 'winds of influence'. Do you agree with the general indications in Figure 9.5? Are there other effects in the data that are not made apparent in this map?

Explore the relationship between distance and flows in this data set, using some of the ideas from the previous exercise. You will need to bear in mind that each airport is duplicated in the file. Look at the data file using `Modify` from the `Data` menu to help you understand how it is structured. Are there any striking differences between the distance effects seen here and those seen in the patient flow example? Speculate on the reasons for such differences.

Exercise 9.3

Open the file 'Dutch migration'. In this file the data items named after provinces each represent the flows from that province to each of the other provinces. So, for example, a choropleth map of the data item 'Drenthe' shows the flows from 'Drenthe' to each other province. Similarly a choropleth map of 'All provinces' shows the total flow from all provinces to each of the other provinces; in other words, the total inflow to each of those provinces. First map the flows from a few of the provinces to get a general impression of any obvious differences in the patterns. What difference, for example, do you observe between the pattern of flows from Drenthe and that from Zuid-Holland?

Next, experiment with using the parameters of the doubly constrained gravity model given in the text, to predict flows from Zuid-Holland (location 11 in the file) to each of the other provinces. The estimated origin and destination parameters given in the text are included in the file as the data items 'D. const. org. parameters' and 'D. const. dest. parameters'. These are, respectively, data items 14 and 15 in the file. Bearing in mind that the distances calculated by INFO-MAP in this example will be in tens of metres and that the model predicts flows in thousands of individuals a suitable formula would be:

```
zuidflows=1000*{14}[11]*{15}*exp(-.000241*dist([11]))
```

Assuming that this new data item is number 17 in the file, convince yourself that your total predicted flows from Zuid-Holland are approximately equal to the total observed (they won't be exactly because we have only reported parameters to limited significant figures in the text). Do this by calculating two worksheet items and comparing them, for example:

```
work[1]=sum({11})
work[2]=sum({17})
```

Next, calculate and map the residuals, that is the differences between the predicted and observed flows from Zuid-Holland. Are there any striking patterns in the residuals? Can you think of factors not included in the model which might explain these? Plot the residuals against the population of each province. What do you observe?

Repeat this exercise for the flows from Utrecht. Are there any major similarities or differences in the pattern of residuals?

Consider how you would predict what flows would result in the Dutch 'system' if, in future years, migration out of Zuid-Holland increased by 50 per cent and that into Zuid-Holland decreased by 50 per cent.

Exercise 9.4

Open the file 'Accounts in Britain', which relates to the study described in the text on the 'attractiveness' of shopping centres for siting bank branches. The data item 'D. const. dest. parameters' contains the $\hat{\beta}_j$ values for those 298 shopping centres with the 'largest' such values out of the 600 shopping centres where the organisation was represented (we have called it 'B&G' although this in no way relates to its real name!). As discussed in the text these were derived by fitting a doubly constrained gravity model to 'flows' of accounts between all existing branches and 8500 postal districts. We cannot usefully calculate any model predictions using this data set, since only a subset of 'destinations' are included. However, as described in the text the main motivation here was one of exploring (and possibly modelling) the $\hat{\beta}_j$ values so as to be able to predict the 'attractiveness' of shopping centres where B&G was not already represented. We have included some possible exogenous measures.

First, `Choropleth map` the variable B&G accounts to show the existing pattern of B&G business (recall this relates to only one particular type of account). You could calculate the ratio of B&G account business to the variable 'Estimated total accounts', and choropleth map the result to show the general pattern of the B&G market share. What general pattern do you observe?

Next, `Dot map` the 'D. const. dest. parameters' variable. These values are given on a scale such that the parameter values for *all* the original 600 shopping centres (of which these 298 are a subset) multiply to unity. Recall that it is only relative comparisons and not absolute values of $\hat{\beta}_j$ which are important. Because of the multiplicative way in which such values enter into the gravity model, it makes sense to compare them on a scale where the mean of their logarithms is zero. Explore variations in the 298 $\hat{\beta}_j$ values provided. Notice there are some very extreme values. Does there seem to be any spatial explanation for where these are occurring and does this lead you to suspect effects that were perhaps not correctly accounted for in the gravity model from which these estimates were derived?

Use `Profile location` with the option 'descending order' to identify these 'outliers' and create a new variable which excludes them by using an appropriate `if()` statement to set their values to 'missing'. Now `Scatter plot` the result against the general 'geodemographic score' provided for each of the shopping centres and alternatively the 'competitor ratio'. Does there seem to be a relationship that would enable the 'attractiveness' values to be modelled in terms of such variables? Use the `north()` and `east()` functions to create new variables which enable you to explore for possible north–south or east–west trends in the 'attractiveness' parameter values.

Now use `Zoom/Restrict` from the `File` menu to zoom into a rectangular area which includes just those shopping centres in the Greater London area. (Use a raster zoom.) `Dot map` the 'D. const. dest. parameters' in this area. Is there an obvious spatial pattern in their values? Are there any better relationships using just these shopping centres between the 'attractiveness' estimates and the general 'geodemographic score' or the 'competitor ratio'?

Suggest some other exogenous variables that it might be worth attempting to use to model the estimates of the 'attractiveness' of the shopping centres in this area? Would some of these involve their proximity to other facilities? If so, which types of facilities suggest themselves as possible candidates?

Exercise 9.5

Return to the 'Dutch migration' data used in an earlier exercise. Now experiment with using the parameters of the singly constrained gravity model given in the text to predict flows from Zuid-Holland (location 11 in the file) to each of the other provinces. The estimated origin parameters given in the text are included in the file as the data items 'S. const. org. parameters'. This is data item 16 in the file. Recall also that the population of each province is variable 13. Bearing in mind, as before, that the distances calculated by INFO-MAP in this example are in tens of metres and that the model predicts flows in thousands of individuals a suitable formula is:

```
fromzuid=1000*{16}[11]*({13}/1000)**(0.4846)
          *exp(-.000241*dist([11]))
```

Assuming that this new data item is number 17 in the file, convince yourself, as in an earlier exercise, that your total predicted flows from Zuid-Holland are approximately equal to the total observed (again this won't be exact because we have only reported parameters to limited significant figures in the text).

Next calculate the flows to, rather than from, Zuid-Holland under this model, that is:

```
tozuid=1000*{16}*({13}[11]/1000)**(0.4846)
          *exp(-.000241*dist([11]))
```

Total these flows into a worksheet item and compare this total with the total observed flows to Zuid-Holland. You may calculate the latter by first zeroing a worksheet item and then accumulating over data items as follows:

```
work[1]=0

i=1,11:work[1]=work[1]+{i}[11]
```

Note that the predicted and observed total flow to Zuid-Holland do not agree under this model.

Now suppose you wish to calculate the total flow to each province from all other provinces, using this model. You could do calculations similar to those you have used for flows to Zuid-Holland, for each of the other provinces in turn, summing the results and putting these totals into a new variable. However, INFO-MAP provides a special function `flow()` to do this for you. In general, a statement such as `flow(0.05,0.02,{1},{2})` will return the total flow to locations which have non-missing values of data item 2 and with relative attractiveness specified by the values given in data item 2, where 'demand' arises at locations identified by non-missing values of data item 1 and at a rate of 0.05 × the values given in data item 1. It is assumed that distance decay will be exponential with a negative parameter, whose absolute value is in this case 0.02. In other words, this function calculates the total flows to locations arising from a singly constrained gravity model, where you supply the totals that have to flow, the relative 'attractiveness' of the destinations and the distance deterrence parameter.

In our case the totals that have to flow from each province are given by the sum of the data items named after each province. We may calculate these as a new data item:

```
i=1,11,totalout[i]=sum({i})
```

We assume that this new variable is number 19 in the file. We now need to calculate a set of relative 'attractiveness' measures for each of the provinces. According to our model this is just:

```
attract=({13}/1000)**(0.4846)
```

Assuming this is variable number 20, we are now in a position to calculate the required total predicted inflow to each province as:

```
predin=flow(1.0,0.000241,{19},{20})
```

You can convince yourself of the result by comparing the 11th element of this new variable with the total predicted flows to Zuid-Holland that you calculated previously in a more direct fashion. These should be almost the same, although they may differ slightly because of rounding errors. Note that the `flow()` function does not use the origin parameters of the model at all. It can derive these by knowing that all the demand specified has to be satisfied.

Calculate the residuals between the predicted total inflow to each province and that observed (the variable 'All provinces'). Map these residuals. What do they tell you about the overall performance of the model?

We can also use the `flow()` function to apply the unconstrained model given in the text. First we calculate the total outflow in each province predicted by the model as:

```
upredout=8.5*({13}/1000)**(0.6965)
```

We use 8.5 rather than 0.0085 here because we want flows in thousands. We assume the result is variable 22. We now need the relative 'attractiveness' of provinces under the unconstrained model because it is slightly different to that used earlier for the singly constrained model. We do this by:

```
uattract=({13}/1000)**(0.4654)
```

Assuming this is variable number 23, we now calculate the required total predicted inflow to each province as:

```
upredin=flow(1.0,0.000242,{22},{23})
```

Calculate the residuals between the predicted total inflow to each province and that observed under this model. Map these residuals, compare them with those obtained under the singly constrained model. Are there any significant differences in the overall picture of predictions?

Postscript

In overall summary then, what have we tried to do in this book on interactive spatial data analysis? In terms of *data* we subdivided data types into four classes: those concerned with the locations of point events; with attributes measured at sampled point locations; with measurements made over a set of areal units; and with flow data arising in geographical space. As far as *spatial analysis* is concerned, we made a distinction, though we emphasised it was a fluid one, between methods that were visualisation tools, those which were exploratory devices, and those which were more formal model-based approaches. Lastly, in terms of the *interactive* nature of analysis we stressed the importance of being able to move easily and naturally between different views of the data, and from one mode of analysis to another.

What then does the future hold for interactive spatial data analysis? In terms of the first word in the book's title, we anticipate major changes. We trust that the software included with the book will prove useful as a pedagogic device; but it is already dated within the context of developing software environments. As we have hinted in the last chapter, new tools for interactive optimisation and location-allocation analysis are being developed. At the same time, as discussed in Chapter 2, proprietary GIS are being revised to incorporate enhanced functionality for spatial analysis. Doubtless, new methods of data analysis will also continue to be developed by statisticians, geographers, and environmental scientists. The explosion in the volume of data being made available for research and policy making, most obviously from satellites but from other sources too, will also ensure a healthy future for spatial data analysis.

We trust that this book will prove to be a useful introductory source to would-be spatial analysts in that future. Above all, we hope that we have succeeded in giving an overview of, and a foundation in, the subject. We also hope that we have interested you enough to make you want to study it more deeply. We can assure you that there is plenty of spatial data analysis left to discover. If you have followed this introduction, then you should be in a good position to go on in the subject and find out that some of what we have said has been over-simplified and that in other areas we have side-stepped some difficult issues!

Our trip has been a fairly long one. We hope that you have found it worthwhile, and that along the way we have managed, in the language of our last chapter, to maximise intellectual benefits and minimise emotional costs!

Appendix

INFO-MAP User's Guide

Contents

A.1 Overview

A.1.1 About INFO-MAP

The software package INFO-MAP discussed in this *User's Guide* was especially written by one of the authors to accompany this book. It was designed to perform a limited range of statistical analyses on small spatial data sets, using a minimal PC hardware configuration. It runs on IBM-compatible PCs under MS-DOS (not Windows) using a standard EGA or VGA graphics adapter and requires 3 Mbytes of disk space to install and a minimum of 500 Kbytes of RAM to run. It supports a Microsoft (or compatible) mouse which is useful, although not essential. Interfaces and drivers are provided to support use of a digitising tablet, scanner or printer; but only a very limited range of these are catered for. The software package is distributed with *Interactive Spatial Data Analysis*, as an educational aid; it is not intended to be a polished commercial product and no support is provided for the package. Versions for other hardware configurations or operating systems are not available, nor are interfaces or drivers for a wider range of peripheral devices. Further releases are not envisaged.

A.1.2 About the *User's Guide*

This Appendix is not intended either to provide tuition on how to analyse data with INFO-MAP, or to explain the statistical methods which are implemented in the package. The main text of this book is concerned with the latter and the former is addressed in the computer exercises provided at the end of each of its chapters. This Appendix simply provides a short overview of the package, notes on installation, basic keyboard and menu use, also the problems that may arise in installing or running the package. Given this basic information we hope the user should be able to get started with INFO-MAP; it is not a complex package and is not difficult to use.

A second document, the *User's Reference*, provides notes on configuration and a detailed explanation of all INFO-MAP menus and options. It is included in a format suitable for laser printing, on the distribution diskette for INFO-MAP and is automatically copied to the sub-directory 'util' of the directory into which INFO-MAP is installed during the installation process. Details on how to print this file may be found in the later section 'Installation'.

The *User's Guide* and *User's Reference* are the only documentation provided for INFO-MAP. The package does not incorporate an 'on-line' help facility; there is a 'help' option, but this only provides brief general notes on the package and issues such as setup and device configuration.

How do we suggest that you use the *User's Guide*? We would advise all users to read the remainder of this first section before installing or trying to use the package. It is concerned with an overview of the package and the way in which data sets are structured, and the general framework under which its various functions are arranged. You may then wish to install the package and will need to refer to the section on 'Installation' for instructions on how to do this. If things go wrong in the installation or you cannot get the package to work, you will need to refer to relevant parts of the section on 'Trouble-shooting'. In the first instance we would advise you not to bother too much about attempting to configure INFO-MAP defaults, or trying to support devices such as a printer; you can always come back to these later. However, if you have one, you should try to get your mouse working with INFO-MAP, because this will make it easier to use the package. Assuming you can run the package we next suggest you read the section on 'Getting Started' in conjunction with running the software. This should introduce you to the basic 'look and feel' of INFO-MAP.

You should then print the *User's Reference* (see the section on 'Installation'), and refer to the relevant chapters in that document as and when required. In particular, when you have some experience with using the package you can consult the 'Configuration' chapter in the *User's Reference* for details on changing defaults, setting up a printer, and so on.

A.1.3 Overview of INFO-MAP

INFO-MAP is a package which allows the user to enter small spatial data sets into a computer and to explore and analyse them. By 'spatial data set' we mean not only data values, but also their relative spatial configuration and a 'picture' of the 'study region' within which they lie. By 'exploring' such a data set we mean creating various kinds of 'maps' or plots of the data values, or transformations of them, and carrying out a variety of statistical analyses on the data values, their spatial configuration, or combinations of both of these.

Each data set in INFO-MAP consists of at least two components; first, a *base file*, and, second, one or more *data files* which are associated with that *base file*. Several example data sets of this type are automatically installed with the package and there are functions within the package for creating new ones.

Base files fulfil two functions within the package. Firstly, they contain a grid referenced screen image, showing a simple outline of the 'study region' for a particular spatial data set and any relevant sub-divisions of this, such as administrative zones or districts. We refer to this as a *boundary*. Secondly, they contain spatial coordinates for a set of points within this 'study region' where data values have been observed. We refer to these as *locations*.

The observed data values are contained in separate *data files*, each of which is associated with a particular *base file*. They are best thought of as a 'spreadsheet' in which rows correspond to the *locations* specified in their associated *base file* and columns correspond to different data items (attributes or variables) recorded at each of these *locations*.

A final optional component may also be present in an INFO-MAP data set—an *overlay file*. *Overlay files*, like *data files*, are each associated with a *base file*. If present they indicate additional contextual information for the 'study region', such as topographic features, road networks or positions of various types of facilities. However, this information is simply in the form of a graphic image which may be 'superimposed' on the *boundary* contained in the *base file*. *Overlay files* are 'dumb' and only of use visually; they do not contain any quantitative information which may be extracted and used in numerical analysis.

Apart from a small number of options concerned with the setting of defaults or the configuration of devices such as printers, all menus and options within INFO-MAP relate to the three basic components of INFO-MAP data sets discussed above (*base files*, *data files* or *overlay files*), or relate to combining two or more of these in order to visualise or analyse the spatial data set. In general, INFO-MAP functions fall into four groups, each relating to a separate module within the package.

Firstly, the package provides a variety of ways to create or modify the two components of a *base file*, the *boundary* and the *locations*. All options of this type are contained in the *base file* module. This allows *boundaries* to be drawn or edited on the screen, using a mouse or graphic tablet, or entered directly from hard copy using a scanner. Alternatively, *boundaries* may be 'imported' from ASCII files of polygon coordinates, or from raster 'bitmaps' of graphic images. The *base file* module also allows *locations* to be defined and named using the mouse and keyboard, or 'imported' from ASCII files of grid coordinates. Related sets of functions are also provided in the module for 'exporting' either *boundaries* or *locations* from INFO-MAP.

Secondly, users are provided with various facilities for creating or editing *data files*. These options are contained in the *data file* module. Such files may either be created on screen using a data editor, or imported from ASCII files of data records. In the latter case, records must either be in the same order as the *locations* specified in the *base file*, or contain suitable grid references which allow them to be matched to those locations. Again, related functions are also provided for 'exporting' *data files* from INFO-MAP.

Thirdly, facilities are provided to create and edit *overlay files*. These options are contained in the *overlay file* module. This allows *overlay files* to be drawn with a mouse or graphics tablet, or 'imported' from raster 'bitmaps'. They may also be 'exported' in a similar format.

Finally, and most importantly, given a particular *base file*, *data file* (and, optionally, an *overlay file*), the rest of INFO-MAP is given over to providing options for visualising and analysing the data set to which these relate. These options are provided in the *mapping & analysis* module. If you are only interested in using INFO-MAP to explore the example data sets provided, then these are the only kind of options that need concern you.

Broadly, they include options to create 'dot', 'proportional symbol' or 'choropleth' maps of data values, with a range of facilities for altering the map style (colour, shading, and so on); options to interpolate data values in the form of simple contour or 'kernel' maps; options to produce scatter plots or frequency distributions of data values, and options to perform various other spatial analyses, such as 'variograms', 'correlograms', '*K* functions' and so on.

In particular, new variables may be created from those that already exist within a *data file*, by using 'spreadsheet' type formulae. Such newly created variables may be mapped or used in an entirely equivalent way to the original data items. The formulae for creating new variables can include standard mathematical operations and logical functions; also a range of basic statistical functions, including means, variances, medians, ranks, and so on. Functions to perform a limited number of more sophisticated analyses are also provided, such as multiple regression, principal components or cluster analysis. Consequently, INFO-MAP may be used, for example, to create a new variable containing the residuals from a regression, or the scores on a particular principal component and then to visualise these results in map form.

Further, INFO-MAP formulae provide a range of functions which allow the user to access the spatial relationships between data values. Functions are included to calculate the areas, adjacencies or perimeters of any closed polygons or zones in the *boundary* component of the *base file*; or to access the coordinates of, distances between, or nearest neighbours of, the *locations* defined in the *base file*, and so on. These spatial functions may be used in conjunction with any of the other statistical, mathematical or logical functions and thus provide the potential to incorporate more directly the spatial configuration of data values into any analyses.

Each of the four modules of INFO-MAP discussed above has its own menu bar running across the top of the screen, from which the options within that module may be selected. The basic design of the package is pictured in Figure A.1.

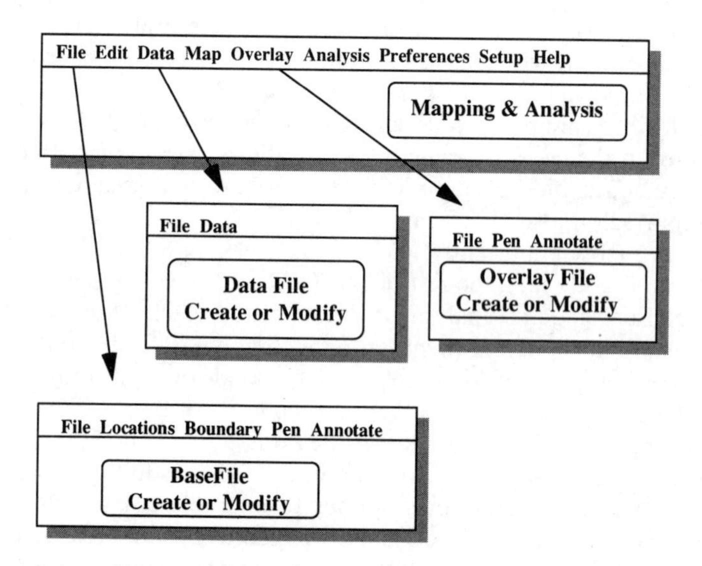

Fig. A.1 Overall structure of INFO-MAP

When you first run INFO-MAP you are in the *mapping & analysis* module. This is the top level of INFO-MAP. The three other modules are invoked from here by either the `New` or `Modify` options in respectively the `File`, `Data` or `Overlay` menus at this level. When you exit from any of these three modules, you always return to the *mapping & analysis* module. Exit from INFO-MAP is always from this top level module, via the `Exit INFO-MAP` option within the `File` menu.

In moving around the various modules in INFO-MAP it is important to bear in mind the way in which INFO-MAP data sets are structured. In particular, remember that there is a hierarchy of files involved. When using INFO-MAP to view or analyse a data set, the *base file* is always 'opened' first, followed by an associated *data file*. Similarly, when entering a new data set, the *base file* must always be created first, followed by one or more associated *data files*. Likewise, you cannot create an *overlay file* before its corresponding *base file* has been created. This hierarchy also implies that when you make changes to a *base file* these changes will potentially affect all *data files* (or *overlay files*) associated with it. Remember that everything that is spatial about a data set is stored in the *base file*.

A.1.4 What INFO-MAP will not do

Overall, then, INFO-MAP is a package to assist you in putting together a small spatial data set, 'picturing' this data set in various ways and going on to seek relationships of interest. We hope that it will enable you to create interesting pictures of spatial data quickly and easily.

However, do not expect too much from the package. It is not designed for computer cartography and will not allow the creation, printing, storage, or manipulation of detailed maps. Neither does it have any pretensions to be a Geographical Information System; there are no facilities provided for maintaining a spatial database of different kinds of spatial object, for buffering and overlay, or for other important GIS operations. Neither is INFO-MAP intended to be a fully comprehensive package for spatial statistics. There are many forms of spatial analysis which it is not possible to perform in INFO-MAP and where techniques are implemented only simple variants are available. There are also limits on the size of data set that can be accommodated. INFO-MAP is simply a package limited to illustrating certain of the ideas discussed in the main text of this book.

A.2 Installation

A.2.1 Distribution diskette

INFO-MAP is provided on single 3.5″ high density (1.4 Mbyte) diskette. This contains the program, drivers, fonts and *User's Reference*, as well as the example data sets. The files on this diskette are held in a compressed format and must be installed to your hard disk using a special 'install' program, also included on the disk. The installation diskette may be used to install more than one copy of INFO-MAP; but **be warned** that during the first such installation, you will be asked to supply an owner's name to appear whenever that copy of INFO-MAP is executed. This cannot be subsequently altered and will apply for all future installations from the same diskette. This has been implemented to provide you with some minor form of copy protection for your own (or your organisation's) copy of INFO-MAP.

A.2.2 Hardware requirements

To run INFO-MAP you need an IBM PC (or compatible) with at least a 286 processor running under DOS, a hard disk and an EGA or VGA card and colour monitor. When uncompressed, the files involved will occupy approximately 3 Mbytes. Unless at least this amount of space is available on your hard disk, the 'install' program will refuse to install the software. If you wish to install INFO-MAP to a network drive on a remote server then this is perfectly possible, but note that steps have to be taken subsequent to the standard installation procedure described below (see 'Configuration' in the *User's Reference*). Use of the package is easier if you have a Microsoft mouse (or compatible), but this is not essential. If you do have such a mouse, then to use it with INFO-MAP an appropriate mouse driver will need to be run before INFO-MAP is executed (see 'Configuration' in the *User's Reference*). This consideration does not affect the standard installation procedure for INFO-MAP as described below. Optional additional devices, of which INFO-MAP supports only a limited range, are a colour or grey-scale printer, a scanner and a digitising tablet (see 'Configuration' in the *User's Reference*). Again the standard installation procedure for INFO-MAP applies whether or not you wish to use such devices.

A.2.3 Memory requirements

INFO-MAP is a DOS program, not a 'protected mode' or Windows application. It therefore runs within the 640 Kbytes of conventional DOS memory. It will not use extended memory and although it will use up to 256 Kbytes of expanded memory (if you have an appropriate memory manager and this is installed on your machine; see 'Configuration' in the *User's Reference*) this option will only marginally reduce the base memory requirement for INFO-MAP, which is approximately 550 Kbytes of conventional memory. If you do not have this much free memory available from DOS, then you may have to remove some of the memory-resident applications loaded when you 'boot' your machine before you can run INFO-MAP (see the later section 'Trouble-shooting'). Whether or not you will eventually need to do this does not affect the standard installation procedure for INFO-MAP described below.

Note that because it uses EGA graphics and is not a Windows application, INFO-MAP will only run 'full screen' as a 'DOS applicaton' under Windows. You may also experience memory problems attempting to run INFO-MAP this way, since there is likely to be less conventional memory available to DOS applications running within Windows that those running 'stand alone'.

A.2.4 Installation instructions

INFO-MAP should be installed directly from DOS and not from 'DOS prompt' within Windows. The following steps should be followed to install the INFO-MAP package to your hard disk:

- Insert the Installation Diskette into a floppy disk drive
- Make that floppy drive the current drive, by typing `'A:'` or `'B:'` as appropriate
- Type `'install'`.

The 'install' program should then run and display an installation screen presenting the following options:

- **Install Directory**—the hard disk drive and directory in which INFO-MAP is to be installed. The default is 'c:\infomap', but this may be changed to any valid directory name of the user's choice. The directory does not have to exist already; it will be created by the 'install' program if it does not. If you are installing INFO-MAP to a network drive on a remote server, you must have write permission for the directory you specify. Alternatively, since some steps in addition to the installation procedure are necessary to run INFO-MAP from a network server, it may be better to install to your local drive in the first instance and then copy the installed files to the network drive later (see 'Configuration' in the *User's Reference*).
- **Registered Owner**—A text string of up to 20 characters, which will appear whenever this copy of INFO-MAP is executed. There is no default for this option and entry is mandatory. If the name you wish to use exceeds the number of characters permitted, then use the 'owner's organisation' option (see below) for a second line of your name, instead of using it to specify the name of your organisation.

  Take some care over your choice of name. Each installation diskette allows this name to be specified only once—the first time that you install a copy of INFO-MAP. The name cannot subsequently be changed and if you install the software a second time from the same installation disk, the 'install' program will use the same name as that entered on the first occasion.

  Note that when initially installed the version of INFO-MAP distributed with this book provides the user with a 'restricted registration'. All options are fully functional except permanent modification of existing *base files* or the saving of new *base files*. If you wish to take advantage of these particular INFO-MAP options in order to create your own spatial data sets, then you will neen to obtain a registration key unique to the serial number of your copy, so enabling you to convert your copy to 'full registration' and overriding the restriction. Details of how to do this are given by using the `About registration` option in the `Help` menu (see the later section 'Getting Started').

- **Owner's Organisation**—A text string of up to 20 characters, which will appear in addition to the 'registered owner' whenever this copy of INFO-MAP is executed. There is no default and entry is mandatory. This is simply a second line of text for specifying the 'registered owner' and the same comments apply.

Each of the above three options is entered by using ⟨TAB⟩ to position the text cursor on the option and then entering the required text. At any time the installation process may be aborted by using the ⟨ESC⟩ key. When the options are as desired, the user should ⟨TAB⟩ to the last option and then press the ⟨CR⟩ key, to commence installation.

The installation software will print progress messages as various groups of files are copied to the hard disk.

A.2.5 What the install program does

The installation software will perform the following tasks:

- Create (if necessary) the directory that you specified for the INFO-MAP installation (subsequently referred to as the INFO-MAP 'root' directory) and copy the file 'infomap.exe' to this directory overwriting any existing file of this name.
- Create (if necessary) a sub-directory of the INFO-MAP 'root' directory named 'util' and copy files 'map.dev', 'map.hlp', 'map.fnt', 'map*.prn', 'map*.scn', 'map*.com' and 'map*.fnt' to this directory overwriting any existing files with these names. Of these files the first three are necessary for INFO-MAP to run; the rest relate to device drivers or additional fonts which you may require in future depending upon whether or not you wish to use a printer, scanner or digitising tablet.
- Copy files 'userref.*' into the sub-directory 'util' of the INFO-MAP 'root' directory. These are copies of the INFO-MAP *User's Reference*, in various formats, including typeset versions suitable for direct laser printing. The *User's Reference* contains a detailed reference to INFO-MAP functions and configuration. It should be printed after installation as described in a later section.

- Create (if necessary) a sub-directory of the INFO-MAP 'root' directory named 'data' and copy files '*.bdy', '*.dta', '*.oly' and '*.txt' to this directory overwriting any existing files with these names. These are the example data sets referred to in the book. None of these are essential for INFO-MAP to run.

If the installation program is successful, it should print a message to that effect, indicate how to run the program, change your current directory to the INFO-MAP 'root' directory and exit. INFO-MAP should then run as described in the later section 'Getting started'. If things go wrong refer to the later section 'Trouble-shooting'.

A.2.6 Printing the *User's Reference*

The INFO-MAP *User's Reference* provides a detailed explanation of all INFO-MAP options and how to configure the package in various ways. You will need this document if you intend to use INFO-MAP seriously.

As described in the previous section, during installation a typeset copy of the *User's Reference*, in a format suitable for printing on a laser printer, will have been copied to the sub-directory 'util' of the directory into which INFO-MAP was installed. Two formats are supplied. The file 'userref.ps' may be printed on any PostScript laser printer. The file 'userref.hp' is in a format suitable for Hewlett Packard laser printers. Neither file requires conversion or translation, it is already in the form which can be directly interpreted by any laser printer designed to operate in the correspoding mode and should be submitted to the printer using a standard print spooler.

If you do not have access to a PostScript or HP format laser printer, the text of the *User's Reference* will also be found in the same sub-directory contained in the file 'userref.rtf'. This is a version of the document in 'rich text format'. Most commonly used word processors will accept documents in this format. You could therefore import the document into your favourite word processor and print it in that way, although printing the PostScript or HP versions is much to be preferred if at all possible.

After you have printed the *User's Reference*, you may wish to conserve space on your disk by deleting all files 'userref.*' in the 'util' sub-directory, or alternatively deleting those in formats which are not required.

A.3 Getting started

A.3.1 Running INFO-MAP

In order to run INFO-MAP you simply change your current directory to the directory in which the INFO-MAP program file 'infomap.exe' resides and type `'infomap'`. You need to make this INFO-MAP 'root' directory your current directory before running the package because INFO-MAP is too large to load into memory all at once and needs to be able to find its own program file in order to read different sections of code, depending on which options you select whilst running. It assumes the program file to be in the current directory at start up time and will fail to load if this is not the case. If you find this requirement inconvenient and don't wish to have to change directory before running INFO-MAP, then there are ways around it (see 'Configuration' in the *User's Reference*). The simplest way is simply to create a '.bat' file in an appropriate directory in your default DOS 'path' and get this file to change directory to the INFO-MAP directory and then run INFO-MAP. Note that if you are running INFO-MAP from a remote server on a network, the procedure for startup is different (see 'Configuration' in the *User's Reference*).

If the program loads correctly, then a copyright 'logo' containing the 'registered owner's' name will appear for a few seconds and then clear, leaving a blank screen except for a menu bar along the top of the screen and a message area at the foot. Pressing ⟨CR⟩ whilst the 'logo' is displayed will cause it to disappear immediately, rather than being displayed for the default number of seconds. You should then be able to use the package as described in the following sections. If the program will not load, or loads but displays an error message, then see the later section 'Trouble-shooting'.

A.3.2 Using the keyboard and mouse

INFO-MAP is controlled by using a mouse, or the keyboard, or a mixture of the two, in much the same way as most other PC software. If a Microsoft mouse driver (or compatible) is detected by INFO-MAP when it loads, then you should see the usual mouse cursor on the screen and this should respond to your mouse movements. If you have a mouse, but the cursor is not displayed in INFO-MAP or it appears distorted, or doesn't respond, then consult the later section 'Trouble-shooting'. If at some future time you use a digitising tablet with INFO-MAP, then the digitising 'puck' or 'pen' will replace the mouse as the pointing device and you should consult the chapter 'Configuration' in the *User's Reference* for details of operation.

As far as general use of a mouse in INFO-MAP is concerned (and we strongly suggest you use one if possible) 'double clicking' the left mouse button is the basic operation for initiating or terminating an action throughout INFO-MAP. INFO-MAP only registers 'single clicks' of the left mouse button in menu selection, elsewhere you always have to make actions more explicit with a 'double click'. INFO-MAP always responds to a 'single click' of the right mouse button, which signifies 'aborting' or 'escaping' from an action throughout the system. For some operations you may be required to 'drag' or 'draw' with the mouse. In these cases you move the mouse whilst at the same time holding down the left button. In one or two options within INFO-MAP a distinction is made between a 'double click' of the left button and the same action whilst holding down either the ⟨SHFT⟩ key on the keyboard, or alternatively whilst holding down the ⟨CTRL⟩ key. It is possible to reverse the action of the mouse buttons in INFO-MAP if you prefer them right–left rather than the standard left–right (see 'Configuration' chapter in the *User's Reference*).

Keyboard use is largely self explanatory. Cursor keys mimic mouse movement vertically or horizontally, ⟨CR⟩ acts like a 'double click' of the left mouse button, ⟨ESC⟩ like a 'single click' of the right button. ⟨Home⟩, ⟨End⟩, ⟨Pg Up⟩ and ⟨Pg Dn⟩ keys (those on the numeric keypad only) may be used to mimic diagonal mouse movements where this is appropriate, at other times (scrolling in menus) they fulfil their usual functions. To mimic 'dragging' or 'drawing' from the keyboard it is necessary to hold down the ⟨SHFT⟩ key (or set ⟨NumLock⟩ to on) whilst at the same time using the cursor keys (those on the numeric keypad only). In general ⟨TAB⟩ or ⟨SHFT⟩ + ⟨TAB⟩ skip respectively forwards or backwards between fields when making multiple on-screen entries. ⟨INS⟩ toggles between text insertion and text overwrite when altering an existing text input field. ⟨DEL⟩ deletes the character to the right in text editing and ⟨BKSP⟩ deletes the character to the left.

A.3.3 Choosing options in INFO-MAP

Given the basic mouse and keyboard operations described above, INFO-MAP is then largely menu-driven. Often menus are automatically generated depending on previous options that you have selected; but at other times the menus available are displayed in the form of a menu bar running across the top of the screen. For example, when you first start the package you will be presented with the following menu bar:

File	Edit	Data	Map	Overlay	Analysis	Preferences	Setup	Help

You may activate menus on the menu bar either by pointing at the option with the mouse and 'clicking', or from the keyboard by pressing the ⟨ALT⟩ key together with the first letter of the option you require. For example ⟨ALT⟩ + ⟨F⟩ would cause the 'file' menu to 'drop down' in the above case. Once you have activated one menu you may move to the next menu to the right by using the ⟨TAB⟩ key, or that to the left using ⟨SHFT⟩ + ⟨TAB⟩. To 'close' a menu without making a selection use the ⟨ESC⟩ key

Using the keyboard, an item is selected from a menu by using the cursor keys to highlight the required item and then pressing the ⟨CR⟩ key. In some cases the items in a menu may exceed the length of the menu 'window', in which case use of the cursor keys when at the top or bottom of the menu will cause the items to scroll. The ⟨Pg Up⟩, ⟨Pg Dn⟩, ⟨Home⟩ and ⟨End⟩ keys can also be used to move around the menu.

When using a mouse in menu selection, pointing to an item and 'clicking' the left button will cause that item to be selected. 'Dragging' the mouse (holding the left button and moving) allows scrolling through items, whilst 'double clicking' with the left button above the first item or below the last item displayed, is the equivalent of using the ⟨Pg Up⟩ or ⟨Pg Dn⟩ keys respectively.

Apart from menus, INFO-MAP often displays 'windows' containing prompts to which the user is expected to provide some input. In these cases the mouse is inactive, except in so much as 'double-clicking' the left button is equivalent to the ⟨CR⟩ key and 'single clicking' the right button is equivalent to the ⟨ESC⟩ key. Movement forward between prompts is effected with the ⟨TAB⟩ key or backwards with ⟨SHFT⟩ + ⟨TAB⟩. ⟨CR⟩ at the last prompt in the 'window' signifies that the user has finished making entries and wishes to proceed with those entries. ⟨ESC⟩ as usual will 'abort' the option and close the 'window'.

A.3.4 The `Help` menu

Note that the `Help` menu in INFO-MAP does not provide guidance on how to use any of the functions in INFO-MAP. It is not intended to be an 'on-line' help facility but merely offers a series of 'about' options which provide the user with some general background information on the package and issues such as the mouse, printers, digitising tablets, scanners and the like.

A.3.5 The working directory

INFO-MAP always looks for data sets in its current 'working directory'. When you attempt to 'open' files only those in the current 'working directory' will be displayed. Similarly, if you save files they will be placed in the current 'working directory' by default. If you attempt to 'import' files they will be looked for in the current 'working directory' unless you specify a full DOS path as part of the file name you supply to the option concerned. At any time you can change the current 'working directory' by using the option `Working directory` in the `Preferences` menu. The 'in-built' INFO-MAP default for the 'working directory' is the sub-directory 'data' of the directory where the 'infomap.exe' file resides; this is where the example data sets are placed during the installation of INFO-MAP. The option `Save settings` from the `Preferences` menu can be used to save the current 'working directory' (along with a number of other current settings discussed in 'Configuration' in the *User's Reference*) to be the default for subsequent runs of INFO-MAP.

A.3.6 Preferences and device setup

Several settings such as colour, working directory, scaling type (class interval selection), number of scaling intervals and so on, have default values when INFO-MAP is executed. Most of these defaults can be changed by using the option `Save settings` from the `Preferences` menu, which saves their current settings to be the defaults for subsequent runs of INFO-MAP. Similar concepts apply if you use the options within the `Setup` menu to install a printer, scanner or digitising tablet. All these various settings are described in the section 'Configuration' in the *User's Reference*. Their values are stored in a file 'infomap.ini' which is created by INFO-MAP in the directory from which the program was executed. Normally this will be the same directory as that in which the 'infomap.exe' file resides. However, if INFO-MAP is being used from a remote network server (see 'Configuration' in the *User's Reference*) INFO-MAP is always executed from a directory on the local drive and so 'infomap.ini' resides locally. The file 'infomap.ini' is optional—if it exists in the directory from which the program is executed then INFO-MAP will use the settings specified in it; otherwise 'built-in' defaults will be used.

A.3.7 Where next?

Assuming you have succeeded in running INFO-MAP and read over the basic use of mouse and keyboard and other ideas as described in the above sections, you may wish to do something simple with the software and gain a quick insight into its general working, without having to immediately consult the *User's Reference* and struggle through the detailed descriptions of all the options given there. At this stage you might like to refer to Chapter 1 of the main text and attempt some exercises. These will give you a good general idea of how the package works.

You should have printed the *User's Reference* as described earlier. The first chapter in that document explains the structure of the software in more detail and the subsequent chapters explain all of the INFO-MAP menus and options in detail. If you want to export the example data sets into ASCII files to use in other packages you will need to refer to the option `Export` under the `Data` menu in the chapter 'Creating and modifying data files' in the *User's Reference*. If you want to create your own spatial data sets in INFO-MAP, then there is an exercise relating to this in Chapter 2 of the main text of this book. You will also need to make sure you have full 'registration' for your copy of INFO-MAP. Details of how to obtain this are given in the `About registration` option in the `Help` menu. Finally you will probably also need to consult details given in the *User's Reference*, such as the chapters 'Creating and modifying base files' and 'Creating and modifying data files'.

A.4 Trouble-shooting

A.4.1 INFO-MAP problems

Various problems may arise in running INFO-MAP. Firstly the program may not run at all. Secondly it may run but fail whilst performing some particular function, display an error message and possibly terminate. This section discusses some possible problems of this type that you may encounter.

There is of course a third class of problem that you may encounter. Your installation diskette may be corrupted for some reason and so you may not be able to install the package at all. Alternatively, INFO-MAP may run but 'crash' in an uncontrolled way because of an error or 'bug' in the code which we have overlooked. Obviously, we hope this third type of problem will not arise, but if it does then all we can do here is to apologise. INFO-MAP is distributed as an educational aid and not as a polished commercial product; we cannot provide any support for the package and further releases or 'bug-fixes' are not envisaged.

A.4.2 INFO-MAP will not run at all

The most likely reason for this is that the program is too big to fit into the conventional memory available on your machine. INFO-MAP is restricted to run within the 640 Kbytes of conventional DOS memory. If the amount of free conventional RAM at the time of attempting to execute INFO-MAP is much below 480 Kbytes then INFO-MAP may not execute at all. The only way around this problem is to free additional RAM by removing memory-resident software or large device drivers unconnected with INFO-MAP. This is normally done by editing either the files 'config.sys' or 'autoexec.bat' to prevent the loading of such drivers when your machine 'boots' and then 're-booting' the machine.

A second possible reason why INFO-MAP will not run is because you are executing the program without making the INFO-MAP directory your 'current' directory. If you wish to run INFO-MAP from elsewhere you must include the INFO-MAP directory in your DOS path, *and* set a DOS environment variable 'MAPDIR' to the path name of the directory where 'infomap.exe' resides (see 'Configuration' in the *User's Reference*).

A third possible reason why INFO-MAP will not run is because INFO-MAP is trying to read an 'infomap.ini' file which is corrupt or invalid in some other way. INFO-MAP always attempts to read any 'infomap.ini' file which is found in the 'current' directory from which the program is executed. It does this in order to determine default user settings (see 'Configuration' in the *User's Reference*). If INFO-MAP fails to execute then it is possible that such a file may have become corrupted in some way. 'infomap.ini' is an optional file of user preferences and not required for INFO-MAP to run, so if you suspect this is the problem delete any such 'infomap.ini' file from DOS and try running INFO-MAP again.

A.4.3 Dynamic memory overflow

In addition to the 480 Kbytes INFO-MAP requires to load, it needs to allocate additional memory dynamically during execution. Between 64 and 100 Kbytes of such dynamic memory should be adequate for most purposes, and hence we specify the base memory requirement for INFO-MAP as approximately 550 Kbytes overall.

You can find out how much memory is available for INFO-MAP dynamic allocation at any time by using the `About memory` option in the `Help` menu. If it is small on your machine then INFO-MAP may fail with the error message 'Dynamic Memory Overflow' when attempting to allocate dynamic memory. One way around this problem is to free additional RAM by removing memory-resident software or large device drivers unconnected with INFO-MAP. Another is to make expanded memory available to INFO-MAP. This will not affect the minimum amount of memory required to execute the program which will remain as above at about 480 Kbytes, but it will allow INFO-MAP to use expanded rather than conventional memory for some dynamic allocation, so reducing the overall memory requirement from 550 Kbytes to around 500 Kbytes. (See 'Configuration' in the *User's Reference* for a discussion of using expanded memory with INFO-MAP.)

A4.4 MAPDIR environment variable not set or invalid

INFO-MAP is too large to load into memory all at once and needs to be able to find its own program file 'infomap.exe' in order to read different sections of code, depending on the options you select when running. By default it assumes 'infomap.exe' to be in the 'current' directory from which it is executed. Alternatively, it expects to find a DOS environment variable 'MAPDIR' set to the path name of the directory where 'infomap.exe' resides *and* this directory to be also included in the DOS path. If one of these is not the case then INFO-MAP will fail to load. The error 'MAPDIR environment variable not set' relates to the case where 'infomap.exe' is not found in the 'current' directory at time of execution and no 'MAPDIR' environment variable has been specified. The error 'Invalid MAPDIR environment variable' relates to the case where a 'MAPDIR' environment variable has been detected but 'infomap.exe' is not found in the directory which it specifies.

A.4.5 Graphics driver not found or invalid

INFO-MAP requires to find and read the graphics driver file 'map.dev' in order to run. It expects to find this file in the sub-directory 'util' of the directory in which the 'infomap.exe' file resides. If the file is not found then the program will not run and the error 'Graphics driver not found' is reported. If such a file is found but is corrupt or not a valid graphics driver then 'Invalid graphics driver' is reported.

A.4.6 Font file not found

In order to run, INFO-MAP requires to find and read the font file 'map.fnt'. It expects to find this file in the sub-directory 'util' of the directory in which the 'infomap.exe' file resides. If the file is not found then the program will not run and the error 'Font file not found' is reported.

A.4.7 Too many files open at one time

There are two reasons why this error may occur. Firstly, INFO-MAP cannot find enough DOS file handles available to open necessary files. In this case you need to increase the number of file handles available in DOS. This is done by use of the DOS command 'files=' in your 'config.sys' file. See your DOS documentation for details. Secondly, the error may arise because of lack of enough free memory for allocation of file buffers. In this case you need to free additional RAM by removing memory-resident software or large device drivers unconnected with INFO-MAP, before running INFO-MAP.

A.4.8 Other problems

The sections above have covered errors that cause INFO-MAP to terminate in a controlled way. There are of course other problems that may arise. One possibility is the program 'crashing' in an uncontrolled way. As said in the introduction to this chapter, we apologise, but are unable to help if this happens.

Another possibility is problems which do not cause the program to terminate. For example, things may go wrong when trying to get a mouse, printer, scanner or digitising tablet to work correctly with INFO-MAP. This area is fully discussed in the chapter 'Configuration' in the *User's Reference*. However, we should acknowledge that the very large variation in such devices means that there will be devices which should in theory work with INFO-MAP but which will not in practice. Again, we apologise but are unable to offer further support.

Another area which may cause problems is inability to read *base files*, *data files* or *overlay files* because they have become corrupted in some way. If INFO-MAP fails when you try to 'open' a file in some 'working directory' you could try moving all files out of this directory and moving them back in groups, running INFO-MAP each time until you track down which file is causing the problem. It is also worth warning the user here against renaming '.bdy' files from DOS. The link between *base files* and *data files* and between *base files* and *overlay files* in INFO-MAP is made by incorporating the '.bdy' file name of the *base file* to which they are associated into the header of both *data files* and *overlay files*. Renaming a '.bdy' file from DOS therefore breaks the link. Renaming of '.dta' or '.oly' files from DOS does not produce a problem.

Finally, there may be a range of additional problems that you may encounter. All we can suggest is that you read the section of the *User's Reference* which is relevant to the area in which the problem occurs to see if you can track down what is going wrong, or find an acceptable 'work around' if INFO-MAP is simply not functioning as it should be.

INDEX